BRITISH MEDICAL BULLETIN

Impact of environmental pollution on health

Balancing risk

Scientific Editors: David J. Briggs, Michael Joffe, Paul Elliott

1 Environmental pollution and the global burden of disease
   David Briggs

25 The impact of environmental pollution on congenital anomalies
   Helen Dolk and Martine Vrijheid

47 Infertility and environmental pollutants
   Michael Joffe

71 Contribution of environmental factors to cancer risk
   Paolo Boffetta and Fredrik Nyberg

95 Air pollution and infection in respiratory illness
   Anoop J Chauhan and Sebastian L Johnston

113 Evaluating evidence on environmental health risks
   Lesley Rushton and Paul Elliott

129 Environmental effects and skin disease
   JSC English, RS Dawe and J Ferguson

143 Ambient air pollution and health
   Klea Katsouyanni

157 Electromagnetic radiation
   Anders Ahlbom and Maria Feychting

167 Hazards of heavy metal contamination
   Lars Järup

183 Health hazards and waste management
   Lesley Rushton

199 Contaminants in drinking water
   John Fawell and Mark J Nieuwenhuijsen

209 Indoor air pollution: a global health concern
   Junfeng (Jim) Zhang and Kirk R Smith

227 Asthma: environmental and occupational factors
   Paul Cullinan and Anthony Newman Taylor

243 Noise pollution: non-auditory effects on health
   Stephen A Stansfeld and Mark P Matheson

259 Risks associated with ionizing radiation
   MP Little

277 Index

http://www.bmb.oupjournals.org

Acknowledgements

The planning committee for this issue of the British Medical Bulletin included
David J. Briggs, Michael Joffe, Paul Elliott, John Fawell, Lesley Rushton,
Mark Nieuwenhuijsen, Bert Brunekreef and Goran Pershagen.

The British Council and Oxford University Press are most grateful to them
for their expert help and advise and for the valuable work of the
Scientific Editors in completing this volume.

Material Disclaimer

The opinions expressed in the British Medical Bulletin are those of the authors and contributors, and do not necessarily reflect those of The British Council, the editors, the editorial board, Oxford University Press or the organization to which the authors are affiliated.

Drug Disclaimer

The mention of trade names, commercial products or organizations in the British Medical Bulletin does not imply endorsement by The British Council, the editors, the editorial board, Oxford University Press or the organization to which the authors are affiliated. The editors and publishers have taken all reasonable precautions to verify drug names and doses, the results of experimental work and clinical findings published in the Journal. The ultimate responsibility for the use and dosage of drugs mentioned in the Journal and in interpretation of published material lies with the medical practitioner, and the editors and publishers cannot accept liability for damages arising from any errors or ommissions in the Journal. Please inform the editors of any errors.

Environmental pollution and the global burden of disease

David Briggs

Small Area Health Statistics Unit, Department of Epidemiology and Public Health, Imperial College, London, UK

Exposures to environmental pollution remain a major source of health risk throughout the world, though risks are generally higher in developing countries, where poverty, lack of investment in modern technology and weak environmental legislation combine to cause high pollution levels. Associations between environmental pollution and health outcome are, however, complex and often poorly characterized. Levels of exposure, for example, are often uncertain or unknown as a result of the lack of detailed monitoring and inevitable variations within any population group. Exposures may occur *via* a range of pathways and exposure processes. Individual pollutants may be implicated in a wide range of health effects, whereas few diseases are directly attributable to single pollutants. Long latency times, the effects of cumulative exposures, and multiple exposures to different pollutants which might act synergistically all create difficulties in unravelling associations between environmental pollution and health. Nevertheless, in recent years, several attempts have been made to assess the global burden of disease as a result of environmental pollution, either in terms of mortality or disability-adjusted life years (DALYs). About 8–9% of the total disease burden may be attributed to pollution, but considerably more in developing countries. Unsafe water, poor sanitation and poor hygiene are seen to be the major sources of exposure, along with indoor air pollution.

Introduction

Correspondence to:
David Briggs, Small Area Health Statistics Unit, Department of Epidemiology and Public Health, Imperial College, London, UK. E-mail: d.briggs@imperial.ac.uk

Despite the major efforts that have been made over recent years to clean up the environment, pollution remains a major problem and poses continuing risks to health. The problems are undoubtedly greatest in the developing world, where traditional sources of pollution such as industrial emissions, poor sanitation, inadequate waste management, contaminated water supplies and exposures to indoor air pollution from biomass fuels affect large numbers of people. Even in developed countries, however, environmental pollution persists, most especially amongst poorer sectors of society[1,2]. In recent decades, too, a wide range of modern pollutants have emerged—not least, those associated with road traffic and the use of modern chemicals in the home, in food, for water treatment and for pest control. Most of these pollutants are rarely present in excessively large concentrations, so effects on health are usually far from immediate or obvious. As Taubes[3] has noted, few of the problems of environmental exposure that concern us today imply large relative risks. Detecting small effects against a background of variability in exposure and human susceptibility, and measurement error, poses severe scientific challenges.

The progressively larger number of people exposed to environmental pollution (if only as a result of growing population numbers and increasing urbanization) nevertheless means that even small increases in relative risk can add up to major public health concerns. The emergence of new sources of exposure and new risk factors, some of them—such as endocrine disruptors—with the capacity to have lifelong implications for health, also means that there is a continuing need for both vigilance and action. As the impact of human activities and issues of environmental health become increasingly global in scale and extent, the need to recognize and to address the health risks associated with environmental pollution becomes even more urgent. Effective action, however, requires an understanding not only of the magnitude of the problem, but also its causes and underlying processes, for only then can intervention be targeted at where it is most needed and likely to have greatest effect. As background to the other chapters in this volume, therefore, this chapter discusses the nature of the link between environmental pollution and health and considers the contribution of environmental pollution to the global burden of disease.

Links between environmental pollution and health

Environmental pollution can be simply, if somewhat generally, defined as the presence in the environment of an agent which is potentially damaging to either the environment or human health. As such, pollutants take many forms. They include not only chemicals, but also organisms and biological materials, as well as energy in its various forms (*e.g.* noise, radiation, heat). The number of potential pollutants is therefore essentially countless. There are, for example, some 30,000 chemicals in common use today, any one of which may be released into the environment during processing or use. Fewer than 1% of these have been subject to a detailed assessment in terms of their toxicity and health risks[4]. The number of biological pollutants is truly unquantifiable. They include not only living and viable organisms, such as bacteria, but also the vast array of endotoxins that can be released from the protoplasm of organisms after death. There is, therefore, no shortage of potential environmental risks to health. What is lacking, for the most part, is an understanding of the nature and mechanisms of these risks.

The source–effect chain

The link between pollution and health is both a complex and contingent process. For pollutants to have an effect on health, susceptible individuals must receive doses of the pollutant, or its decomposition products, sufficient to trigger detectable symptoms. For this to occur, these individuals must have been exposed to the pollutant, often over relatively long periods of time or on repeated occasions. Such exposures require that the susceptible individuals and pollutants shared the same environments at the same time. For this to happen, the pollutants must not only be released into the environment, but then be dispersed through it in media used by, or accessible to, humans. Health consequences of environmental pollution are thus far from inevitable, even for pollutants that are inherently toxic; they depend on the coincidence of both the emission and dispersion processes that determine where and when the pollutant occurs in the environment, and the human behaviours that determine where and when they occupy those same locations.

The whole process can simply be represented as a causal chain, from source to effect (Fig. 1). As this indicates, most pollutants are of human origin. They derive from human activities such as industry, energy production and use, transport, domestic activities, waste disposal, agriculture and recreation. In some cases, however, natural sources of pollution may also be significant. Radon, released through the decay of radioactive materials in the Earth's crust, arsenic released into groundwaters from natural rock sources, heavy metals accumulating in soils and sediments derived from ore-bearing rocks, and particulates and sulphur dioxides released by wildfires or volcanic activity are all examples.

Release from these various sources occurs in a wide range of ways, and to a range of different environmental media, including the atmosphere, surface waters, groundwaters and soil (Fig. 2). Estimates of emission by source and environmental medium are inevitably only approximate, for they can rarely be measured directly. Instead, most emissions inventories derive from some form of modelling, either based on emission factors for different processes or source activities[5] or on input–output models (*i.e.* by calculating the difference between quantities of the material input into the process and quantities contained in the final product).

Atmospheric emissions

Emissions to the atmosphere tend to be more closely modelled and measured, and more generally reported, than those to other media, partly

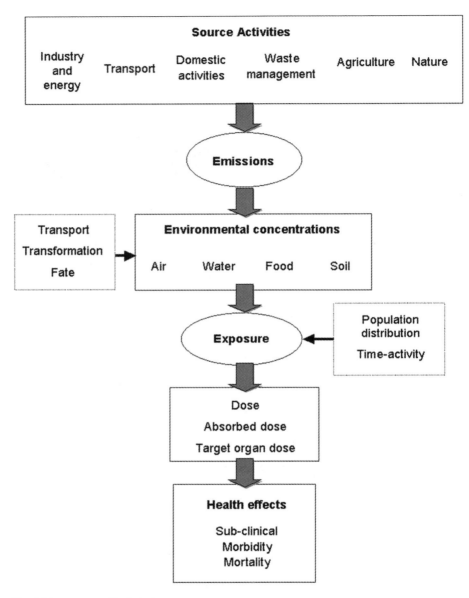

Fig. 1 The source–effect chain.

because of their greater importance for environmental pollution and health (emissions to the atmosphere tend to be more readily discernible and to spread more widely through the environment), and partly because of the existence of better established policy and regulation. Figure 3 shows the main sources of emissions of selected pollutants in the European Union. As this shows, combustion represents one of the most important emission processes for many pollutants, not only from industrial sources, but also

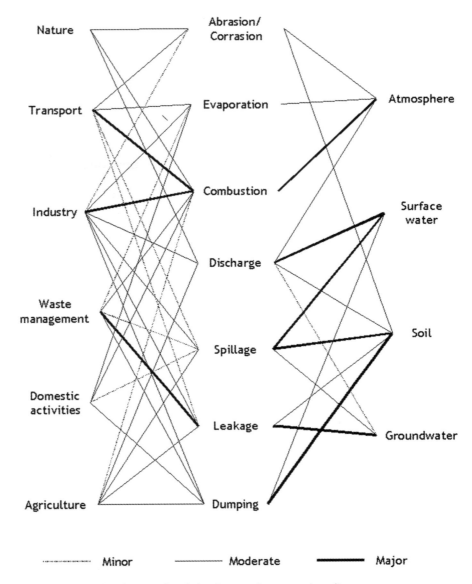

Fig. 2 Sources and pathways of emission into environmental media.

from low-level sources such as motorized vehicles and domestic chimneys, as well as indoor sources such as heating and cooking in the home or workplace. Emissions from industrial combustion or waste incineration tend to be released from relatively tall stacks, and often at high temperature, with the result that they are dispersed widely within the atmosphere. Emissions from low-level sources such as road vehicles and low-temperature combustion sources such as domestic heating, in contrast, tend to be much less widely dispersed. As a result, they contribute to local pollution

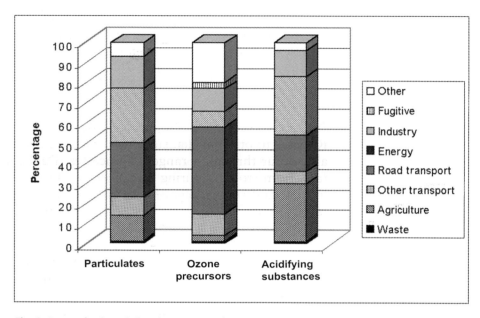

Fig. 3 Atmospheric emissions by source in the European Union (EU15). Note: Ozone precursors include methane, NMVOCs, NO_x and CO; acidifying substances include NO_2, SO_2, NH_3. Source: European Environment Agency.

hotspots and create steep pollution gradients in the environment. In urban environments, for example, traffic-related pollutants such as nitrogen dioxide and carbon monoxide typically show order-of-magnitude variations in concentration over length-scales of tens to a few hundred metres[6]. Evaporation and leakage are also important emission processes contributing to local variations in environmental pollution. In the UK, releases from filling stations account for *ca.* 1.8% of benzene emissions; leakages from gas pipelines contribute *ca.* 13.7% of methane emissions to the atmosphere; evaporation and leakage of solvents during processing and use produce *ca.* 40% of atmospheric emissions of non-methane volatile organic compounds (NMVOCs)[7]. In addition, abrasion, corrasion and corrosion release significant quantities of emissions to the atmosphere. Wear and tear of catalytic converters during operation is a major source of platinum emissions[8], for example, whereas tyre wear (corrasion) and road wear (abrasion) account for about 16% of particulate emissions from road transport and almost 97% of zinc emissions from road transport[7]—and perhaps more where studded tyres are used[9].

These fugitive and local emissions are often overlooked in epidemiological and other studies that use modelling techniques to estimate exposures, but they can be extremely important, both because they are frequently responsible for the highest concentrations of environmental pollution, and because—unlike high-level emissions—they remain close to source and show marked dilution gradients with distance from source. In many cases, they may therefore be the real sources of variation being considered when distance is used as a proxy for exposure around point industrial sources in epidemiological studies.

Emissions to surface water, groundwater and soil

Releases to other media, such as surface waters, groundwaters and soil, also occur through a range of processes. Deliberate discharge, spillage (*e.g.* from storage, during transport, or during processing and usage), leakage and runoff (*e.g.* of agricultural chemicals) are all important in terms of aqueous pollutants. Legal limits for discharges to streams are set for many industries, aimed at keeping levels of contamination within accepted limits. Illegal discharges, or accidental spillage, however, sometimes occur and accounted for the majority of reported surface water pollution incidents in the UK in 2001, for which the cause is known[10]. Dumping (both legally in landfill sites and illegally) represents a major source of emission of solid wastes, though final release into the wider environment may only occur when these materials decompose or break up. Landfill sites may thus be responsible for emissions of a wide range of pollutants, *via* different pathways, especially when these sites are inadequately sealed or poorly maintained[11]. The contribution of informal and illegal dumping to environmental pollution is, inevitably, only poorly known.

Environmental fate

Once released into the environment, pollutants may be transported *via* many different processes and pathways, often moving from one medium to another, and undergoing a wide range of modifications in the process. Chemical reactions, physical abrasion, sorting by size or mass and deposition all change the composition of the pollutants and alter the pollution mix. Dilution occurs as pollutants spread outward into a wider volume of space; concentration may occur as pollutants accumulate in local 'sinks' or in the bodies of organisms, as they pass along the food chain.

In general, these processes tend to result in some degree of distance-decay in environmental concentrations, if only because the opportunity for dilution, decomposition and deposition increases with increasing distance of transport. It is largely on this basis that distance is often used as a surrogate for exposure in many epidemiological studies. The realities of environmental patterns of pollution are, however, often much more complex than these simple distance-based models imply. They also vary greatly between different pollutants and environmental media, because of the different transportational behaviours that are involved. In addition, dispersion processes and resulting pollution concentration fields may vary substantially depending on the prevailing (*e.g.* meteorological) conditions at the time. Patterns of atmospheric dispersion, for example, differ not only in relation to windspeed and direction but also atmospheric stability (*e.g.* between stable and unstable weather conditions, or when there is a temperature inversion)[6]. Movement of many pollutants through soils occurs mainly as mass flow in water passing through larger pore spaces and fissures: the irregular distribution of these within highly structured soils means that dispersion often follows highly discrete pathways[12,13]. Gaseous pollutants may follow similar preferred pathways. Releases from landfill sites may thus travel relatively long distances in the soil or bedrock, before emerging at the surface, where they can cause local hazards including explosions[14]. Radon shows the same discrete and complex pattern, such that concentrations may vary by orders of magnitude from one home to another in the same district[15,16]. Modelling these locally variable pathways poses severe challenges.

To a large extent, the increased opportunity for mixing means that dispersion of pollutants in surface and groundwaters is more regular, leading to more uniform patterns of contamination, at regional scales. In developed countries, also, considerable water mixing often occurs during treatment and distribution, so that water quality is relatively uniform across large areas and populations. Local variations may occur, however, because of contamination within the distribution system or differences in the length of the network, and thus in the time available for contamination and decomposition of the disinfectants incorporated at treatment[17]. In developing countries, especially, considerable variations may also occur between waters in shallow wells, particularly where these are affected by local pollution sources, such as badly sited latrines or agricultural activities. Again, this makes exposure assessment difficult, without the ability to collect data on water quality for individual wells.

Similar difficulties occur in tracking and modelling transport of pollutants in the food chain. Whilst the general pathways followed by pollutants are often clear in natural (and some farmed) food chains, in that persistent compounds tend to accumulate as they pass from one trophic level to another, the detailed patterns of contamination are often far more complex. Many animals have very restricted feeding behaviours: even in areas of open grazing land, for example, sheep tend to focus on distinct home ranges from which they rarely stray[18]. As a result, marked variations may occur in contaminant uptake by livestock, even over short distances, as illustrated by patterns of contamination from the Chernobyl incident in the UK[19]. Significant accumulation of these contaminants in humans likewise tends to occur only where small groups of individuals rely on local food sources. On the other hand, in many modern food supply systems, industrial-scale processing and distribution operations mean that foodstuffs often travel large distances before consumption and are drawn from far-flung sources. In the UK, as in most developed countries, therefore, the average distance travelled by foodstuffs before consumption has increased markedly, from an average of about 82 km in 1978 to 346 km in 1998[20]. In the light of these changes, several attempts have been made in recent years to calculate the distance travelled by ingredients in common food products or meals (so called 'food-miles'). In Iowa, USA, for example, ingredients for a standard meal of stir-fry and salad were estimated to have been transported 20,000 km[21]; in the UK, Sustain, a pressure group on food and agriculture, estimated that ingredients for a traditional turkey dinner had been transported some 38,620 km[22]! Apart from implications for increased energy consumption and environmental pollution, such extended distribution networks clearly mean that it can be difficult to track and control potential contamination between source and consumption.

By whatever pathways and processes pollutants pass through the environment, four related factors are especially important in determining the potential for exposure and health effects: their persistence, their mobility, their decomposition products and their toxicity. The problems associated with the release of persistent pollutants into the environment were highlighted many years ago with recognition of the global extent of contamination, and wide-ranging environmental and health effects, caused by DDT and other organochlorine pesticides[23]. The story is in many ways now being repeated in relation to chlorofluorocarbons and other atmospheric pollutants that act as greenhouse gases or scavengers of stratospheric ozone[24], and perhaps also in relation to endocrine disruptors[25]. Persistence, however, is not necessarily the most important issue, for where they persist in inert yet inaccessible forms, pollutants may pose relatively limited risks. Thus, whereas inorganic mercury is persistent, it is less toxic and less readily bioavailable than methyl mercury, to which it is naturally converted through chemical reactions and the action of soil and aquatic microorganisms[26,27]. Equally, many solid wastes represent little risk to health so long as they remain in their original form. The problems in these cases often come when decomposition occurs, either because the decomposition products are inherently more toxic or because they are more mobile, and thus are more likely to result in human exposure.

Exposure and dose

Whilst the potential for health impairment initially depends upon the existence and concentrations of pollutants in the environment, for health effects to occur exposures must take place that lead to a dose sufficient to have adverse health consequences. Exposure in this context is defined as the contact between a hazardous agent (in this case a pollutant) and an organism. Dose refers to the quantity of the substance in the body. The absorbed dose refers to the amount of the substance entering the body as a whole; target organ dose refers to the amount reaching the specific organs that are affected.

Exposure can take place in many different ways. Three main forms of exposure are generally recognized: dermal contact, inhalation and ingestion. In some cases, however, it may also be useful to recognize a fourth—injection—for example, when pollutants are transmitted by animal bites or by deliberate injection. In each of these cases, exposure may occur in a range of different environments. Whilst some exposures occur in the outdoor (ambient) environment, most people spend the majority of their time indoors, either at home or at their place of work or learning. Indoor exposures therefore often make up a major proportion of total exposure[28]— although they tend to be rather neglected in many epidemiological studies. Food and drinking water are likewise important routes of exposure for many pollutants.

What determines levels of exposure is consequently not just the distribution of pollution within the environment, but also human behaviours and lifestyles, and thus the sorts of exposure environments in which people spend their time. By the same token, exposure is not only an environmental process, it is also a social, demographic and economic one. Indeed, because of the myriad ways in which socio-economic and demographic factors influence and interact with environmental conditions, exposures, human susceptibility and health outcome, they may often appear to outweigh the effects of the environment *per se* in associations between pollution and health. One expression of these complex interactions is the patterns of what is often called environmental injustice that are seen throughout the world: namely, the tendency for environmental pollution and poverty or other forms of disadvantage to be strongly correlated, such that poorer people tend to live in more polluted environments[29,30]. Whilst the reasons for this association are not fully understood, and may be more subtle than often assumed, the double jeopardy that it represents seems to be generally reflected in terms of health inequalities as well. The problem, however, comes in trying to separate the contributions to these adverse health outcomes from socio-economic and environmental factors—and thus to quantify the attributable effects of pollution.

It also has to be recognized that pollutants rarely occur in isolation; more typically they exist in combination. Exposures are therefore not singular. Instead we are usually exposed to mixes of pollutants, often derived from different sources, some of which may have additive or synergistic effects. Unravelling the effects of individual pollutants from this mix is a challenging problem that has yet to be adequately resolved in many areas of epidemiology.

Health effects: dose–response relationships, latency and attributable risk

One of the underlying tenets of environmental epidemiology is that, for the health effects of interest, a relationship exists between the level of exposure (or dose) and the degree of effect. Effects can, in fact, be represented in two different ways: by the type of effect or by its severity or the probability of its occurrence (often termed the 'response'). In either case, these associations are generally assumed to be broadly linear, such that the effect or response increases with each increment of exposure to a pollutant (Fig. 4A). For many pollutants and many health effects, this assumption seems to hold true at least over a wide range of exposures and responses. Some, however, appear to be characterized by more complex associations. Thresholds may exist, for example, below which no detectable health effects occur (Fig. 4B). At high levels of exposure, responses may weaken, so that the dose–response relationship is essentially curvilinear-convex (Fig. 4C) or S-shaped (Fig. 4D). In a few cases, there is some evidence that ∩- or, more rarely, U-shaped relationships may exist—for example, in relation to solar radiation or vitamin intake. One of the main purposes of epidemiology is to demonstrate and, if possible, quantify these relationships, where they exist.

Just as exposures may be long- or short-term, so health effects can be short-lived (acute) or prolonged (chronic). Health effects may also be delayed to a greater or lesser extent after initial exposure, either because

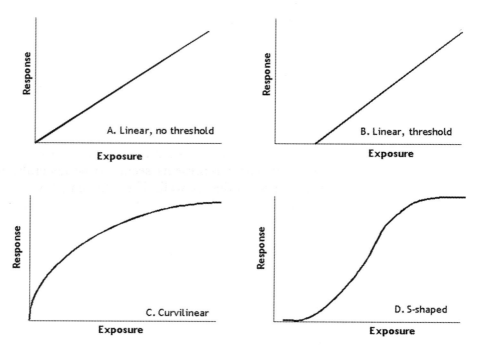

Fig. 4 Common forms of exposure–response relationships.

it takes time for exposures to reach a critical level, or because the disease itself takes time to develop and become apparent (latency). Many acute effects are almost immediate, and have latencies typically of no more than a few minutes to a few days. Many chronic effects, on the other hand, can have latencies of several years—up to 20 years or more, for example, in the case of some cancers and diseases such as asbestosis. Dealing with these latencies is problematic in epidemiological studies, both because they often imply the need for information on past (in some cases long-past) exposures, and because the degree of latency may vary from one individual to another, depending on factors such as the level of exposure, the age at which exposure occurred and pre-existing health status. In some cases, too, a so-called harvesting effect may occur, such that, following a brief increase in disease rates as more vulnerable people are affected, rates of illness fall because there are fewer vulnerable people left (Fig. 5). This, too, can complicate epidemiological studies, for the scale (and even the direction) of the dose–response relationship may vary depending on the length of latency allowed for in the analysis.

Long-term legacies for health may also occur as a result of sensitization and predisposition of individuals to the effects of exposure in early life. Sensitization to house dust mite and other allergens both in the diet and the indoor environment during the first few months of life, for example, appears to increase risks of allergic airway disease later in childhood[31]. Similarly, inverse associations have been found between birth weight and the incidence of a range of diseases including hypertension, type-2 diabetes

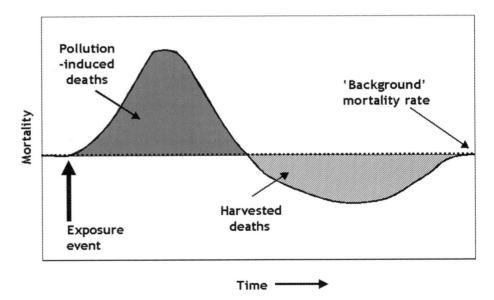

Fig. 5 Harvesting effects.

and cardiovascular disease in adults[32–34]. Environmental exposures, such as air pollution, that contribute to these predisposing conditions may thus have long-term (and in some cases lifelong) implications for health.

Health outcomes may also be more or less specific to exposures to particular pollutants. Very few diseases are, in practice, pollutant-specific. Amongst these are asbestosis and mesothelioma (due to exposures to asbestos) and bagassosis, through exposure to organic dusts (the most common example, of which, is 'Farmer's lung'). Far more commonly, individual health effects may arise as a result of exposures to a number of different risk factors, either individually or in combination, whereas individual exposures can give rise to a range of different health effects. Environmental health is thus characterized by many-to-many relationships; understanding these is, again, a major challenge for epidemiology. Partly for this reason, it is often extremely difficult to assess the health burden attributable to an individual pollutant. Over-estimation may occur due to double-counting (or multiple attribution) of health effects; under-estimation may arise due to the failure to recognize some contributions to the disease burden as a result of masking by other risk factors.

In addition, of course, all epidemiological studies—and other studies that contribute to the establishment of dose–response relationships, such as laboratory experiments and clinical trials—are subject to error and uncertainty. These arise for many different reasons: because of errors in exposure assessment or classification, because of errors in diagnosis or reporting of health outcome, because of inadequate sample size, because of inadequate adjustment for confounding or effect modification by other factors, because of biases in sampling and statistical analysis, and because of the underlying indeterminacy of some of the associations of interest. As a result, dose–response reported by different studies often shows substantial differences, and many separate studies may be needed before a clear pattern of association emerges. Even then, problems may be encountered in deriving reliable dose–response relationships (*e.g.* through some form of meta-analysis), because of inconsistencies in study design (*e.g.* in methods of exposure classification, target populations or specification of health outcome). Most dose–response relationships are thus accompanied by a relatively large degree of uncertainty.

Models of pollutant pathways

As the discussion above has indicated, the relationships between pollution and health are both complex and often indirect. Considerable difficulties are thus encountered in quantifying the associations involved. It is largely for this reason that many of the health effects of environmental pollution are still uncertain, and that problems arise in attempting to attribute health outcome to environmental causes—for example, when trying to confirm or explain apparent spatial clusters in health.

These subtleties and complexities highlight the importance of examining critically any hypothesis about a relationship between a pollutant and an apparent health effect, and of setting such hypotheses within a wider environmental context. Assumptions about simple, singular cause–effect relationships often need to be eschewed; in their place we need to recognize the possibility for multifactorial effects in which single health outcomes are attributable to a wide variety of (possibly inter-related) environmental and other risk factors; and in which individual exposures may contribute to a range of different health effects. The contingent and historical nature of many of these associations also needs to be appreciated: health effects seen now in many cases owe their existence to exposures, sensitization or some process of predisposition far in the past. Because environmental conditions, and even the very nature of the risk factors involved, may change quite considerably over time, uniformitarianist principles may not hold true, *i.e.* the present cannot always be seen as the key to the past.

Against this background, the use of models to conceptualize the possible interplay of different risk factors and exposure pathways, and how they might have evolved over time, represents an important tool for attempts to understand associations between pollution and health. One example is illustrated in Figure 6, which shows possible sources and pathways of exposure of environmental pollution associated with landfill sites. Several important lessons can be drawn from this example. First, it is evident that the pathways of exposure are highly varied and complex. Which is the most important may well differ from one situation to another. The possibility of contributions from each and all of them needs to be allowed. Second, it is evident that landfill sites leave a legacy which may persist long after they are no longer operational. Present-day land use and activity may therefore not account for current exposures. Third, related to this, sources and pathways of exposure change markedly over time—and, indeed, many of the risks associated with what are now landfill sites may well predate the sites themselves (*e.g.* from prior land use). Perhaps it is for this reason that several studies of health risks around landfill sites have found that raised levels of risk existed before the landfill sites were opened[35,36].

The contribution of environmental pollution to the global burden of disease

Estimating the global burden of disease

For all the reasons outlined above, estimating the contribution of environmental pollution to the burden of disease is far from easy. In general, too little is known either about the causal links between environmental

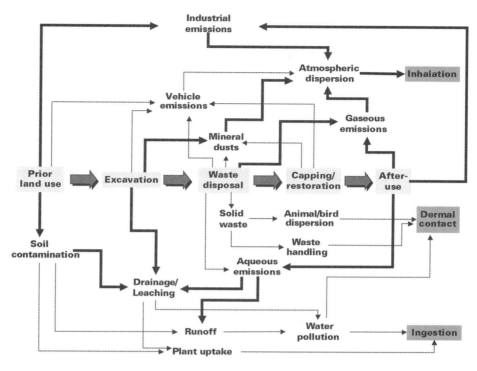

Fig. 6 A model of emission processes and exposure pathways from landfill sites.

pollution and health, or about the levels of exposure across the population, to make reliable assessments of the proportion of disease or mortality attributable to pollution. These difficulties are severe in developed countries, where disease surveillance, reporting of mortality, environmental monitoring and population data are all relatively well established. In most developing countries they become all but insurmountable, because of the generally impoverished state of routine monitoring and reporting. Given that controls on emissions and exposures in the developing world are often limited, it is in these countries that risks from environmental pollution are likely to be greatest. Such uncertainties thus render any attempt to quantify the environmental burden of disease highly approximate at best.

Assessments of the disease burden attributable to different forms and sources of pollution are nevertheless worth the effort. They are needed, for example, to raise awareness about some of the risks associated with environmental pollution, and as a basis for advocacy—to ensure that those most in need have a voice. They are needed to help motivate and prioritize action to protect human health, and to evaluate and monitor the success of interventions. They provide the foundation, therefore, for extremely powerful indicators for policy support, and a means of pricking the global conscience about inequalities in health.

Over recent years, therefore, many attempts have been made to assess the health status of the population, both nationally and globally, and to deduce the contribution made by pollution and other environmental factors. In Europe, for example, more than 50 national environmental health action plans have been developed, following the Helsinki Conference in June 1994, setting out strategies to tackle problems of environmental health[37]. Although these differ substantially in terms of their content and scope, many have involved attempts to make formal assessments of the disease burden attributable to different environmental hazards, and to rank these in terms of their public health significance[38,39]. Various methods were used for this purpose, though most relied on some form of expert judgement, informed where available by quantitative data on mortality or disease rates. Whatever the weaknesses of these assessments, their practical importance is evident, for they have contributed directly to policy prioritization and development in the countries concerned.

The same need has arisen to support the development of environmental health indicators. Since the early 1990s, largely motivated by WHO, increasing attention has been given to constructing indicators on environmental health at all levels from the local to the global scale[40–43], and a number of indicator sets have been created (and to a lesser extent used)[44–46]. Environmental pollution is, inevitably, a major focus of concern in these indicator sets. By definition, also, environmental health indicators provide measures that link environmental hazards and health effect[41]. As such they depend upon an understanding of the association between pollution and health, either in the form of what have been called 'exposure-side indicators', which use information on exposures to imply degrees of health risk, or 'health-side indicators', which use information on health outcome to suggest attributable effects[41,43]. In both contexts, the ability to make at least semi-quantitative interpretations of the link between pollution and health, and thus to assess the contribution to the burden of disease, is assumed.

Mortality

The most explicit attempts to quantify these links, however, have come in recent years through work to estimate the global (and to some extent regional) burden of disease. Earlier efforts in this direction were targeted specifically at making broad-scale enumerations of the total disease burden across the world[47,48]. The traditional measure used for such assessments was mortality, both because data on deaths tended to be more reliable and widely available, and because mortality is directly comparable in terms of health outcome, unlike morbidity which implies differences in severity of effect. Even so, results from the various efforts differed somewhat, largely because of the ways in which gaps and uncertainties in the available data were dealt with[49]. Overall, however, cardiovascular diseases were seen to be the major cause of mortality, accounting for between 19% (based on the World Health Survey estimate) and 28% (based on Murray and Lopez's Global burden of disease study[44]) of total deaths worldwide. Cancer (an estimated 12% in each case), acute respiratory diseases (8.1, 8.7%, respectively), unintentional injuries (5.7, 6.4%), diarrhoeal diseases (*ca.* 5.8%), chronic respiratory diseases (*ca.* 5.7%) and perinatal conditions (6.2, 4.8%) were other major killers.

Years of life lost and disability adjusted life years

Crude estimates of the number and proportion of deaths due to different diseases, of this nature, obviously give only a distorted picture of the true burden of disease, for they take no account of the age of death or the duration of any preceding illness and disability, nor the amount of suffering involved. In an attempt to redress this, Murray and Lopez[48] also computed estimates of the 'years of life lost' (YLL) and 'disability adjusted life years' (DALYs). Years of life lost are estimated as the difference between age at death and the life expectancy in the absence of the disease, based on an advanced developed country (82.5 years for women and 80 years for men at that time). DALYs also incorporate an allowance for the number of years lived with a disability due to disease or injury, weighted according to its severity (based on expert assessments of the relative impact of some 500 different conditions and disease sequelae). The years of disability or life lost are also discounted according to the age of onset (since it is assumed that future years of life lost contribute less to the burden of disease than current ones).

Results of these calculations are summarized and discussed by WHO[49]. Estimates of YYL and DALYs provide a somewhat different ranking of disease compared to crude mortality, since they give additional weight to early-onset diseases and chronic illness. Cardiovascular diseases are thus seen as somewhat less important (making up *ca.* 13% of YYL and 9.7% of DALYs). Acute respiratory diseases (12% and 8.5%, respectively), diarrhoeal diseases (10% and 7.2%) and unintentional injuries (9.3% and 11%) all become proportionally more significant. Mental health conditions also figure as a major source of ill health in terms of DALYs, contributing 11% of the total burden of disease worldwide.

Variations in the global burden of disease

By whichever method they are computed, marked variations are evident in the burden of disease between different sectors of the population. Children are seen to be especially at risk—and young children most of all. More than 30% of all deaths for all diseases in the *Global burden of disease study* occurred to children under 15 years of age; in the case of diarrhoeal diseases they accounted for 88% of deaths, and for acute respiratory illness 67% of deaths. Malaria also struck children disproportionately

(82% of deaths), whilst mortality as a result of perinatal diseases and vaccine-preventable diseases was inevitably almost wholly of children. When measured in terms of DALYs, the overwhelming burden of all these diseases falls on children[49].

Similar inequalities occur both socially and geographically. The World Bank, for example, compared mortality rates and DALYs between poor and rich nations in the world[50]. Clear differences were shown. Whereas ischaemic heart disease, for example, was responsible for 23.4% of deaths in the rich countries, it accounted for only 7.3% in the poor countries; malignant neoplasms were responsible for 22.6% of deaths among rich countries, but only 5.6% among the poor. Conversely, respiratory infections and diarrhoeal diseases accounted for 13.4% and 11.3% of deaths, respectively in poor nations, compared with 4.0% and 0.3% in rich countries; for childhood cluster diseases, the proportions were 7.8% and 0.1%, respectively. Similar disparities have been shown in a comparison of 'less developed' and 'more developed' countries by Smith et al[51]. As these examples show, generalizations about the burden of disease thus need to be interpreted with care. Beneath the often stark global figures lie even starker indications of health inequalities that cry out to be addressed.

The environmental burden of disease

Whilst the original estimates of the global burden of disease made by Murray and Lopez and WHO during the mid 1990s were a major step forward in terms of providing comparable data on health status across the world, they gave information only on health outcomes and did not for the most part attempt to attribute these outcomes to specific causes. Smith et al[51] did, however, make an attempt to assess the environmental contribution to the global burden of disease, using Murray and Lopez's data. This suggested that environmental factors accounted for between 25% and 33% of the total burden of disease, but with a disproportionate share of this falling on children under 5 years of age. Diarrhoeal diseases (for which some 90% of DALYs were attributed to the environment), malaria (ca. 88%) and acute respiratory illness (60%) were seen as outcomes for which the environment was especially influential. Murray and Lopez[52] also made preliminary assessments of the relative importance of different risk factors for the global burden of disease, based on their 1990 data. Malnutrition stood out as the most important factor considered, accounting for ca. 11.7% of deaths and 15.9% of DALYs worldwide. Poor water and sanitation was estimated to be responsible for ca. 5.3% of deaths and 6.8% of DALYs, whereas air pollution contributed 1.1% and 0.5%, respectively. Subsequently, a more specific attempt was made by Prüss et al[53] to assess the effects of water, sanitation and hygiene, using a combination of exposure-based risk assessment and outcome-based disease attribution[54]. Based on available information, they estimated that these environmental factors were responsible for ca. 4% of global mortality and 5.7% of total DALYs. These estimates are somewhat lower than those implied by the original Global burden of disease study[52], partly perhaps because of differences in methodology and partly because of a decline in mortality in the intervening years.

All these attempts to partition the global burden of disease by causative risk factor have faced, and admitted, a number of major difficulties. These relate not only to uncertainties in the available data on health outcome, but also to problems of how to attribute any single death to a single cause or risk factor. Two main approaches have been proposed for disease attribution[55]. Categorical attribution assigns each death to a specific disease or risk factor, according to a defined set of rules (e.g. the ICD system). The advantage of this approach is that it is relatively straightforward and consistent, and avoids double-counting; the disadvantage is that it ignores the multi-factorial nature of many diseases and still leaves unresolved the problem of how to define appropriate rules. Counterfactual attribution involves comparing the current level of disease or mortality with that which might be expected to occur in the absence of the risk factor (or at some other reference level). One of the main difficulties with this approach is how to define this reference level. Several possibilities exist: for example, the complete absence of the risk factor, the level of risk in some reference population or area, or the achievable level of risk with current technologies. Each will tend to give a rather different measure of the attributable burden of disease. In this context, another difficulty also arises, i.e. how to assess the likely change in disease burden under the selected scenario, in the absence of empirical data.

Notwithstanding these difficulties, a revised assessment of the global burden of disease has recently been carried out, involving explicit attempts at attribution by risk factor or hazard[56]. A counterfactual approach was used, with the reference level for each disease being defined as that which would occur under conditions of a minimum theoretical exposure distribution (i.e. that which would achieve the lowest population risk, irrespective of whether this is achievable in practice). Assessments were carried out by a series of expert groups, who first undertook a detailed review of relevant literature, and derived estimates of the exposures and relative risks for specific age and gender groups, for each of 14 sub-regions. Based on these data, estimates were then made of the population impact fraction of the disease or death in each region, for each risk factor:

$$PIF = \frac{\int_{x=0}^{m} RR(x)P(x) - \int_{x=0}^{m} RR(x)P'(x)}{\int_{x=0}^{m} RR(x)P(x)}$$

where PIF is the population impact fraction, $RR(x)$ is the relative risk at exposure level x, $P(x)$ is the population distribution of exposure, $P'(x)$ is the counterfactual distribution of exposure and m is the maximum exposure level.

Results from this assessment for a number of environmental risk factors are summarized in Table 1. More detail on the methods and results are available on the WHO website for a number of the disease groups, including chronic obstructive pulmonary disease, malnutrition and injuries[57], and others will be published as available. The sources of environmental and occupational pollution listed in Table 1 account for 8–9% of the total global burden of disease, measured either in terms of mortality or DALYs. Amongst these risk factors, water, sanitation and hygiene and indoor air pollution are seen to be the most important; health effects of outdoor air pollution are comparatively small, although to some extent this may reflect differences in methodology between this and the other expert groups. It is also evident that a range of other sources of pollution, not included in this table, might be implicated in the global burden of disease, such as exposures to ionizing and non-ionizing radiation, food contamination, pesticides, household hazardous chemicals, wastes and other forms of indoor air pollution. The overall burden of disease attributable to pollution, therefore, cannot yet be assessed.

As with previous estimates of the burden of disease, marked variations can also be recognized between different parts of the world. As is to be

Table 1 Global burden of disease (thousand and percent) attributable to selected sources of environmental and occupational pollution

Risk factor	Deaths		DALYs	
	Number	%	Number	%
Total (all risk factors)	55,861		1,455,473	
Water, sanitation and hygiene	1730	3.1	54,158	3.7
Urban outdoor air pollution	799	1.4	6404	0.4
Indoor smoke from solid fuels	1619	2.9	38,539	2.6
Lead	234	0.4	12,926	0.9
Occupational carcinogens	118	0.2	1183	0.1
Occupational airborne particulates	356	0.6	5354	0.4
Occupational noise	0	0.0	4151	0.3
Total (pollution-related)	4856	8.7	122,715	8.4

Source: based on Ezzati et al[56].

expected, the developing countries bear the major proportion of the burden. Problems of unsafe water, sanitation and hygiene, for example, account for an estimated 6.6% of DALYs in Africa, and 4.7% in south-east Asia, compared with 0.5% in Europe. Indoor air pollution accounts for 4.4% of DALYs in Africa and 3.6% in south-east Asia, compared to 0.4% in Europe. In absolute terms the differences are even more stark. The total number of DALYs per head of population attributable to these two risk factors in Africa are 29.1 per thousand for unsafe water, sanitation and hygiene and 19.3 per thousand for indoor air pollution; in south-east Asia they are 12.8 and 9.9 per thousand, respectively; in Europe they are 0.8 and 0.6 per thousand, respectively. Risks attributable to environmental pollution in the developing world are thus 15–35 or more times greater than in developed countries.

Conclusions

The complexities involved in the link between environmental pollution and health, and the uncertainties inherent in the available data on mortality and morbidity, in existing knowledge about the aetiology of diseases, and in environmental information and estimates of exposure, all mean that any attempt to assess the environmental contribution to the global burden of disease is fraught with difficulties. The estimates produced to date must therefore be regarded as no more than order-of-magnitude estimates. Despite these limitations, however, several conclusions seem beyond refute.

The first is that environmental pollution plays a significant role in a number of health outcomes, and in several cases this adds up to a serious public health concern. Water pollution, sanitation and hygiene, indoor air pollution, and to a lesser extent outdoor air pollution and exposures to chemicals in both the indoor and outdoor environment are all important risk factors in this respect. Ionizing and non-ionizing radiation and noise are also causes for concern in many cases.

Secondly, it is clear that the distribution of risks from these factors is not equal across the world. The global burden of disease may be difficult to quantify, but stark contrasts in that burden are evident between the developed and the developing world, between rich and poor, and often between children and adults. The developed world is not risk-free, and development is no panacea for all environmental health ills. On occasions, in fact, the opposite is true: developments, such as increased reliance on road transport, increased use of chemicals in agriculture, and increased proportions of time spent in modern, hermitically sealed buildings surrounded by chemically-based fabrics and furnishings may actually increase exposures and exacerbate health risks. But overall the developing world is far more severely affected by pollution, and in many instances becoming more so, as pressures from development add to traditional sources of exposure and risk.

Thirdly, and perhaps most importantly, many of these risks and health effects are readily avoidable. Rarely does the solution lie in advanced technologies or even expensive drugs. Instead, the need is for preventive action to reduce the emission of pollutants into the environment in the first place—and that is largely achievable with existing know-how. Indeed, in many cases it has already been implemented in many of the richer countries. Science, therefore, certainly has a role to play in addressing these issues. More research is undoubtedly needed on a range of emerging environmental health issues. But the deficit of action that has allowed environmental pollution still to take its toll on health derives not so much from failures in science or technology as from the lack of political will and economic empowerment. It is from that direction that salvation needs ultimately to come for those at the mercy of environmental pollution.

References

1 Samet JM, Dearry A, Eggleston PA *et al*. Urban air pollution and health inequities: A workshop report. *Environ Health Perspect* 2001; **109** (**Suppl. 3**): 357–74
2 Sexton K, Adgate JL. Looking at environmental justice from an environmental health perspective, *J Expos Anal Environ Epidemiol* 2000; **9**: 3–8
3 Taubes G. Epidemiology faces its limits. *Science* 1995; **269**: 164–9
4 Royal Commission on Environmental Pollution. *Chemicals in products: safeguarding the environment and human health. 24th Report of the Royal Commission on Environmental Pollution.* London: Royal Commission on Environmental Pollution, 2003
5 Lindley SJ, Longhurst JWS, Watson AFR, Conlan DE. Procedures for the estimation of regional scale atmospheric emissions—an example from the NW region of England. *Atmos Environ* 1996; **30**: 3079–91
6 Colvile R, Briggs DJ. Dispersion modelling. In: Elliott P, Wakefield JC, Best NG, Briggs DJ (eds) *Spatial Epidemiology. Methods and Applications.* Oxford: Oxford University Press, 2000; 375–92
7 National Atmospheric Emissions Inventory http://www.naei.org.uk/emissions/emissions_2000.php?action=data (date last accessed 29 June 2003)
8 Jarvis KE, Parry SJ, Piper JM. Temporal and spatial studies of autocatalyst-derived platinum, rhodium, palladium and selected vehicle-derived trace element in the environment. *Environ Sci Technol* 2001; **35**: 1031–6
9 Björklund S. Emissioner av partiklar från dieselfordon och vedförbränning. In: Partiklar och hälsa-ett angeläget problem att undersöka. Skandias miljökommision, rapport number 5. Försäkringsaktiebolaget Skandia, 1996, S-103 50
10 Environment Agency http://www.environment-agency.gov.uk/yourenv/eff/pollution/296030/296054/?lang=_e&region= (date last accessed 29 June 2003)
11 Rushton L. Health hazards and waste management. *Br Med Bull* 2003; **68**: 183–198
12 Department of Food, Environment and Rural Affairs. *The Government's Strategic Review of Diffuse Water Pollution from Agriculture in England. Agriculture and Water: A Diffuse Pollution Review.* London: DEFRA, 2001
13 Beven K, Germann P. Macropores and water flow in soils. *Water Resour Res* 1982; **18**: 1311–25

14 ATSDR. *Landfill Gas Primer: An overview for Environmental Health Professionals.* Washington: Agency for Toxic Substances and Disease Registry, 2001
15 WHO. *Air Quality Guidelines for Europe*, 2nd edn. Copenhagen: WHO Regional Office for Europe, 2001
16 Henshaw DL. Radon exposure in the home: its occurrence and possible health effects. *Contemp Phys* 1993; **34**: 31–48
17 Fawell J, Nieuwenhuijsen MJ. Contaminants in drinking water. *Br Med Bull* 2003; **68**: 199–208
18 Fisher A, Matthews L. The social behaviour of sheep. In: Keeling LJ, Gonyou HW (eds) *Social Behaviour in Farm Animals.* Wallingford: CAB International, 2001
19 Beresford NA, Barnett CL, Crout NMJ, Morris CC. Radiocaesium variability within sheep flocks: relationships between the 137Cs activity concentrations of individual ewes within a flock and between ewes and their progeny. *Sci Total Environ* 1996; **177**: 85–96
20 UK Sustainable Development Commission. *State of Sustainable Development in the UK. Preparatory Paper.* London: UK Sustainable Development Commission, 2001
21 Pirog R, van Pelt T, Enshayan K, Cook E. *Food, Fuel and Freeways. An Iowa Perspective on how Far Food Travels, Fuel Usage and Greenhouse Gas Emissions.* Ames, IA: Leopold Centre, Iowa State University, 2001
22 Jones A. *Eating Oil: Food Supply in a Changing Climate.* London: Sustain and Elm Farm Research Centre, 2001
23 Bro-Rasmussen F. Contamination by persistent chemicals in food chain and human health. *Sci Total Environ* 1996; **188**: S45–S60
24 McFarland M, Kaye J. Chlorofluorocarbons and ozone. *Photochem Photobiol* 1992; **55**: 911–29
25 Joffe M. Infertility and environmental pollutants. *Br Med Bull* 2003; **68**: 47–70
26 WHO. *Mercury, Inorganic. Environmental Health Criteria*, vol. **118**. Geneva: World Health Organization, 1991
27 WHO. *Methylmercury. Environmental Health Criteria*, vol. **101**. Geneva: World Health Organization, 1990
28 Zhang J, Smith KR. Indoor air pollution: A global health concern. *Br Med Bull* 2003; **68**: 209–225
29 Morello-Frosch R, Paster M, Porras C, Sadd J. Environmental justice and regional inequality in Southern California: Implications for future research. *Environ Health Perspect* 2002; **110**: 149–53
30 Waller LA, Louis TA, Carlin BP. Environmental justice and statistical summaries of differences in exposure distributions. *J Expos Anal Environ Epidemiol* 1999; **9**: 56–65
31 Halken S. Early sensitisation and development of allergic airway disease—risk factors and predictors. *Paediatr Respir Rev* 2003; **4**: 128–34
32 Leon DA, Lithell HO, Vagero D, Koupilova I, Mohsen R, Berglund L, Lithell UB, McKeigue PM. Reduced fetal growth rate and increased risk of death from ischaemic heart disease: cohort study of 15,000 Swedish men and women born 1915–29. *BMJ* 1998; **317**: 241–5
33 Rich-Edwards JW, Stampfer MJ, Manson JE, Rosner B, Hankinson SE, Colditz GA, Willett WC, Hennekens CH. Birth weight and risk of cardiovascular disease in a cohort of women followed up since 1976. *BMJ* 1997; **315**: 396–400
34 Suzuki T, Minami J, Ohrui M, Ishimitsu T, Matsuoka H. Relationship between birth weight and cardiovascular risk factors in Japanese young adults. *Am J Hypertens* 2000: **13**: 907–13
35 Elliott P, Briggs DJ, Morris S, de Hoogh C, Hurt C, Kold Jensen T, Maitland I, Richardson S, Wakefield J, Jarup L. Risk of adverse birth outcomes in populations living near landfill sites. *BMJ* 2001; **323**: 363–8
36 Fielder HM, Poon-King CM, Palmer SR, Moss N, Coleman G. Assessment of impact on health of residents living near the Nant-y-Gwyddon landfill site: retrospective analysis. *BMJ* 2000; **320**: 19–22
37 Kello D, Haralanova M, Stern RM, Briggs DJ. National environmental health action plans: background and process. In: Briggs DJ, Stern RM, Tinker TL (eds) *Environmental Health for All. Risk Assessment and Risk Communication for National Environmental Health Action Plans.* Dordrecht: Kluwer Academic Publishers, 1999; 3–15
38 Cizcova H, Kazmarova H, Dumitrescu A, Janikowski R. The NEHAP experience in the Czech Republic, Romania and Poland. In: Briggs DJ, Stern RM, Tinker TL (eds) *Environmental Health for All. Risk Assessment and Risk Communication for National Environmental Health Action Plans.* Dordrecht: Kluwer Academic Publishers, 1999; 17–34

39 Victorin K, Hogstedt C, Kyrklund T, Eriksson M. Setting priorities for environmental health risks in Sweden. In: Briggs DJ, Stern RM, Tinker TL (eds) *Environmental Health for All. Risk Assessment and Risk Communication for National Environmental Health Action Plans.* Dordrecht: Kluwer Academic Publishers, 1999; 35–51

40 Corvalán CF, Kjellström T, Smith KR. Health, environment and sustainable development. Identifying links and indicators to promote action. *Epidemiology* 1999; **10**: 656–60

41 Corvalán C, Briggs D, Zielhuis G (eds) *Decision-making in Environmental Health: From Evidence to Action.* London: E & F.N. Spon, 2000

42 Kjellström T, Corvalán C. Framework for the development of environmental health indicators. *World Health Stat Q* 1995; **48**: 144–54

43 Briggs DJ, Wills J. Presenting decision-makers with their choices: environmental health indicators for NEHAPs. In: Briggs DJ, Stern RM, Tinker TL (eds) *Environmental Health for All. Risk Assessment and Risk Communication for National Environmental Health Action Plans.* Dordrecht: Kluwer Academic Publishers, 1999; 187–201

44 WHO. Environmental Health Indicators: Framework and Methodology. 1999, http://www.who.int/environmental_information/Information_resources/on_line_general.htm (date last accessed 30 June 2003)

45 von Shirnding H. *Health in Sustainable Development Planning: The Role of Indicators.* Geneva: WHO, 2003

46 Briggs DJ. *Making A Difference: Indicators For Children's Environmental Health.* Geneva: WHO; In press

47 WHO. *World Health Report.* Geneva: WHO, 1995

48 Murray CJL, Lopez AD (eds) *The Global Burden of Disease.* Cambridge: Harvard School of Public Health (on behalf of the World Health Organization and World Bank), 1996

49 WHO. *Health and Environment in Sustainable Development. Five Years after the Earth Summit.* Geneva: WHO, 1997

50 Gwatkin DR, Guillot M. *The Burden of Disease Among the Global Poor. Current Situation, Future Trends, and Implications for Strategy.* Washington, DC, USA: The World Bank, 1999

51 Smith KR, Corvalán C, Kjellström T. How much global ill health is attributable to environmental factors? *Epidemiology* 1999; **10**: 573–84

52 Murray CJL, Lopez AD. On the comparable quantification of health risks: lessons from the global burden of disease study. *Epidemiology* 1999; **10**: 594–605

53 Prüss A, Kay D, Fetrell L, Bartram J. Estimating the burden of disease from water, sanitation and hygiene at a global level. *Environ Health Perspect* 2002; **110**: 537–42

54 Prüss A, Corvalán C, Pastides H, de Hollander AEM. Methodologic considerations in estimating burden of disease from environmental risk factors at national and global levels. *Int J Occup Environ Health* 2002; **7**: 58–67

55 Mathers CD, Ezzati M, Lopez AD, Murray CJI, Rodgers AD. Causal decomposition of summary measures of population health. In: Murray CJL, Saloman J, Mathers CD, Lopez AD (eds) *Summary Measures of Population Health.* Geneva: WHO, 2002

56 Ezzati M, Lopez AD, Rodgers A, Vander Hoorn S, Murray CJL, the Comparative Risk Assessment Collaborating Group. Selected major risk factors and global and regional burden of disease. *Lancet* 2002; **360**: 1347–60

57 WHO. http://www.who.int/peh/burden/globalestim.htm (date last accessed 30 June 2003)

The impact of environmental pollution on congenital anomalies

Helen Dolk* and **Martine Vrijheid**[†]

University of Ulster, Newtownabbey, UK and [†]International Agency for Research on Cancer, Lyon, France

Major congenital anomalies are diagnosed in 2–4% of births. In this paper we review epidemiological studies that have specifically looked at congenital anomalies as a possible outcome of community exposure to chemical exposures associated with environmental pollution. These include studies of drinking water contaminants (heavy metals and nitrates, chlorinated and aromatic solvents, and chlorination by-products), residence near waste disposal sites and contaminated land, pesticide exposure in agricultural areas, air pollution and industrial pollution sources, food contamination, and disasters involving accidental, negligent or deliberate chemical releases of great magnitude. We conclude that there are relatively few environmental pollution exposures for which we can draw strong conclusions about the potential to cause congenital anomalies and, if so, the chemical constituents implicated, to provide an evidence base for public health and clinical practice. A precautionary approach should be adopted at both community and individual level. In order to prevent congenital anomalies, one must reduce exposure to potential teratogens before pregnancy is recognized (*i.e.* preconceptionally and in the first few weeks of pregnancy). It is a challenge to develop effective strategies for preconceptional care within the primary care framework. Prenatal service providers and counsellors need to be aware of the uncertainties regarding environmental pollution when addressing parental concerns.

Introduction

Correspondence to:
Helen Dolk, Faculty of Life
and Health Sciences,
University of Ulster at
Jordanstown, Shore Road,
Newtownabbey
BT37 0QB, UK.
E-mail: h.dolk@ulster.ac.uk

In this paper, we review epidemiological studies that have specifically looked at congenital anomalies as a possible outcome of community exposure to chemical exposures associated with environmental pollution. The assessment of whether and to what extent environmental pollution causes birth defects in the population also draws on other evidence, principally toxicological data, data from animal studies, detailed exposure data, and human data from those occupationally exposed to high levels of the chemical. This review does not constitute a risk assessment including these sources of evidence.

Congenital anomalies are part of a spectrum of adverse pregnancy outcomes that may be associated with exposure to environmental pollution. This spectrum also includes fetal death, including early spontaneous abortion, low birth weight associated with prematurity or intrauterine growth retardation, and neurodevelopmental effects that can only be detected in later infancy and childhood.

Environmental pollution can in principle cause congenital anomalies through preconceptional mutagenic action (maternal or paternal) or postconceptional teratogenic action (maternal). Preconceptional mutagenic effects may include chromosomal anomalies and syndromes as a result of new mutations. Postconceptional action, the main focus of this paper, depends on the precise timing of exposure: in embryonic and fetal development, each normal developmental process occurs during a specific period of a few days or weeks, and it is during this 'sensitive period' that exposure to a teratogenic agent may lead to an anomaly. Thus, a particular chemical may cause a congenital anomaly after exposure in, say, the sixth week of development, but exposure during the previous or succeeding week may have no effect, or an anatomically distinct effect. Where a child has more than one anomaly ('multiply malformed'), this may be because exposure has covered a number of sensitive periods for different congenital anomalies, or because exposure at one developmental stage has a number of different effects on organogenesis.

The majority of these sensitive periods occur during the first trimester of pregnancy, when most organogenesis occurs. Thus, as much organogenesis occurs before the pregnancy is even recognized, protection of the embryo cannot rely on actions taken once the woman knows that she is pregnant. Relevant preconceptional exposures may also have postconceptional effects, for example if these are indirect (*e.g.* effects on endocrine function) or if the chemical has a long biological half-life in the body (*e.g.* PCBs). The development of the brain remains subject to adverse influences well into the second trimester and beyond[1,2].

Limitations of the epidemiological literature

Registries of congenital anomaly report 2–4% of births with congenital anomaly, depending on inclusion criteria and ascertainment methods. Cardiac defects account for over one-quarter of all cases, limb anomalies one-fifth, chromosomal anomalies and urinary system anomalies each around 15%, central nervous system anomalies including neural tube defects 10% and oral clefts 7%[3]. The subgroups that have been most commonly singled out for specific study are neural tube defects, oral clefts and more recently cardiac defects and it is therefore unsurprising that these subgroups arise most commonly in reports of pollution-related effects.

Congenital anomalies are a heterogeneous group of many individually quite rare conditions, even within the subgroups mentioned above. When looking for associations between an environmental exposure and many individual congenital anomaly types, one can expect some nominally statistically significant associations by chance alone, but on the other hand low statistical power to detect true associations for any one congenital anomaly type. When individual congenital anomaly types are lumped together to increase statistical power, associations specific to one or two congenital anomaly types may be obscured, especially if groupings are not consistent with what is known of aetiologic and pathogenetic heterogeneity, and if the grouping results in diagnostic noise where well defined conditions are mixed in with variably diagnosed conditions. Many studies only have the statistical power to detect rather large increases in risk, and a negative result must be assessed in this context. On the other hand, some researchers do not declare the process of lumping and splitting of congenital anomaly types that lies behind their presentation of results. Grouping different types together only after seeing results on individual conditions has doubtful statistical validity.

Some progress has been made in identifying pathogenetically more homogeneous groups, such as defects related to vascular disruption or defects related to cranial neural crest cells, where different anatomical types can be combined[4]. There is also discussion as to whether isolated defects and multiple malformations are aetiologically distinct, *e.g.* whether an isolated neural tube defect should be grouped together with cases where a neural tube defect is associated with other malformations[5,6]. 'Sequences' are excluded from the category of multiple malformations, where one malformation is a consequence of or closely related to another, *e.g.* spina bifida with hydrocephalus or clubfoot should be analysed as spina bifida only.

Epidemiologic studies address the problem of variation in diagnosis and reporting of more minor anomalies with greater or lesser success. Most registries employ exclusion lists of 'minor anomalies' which although they may be of relevance to environmental exposures are too inconsistently diagnosed and reported to be useful in population studies[3,4]. However, some malformations range from major to minor forms where thresholds for diagnosis may vary and description of severity is often lacking, *e.g.* microphthalmia, microcephaly, polydactyly or syndactyly. Hypospadias has been of particular recent interest in this regard, since data from a number of areas in the world have shown increasing prevalence in the decades preceding the 1980s, along with a range of other male reproductive abnormalities (see next section). Hypospadias ranges from distal to proximal forms, the distal (glanular) forms being much more common, and it has been difficult to determine whether differences in prevalence between populations in time or geography are real or the result of changes in diagnosis and reporting of distal forms[7].

In the last few decades, the increase in prenatal diagnosis of congenital anomaly followed by termination of pregnancy has presented a particular challenge to epidemiologists. During the period 1995–99 for example, EUROCAT data from 32 European regions showed that 53% of spina bifida cases and 33% of Down Syndrome cases were prenatally diagnosed leading to termination of pregnancy[3], averages which range from 0% in some regions to over 75% in other regions. Since these proportions also vary over time and subregionally, it is important to include terminations of pregnancy following prenatal diagnosis in epidemiological studies, often requiring different sources of information for case ascertainment. The practice of prenatal screening has also brought forward the time of diagnosis of a range of internal congenital anomalies (cardiac and urinary system particularly) leading to increases in reported prevalence of these anomalies in recent decades[3], particularly in areas with more intensive screening.

Many embryos and fetuses with a congenital anomaly are lost as spontaneous abortions, and indeed many chromosomal anomalies are never seen at birth as the malformations are not compatible with continuing *in utero* life. Studies of congenital anomalies diagnosed prenatally or at birth focus on 'survivors' and it is possible that environmental exposures may act on the probability of survival of the malformed fetus, rather than causing abnormal morphological development itself. Although this caution is often raised (especially to explain surprising results such as a negative association between exposure and risk of malformation), direct investigation of such an effect is difficult and it has yet to be established that any environmental exposure acts on differential survival of malformed and normal fetuses. Another consequence of the fact that we analyse congenital anomalies only after a period of 'prenatal selection' is that as prenatal screening and diagnosis are carried out earlier in pregnancy, followed by termination in cases of severe anomaly, so registries of congenital anomalies will pick up cases which would otherwise have been lost as unrecorded spontaneous abortions, particularly for conditions such as Down Syndrome where spontaneous fetal death rates are particularly high. This can introduce artefactual differences in prevalence over time and between areas.

Where teratogenic (malformation causing) effects are concerned, it is usually assumed that there is a threshold of exposure below which the exposure is insufficient to overcome the natural regulatory and repair mechanisms during fetal development and therefore will not lead to a major malformation[8]. This contrasts with the stochastic model for cancer initiation, where ever-smaller doses are simply associated with ever-smaller probabilities of effect and a linear dose–response curve is often adequate. Especially where there is evidence from occupational or animal studies that high dose exposures can be teratogenic, the relevant question is whether chronic low-level pollution reaches the threshold for effect. Biologically, it is reasonable to assume that each individual has their own threshold which an exposure would need to exceed to lead to major disturbance of development. Epidemiologically, it is reasonable to assume that individuals have different thresholds, depending on co-existing or previous environmental exposures as well as genetic susceptibility. Epidemiologic studies determine whether the environmental exposure is sufficient to exceed the thresholds of a significant proportion of the population. Under this model, it is important not to engage in the 'averaging' of exposure used for cancer epidemiology, *i.e.* the distribution of exposure dose in the population matters, and duration may have importance independently of total dose. Thus, the effects in a population with uniform medium exposure may differ from the effects in a population with the same average exposure but where some are unexposed and others highly exposed. This is relevant to the type of exposure modelling one might undertake. For example, wind direction may be important in determining which areas near a factory have the highest average exposure (it may be that further residents downwind experience higher average exposure than nearer residents upwind), but maximum exposure may occur on windless days near the factory and affect all those living closest to the source regardless of wind direction.

If one establishes that an environmental pollutant is causing congenital anomalies, then the next relevant question is usually what genetic and other environmental factors interact with exposure such that only the minority of exposed pregnancies are affected (or why only the minority of individual thresholds have been exceeded). Low periconceptional folate status is now a well-established risk factor for neural tube defects and possibly a much wider range of anomalies[9]. A study of trihalomethane exposures in drinking water looked for (but did not find) interactions with MTHFR genotype, a folate related gene, after suggestions that adverse effects of consumption of water with high trihalomethane levels was restricted to women not taking periconceptional folic acid[10]. Other genes of potential interest are those involved in detoxification of xenobiotics. If genetic susceptibility to the effects of the pollutant is rare, classic epidemiological approaches are unlikely to detect an increase in risk at the population level.

Exposure assessment for studies of environmental pollution is often poor, as discussed elsewhere in this volume. Frequently, measures such as residence near a potentially polluting source are taken as surrogates for exposure rather than a more direct individual measure of exposure, and such measures may relate to the time of birth rather than organogenesis. Poor exposure measurement reduces the power of studies to detect true pollution related effects and the size of the observed risks. One particular problem is that environmental pollution often constitutes a mixture of chemicals, and we may not be sure which constituents to measure in relation to their health effects, or whether indeed the interaction of constituents is important. This also means that it is difficult to assess 'similarity' between communities exposed or potentially exposed to environmental pollution, and thus to assess whether the body of evidence relating to any particular type of pollution suggests an impact on congenital anomalies or not. 'Inconsistent' studies may be inconsistent not only because of statistical noise related to low statistical power or different sources of bias, but the fact that they are investigating different exposures either qualitatively or quantitatively. Moreover, the epidemiological method depends on there being relevant variation in exposure in the population and a study may well be 'negative' if the range of exposures in the study population is so narrow that differences in resulting risks are small, even if the exposure is causally related to congenital anomaly risk.

Environmental epidemiological studies are observational, not experimental, and as such open to 'confounding'. A 'confounder' is a factor associated with both the health outcome (congenital anomalies in this case) and the exposure (environmental pollution). Typically, for example, people of lower social status are more highly exposed to pollution, either because they move where housing prices are lowest, have less power or advocacy skills to prevent exposure, have less access to environmental health information, or because aspects of lifestyle associated with greater deprivation (such as ability to buy bottled water) lead to higher exposure. Although information on the extent to which congenital anomaly prevalence is linked to social status is rather limited, current evidence does suggest that more socio-economically deprived groups have higher non-chromosomal congenital anomaly rates[11], and part of this may be explained by nutritional status. Thus we have to take this into account when interpreting an association between, for example, residence near an incinerator and a raised prevalence of congenital anomaly, as a causal effect of incinerator releases. Since, when assessing environmental pollution, we are usually interested in fairly small increases in risk affecting large numbers of exposed people, socio-economic confounding is a particular problem. Chromosomal anomalies such as Down Syndrome are strongly related to maternal age, and since at present average maternal age increases with social status, higher social status is associated with a greater risk of a Down Syndrome affected pregnancy reversing the direction of socio-economic confounding.

There is a tendency for the sceptical to dismiss observational studies on the basis of uncontrolled confounding, but it is difficult to come up with other realistic strong confounding scenarios for congenital anomalies in relation to established risk factors. Maternal age is strongly related to gastroschisis (young maternal age), and to Down Syndrome and related abnormalities (older maternal age), but not to other non-chromosomal anomalies[3]. Occupational risks and drug related risks are unlikely to have a significant effect at population level as the proportion of the maternal population exposed is small. Differences in folate status are possible between groups exposed and unexposed to environmental pollution, though control for differences in socio-economic status may deal with such differences quite well. Smoking is not a well-established risk factor for congenital anomalies, although it may play a role for specific anomalies.

The scientific and public health literature can be divided into papers pursuing an *a priori* defined hypothesis about a particular pollution source, and studies responding to clusters of congenital anomalies with a suspected local environmental cause. Reports in the media of a 'cluster' of birth defects, often associated with suspected local contamination of air or water, are relatively frequent and may result in a public health investigation. A random distribution of cases in space and time is not a regular distribution, and there will be patches of denser concentration of cases. A community may become aware of an aggregation of cases in their area, and seek the nearest reason such as a waste site or power line. The problem has been likened to the 'Texan sharpshooter' who draws his gun and fires at the barn door, and only afterwards goes and draws the target in the middle of the densest cluster of bullet holes. Since random 'clusters' are expected to occur and there are usually few cases for investigation, some argue that the likelihood of finding a common causal factor is so low that it may often be better not to investigate but instead to 'clean up the mess' of the suspected contaminant without demanding causal proof[12]. Others have tried to derive guidelines for deciding which clusters are worth investigating. Nevertheless, distinguishing random clusters from clusters with a true local environmental cause has proved a generally intractable problem.

Sources of pollution and their impact on congenital anomalies

We have based this section on a Medline and BIDS review, limited to publications in the English language. Space does not permit us to describe or fully reference the studies identified, but the full review is available on request to the authors. We chose to focus in this paper on exposures which have been the subject of a larger number of studies or larger studies (multi-community or multi-site studies), and additionally mention some smaller single community studies which have been particularly influential in creating awareness of the potential impact of different types of environmental pollution. In this way we also seek to avoid the problem of publishing bias, where smaller studies are more likely to be published (and therefore reviewed) if they have positive results.

Most of the relevant epidemiological studies are either case–control studies (where a group of cases with congenital anomaly are compared to a group of controls without, seeking to answer whether a greater proportion of cases than controls have a certain risk factor present) or ecological studies (where in each population subgroup, the frequency/intensity of one or more risk factors is measured, as well as the frequency of birth defects, seeking to answer whether population subgroups with higher levels of a risk factor also have a higher proportion of affected births). Especially where environmental pollution is the subject of study, it cannot be assumed that a case–control study provides better evidence than an ecological study. For example, a case–control study may use as its exposure measurement for each study subject zonal drinking water measurements, and differ little from ecological studies based on these same measurements, except where more detailed individual information on confounders is sought. Difficulties in achieving unbiased control selection are avoided by ecological studies based on all births in the population. On the other hand, it should be clear from the discussion above that there are many reasons why reported prevalence of congenital anomalies may differ between areas, other than environmental exposures, and interpretation of ecological studies should recognize this. Consistency of results between different types of study with different types of bias helps interpret association as causation[13].

Studies of large-scale geographical differences or temporal trends in congenital anomaly prevalence are rarely able to overcome difficulties of interpretation related to differences in diagnostic criteria and ascertainment methods between populations. The higher prevalence of neural tube defects in UK and Ireland compared to the rest of Europe, and the strongly declining total prevalence of neural tube defects (including terminations of pregnancy following prenatal diagnosis) during the 1960s to 1980s in UK and Ireland are however well established phenomena[3,14], possibly related to nutritional factors. The reported prevalence of congenital heart disease and some internal urinary system anomalies has been increasing in Europe in the last two decades, but much of this may be explained by earlier and better diagnosis of these conditions as previously mentioned and against this background it would be difficult to discern exposure-related trends. A worldwide increase in the prevalence of gastroschisis, a rarer anomaly, is without explanation[15]. Recent interest in relation to environmental pollution has focused on the possibility that male reproductive abnormalities, hypospadias and cryptorchidism, have been increasing in prevalence in the decades preceding the 1980s, along with a range of other male reproductive system health outcomes[16]. The difficulty in interpreting hypospadias trends has been mentioned above. Although the evidence for the increasing trends in most of the outcomes is disputed, the possibility that all of these apparent increases are related to increasing exposures to endocrine disrupting chemicals in the environment (present in many of the categories of pollution we discuss below such as pesticides, waste releases and industrial releases) is an active and important area of present research.

Drinking water contamination

Inorganic contaminants in drinking water studied in relation to congenital malformation risk include heavy metals (lead, cadmium, arsenic, barium, chromium, mercury, selenium, silver) other elements, nitrates, nitrites, fluoride and water hardness.

Whereas detailed studies have established beyond doubt the neurotoxic effects of pre- and postnatal lead exposure on children, there is much less evidence concerning the risk of congenital anomalies[17,18]. Two recent studies report conflicting results in relation to neural tube defects and lead in the water supply[19,20], and an Italian study reported a positive association between lead pollution emitted by ceramic factories and the prevalence of cardiovascular anomalies, oral clefts and musculoskeletal anomalies[21].

Large differences in the prevalence of neural tube defects in the British Isles gave rise in the 1970s to theories that the hardness of local water supply could be responsible for this difference. Water hardness is a measurement of total calcium and magnesium levels of the water. This theory was tested in several studies[22,23]. Generally, associations between rates of neural tube defects and water hardness reported in ecological studies have not been substantiated by case–control studies.

In Australia, reports of a high incidence of perinatal mortality due to congenital malformations in one district led investigators to study the local water supply of that area. Nitrates were found to be high in the drinking water in this area, especially in groundwater sources. A case–control study[24] reported an increased risk of congenital malformations for those consuming groundwater and for those consuming water with high levels of nitrates. The nitrate association showed a dose–response effect, supportive of a causal interpretation. Risks for central nervous system and musculoskeletal defects especially were raised for mothers consuming water from groundwater sources. Some subsequent studies have been supportive of a potential effect of high nitrate levels on central nervous system defects[25], anencephaly but not spina bifida[26] and cardiac defects[27].

Early reports in the 1950s suggested that fluoridation of water supplies might result in an increase in the frequency of Down Syndrome. A subsequent comparison of overall Down Syndrome rates in fluoridated and non-fluoridated areas in Massachusetts found no evidence for a difference[28]. Analysis of data from 51 American cities also found no difference in maternal age-specific Down Syndrome rates between fluoridated and non-fluoridated areas[29].

Chlorinated and aromatic solvents (trichloroethylene, benzene) have entered drinking water from leaking underground storage tanks, landfill, and other waste disposal facilities; for example in Woburn, Massachusetts, toxic chemicals (industrial solvents, mainly trichloroethylene) from a waste disposal site were detected in municipal drinking water wells. Residents of Woburn reported a cluster of childhood leukaemia. Lagakos *et al*[30] followed up these findings by compiling an exposure score for residential zones in Woburn, using information on what fraction of the water supply in each zone had come from the contaminated wells annually since the start of the wells. Childhood leukaemia incidence, perinatal deaths, congenital anomalies and childhood disorders were studied in relation to the exposure scores. A significant excess of leukaemia was confirmed and the pregnancy outcome survey found associations with eye/ear congenital anomalies and central nervous system/oral cleft/chromosomal anomalies (mostly Down Syndrome).

Chlorination by-products are halogenated solvents, predominantly trihalomethanes (THM): chloroform, bromodichloromethane, dibromochloromethane and bromoform. The evidence is growing that chlorination by-products may be associated with poor pregnancy outcome[31,32]. Studies of congenital anomalies have reported a range of associations between specific measures of chlorination by-products and specific anomalies, including NTD, oral clefts, cardiac anomalies and urinary tract defects but it is not yet clear which of these represent causal associations[27,31,32].

Waste disposal (landfill sites and incinerators) and contaminated land

Routes of exposure to landfill sites may be through drinking water (see above), other contact with contaminated water, releases to air or contaminated soil. The majority of studies evaluating possible health effects in human populations living near landfill sites investigate communities near one specific waste disposal site ('single-site' studies), frequently in response to concerns from the public about reported contamination from the site, or reported clusters of disease[33]. Love Canal, New York State, brought the world's attention to the potential problems of landfill and contaminated land. Large quantities of toxic materials (residues from pesticide production) were dumped at the landfill during the 1930s and 40s, followed by the building of houses and a school on and around the landfill in the 1950s. By 1977 the site was leaking and chemicals were detected in neighbourhood creeks, sewers, soil, and indoor air of houses. Exposure of Love Canal residents, although not well understood, may have occurred *via* inhalation of volatile chemicals in home air or *via* direct contact with soil or surface water The drinking water supply was not contaminated. Chemicals detected at Love Canal were primarily organic solvents, chlorinated hydrocarbons and acids, including benzene, vinyl chloride, PCBs, dioxin, toluene, trichloroethylene and tetrachloroethylene. A subsequent study interviewing parents reported an increase in birth defects[34].

Sosniak et al[35] investigated the risk of adverse pregnancy outcomes for people living within 1 mile of a total of 1281 NPL sites in USA. The risk for low birth weight and other pregnancy outcomes (infant and fetal death, prematurity, and congenital anomaly) was not associated with living near a site after taking into account a large number of potential confounding factors, including socio-economic variables, collected through questionnaires. However, only around 63% of women originally sampled for the study returned the questionnaire and were included in the study. Also, it is unclear how congenital anomalies were defined and no subgroups of malformations were studied.

Geschwind et al[36] investigated the risk of congenital malformations near 590 hazardous waste sites in New York State. A 12% increase in congenital malformations was found for people living within 1 mile of a site. For malformations of the nervous system, musculoskeletal system, and integument (skin, hair and nails), higher risks were found. Some associations between specific malformation types and types of waste were evaluated, and found to be significant. A dose–response relationship (higher risks with higher exposure) was reported between estimated hazard potential of the site and risk of malformation, adding support to a possible causal relationship. However, a follow-up study of Geschwind's findings found no relation between two selected types of malformations (central nervous system and musculoskeletal) and living near a hazardous waste disposal site[37]. The study did report an increased risk of central nervous system defects for those living near solvent or metal emitting industrial facilities. Subjects for the first 2 years of this study were also included in Geschwind's study, and two more years were added. Marshall et al's[37] attempts to improve the exposure measurement in the first study by assessing the probability of specific contaminant–pathway combinations in 25 sectors of the 1-mile exposure zones were limited by small numbers of cases in each exposure subgroup.

A study by Croen et al[38] in California based exposure measurement on both residence in a census tract containing a waste site and on distance of residence from a site. Three specific types of birth defects: neural tube defects, heart defects and oral clefts were studied. Little or no increase in the risk was found using either measure of exposure. Risk of neural tube (two-fold) and heart defects (four-fold) were increased for maternal residence within a quarter mile of a site although numbers of cases and controls were too small (between two and eight) for these risk estimates to reach statistical significance. Births were ascertained from non-military base hospitals only and the authors point out that the increased risk of NTD may have resulted from military-base residents with NTD-affected pregnancies being more likely to deliver in non-military hospitals than military-base residents with unaffected pregnancies. A subsequent study focusing on ethnic minority infants in California found some small increases in risk of congenital anomaly, although not statistically significant[39].

A European multi-site study reported a 33% increase in risk of all non-chromosomal birth defects combined for residents living within 3 km of 21 hazardous waste landfill sites in 10 European regions[40]. Neural tube defects and specific heart defects showed statistically significant increases in risk. Socio-economic confounding did not readily explain the results. A second part of the study reported a similar increase in risk of chromosomal anomalies (OR 1.41, 95% CI 1.00–1.99)[41]. The study included both open and closed sites that ranged from uncontrolled dumps to relatively modern controlled operations. This disparity makes it difficult at this stage to conclude, if indeed the association is causal, whether risks are related to landfill sites in general or whether specific types of sites may be posing the risks. There was little indication that risk of congenital anomaly was associated with an agreed ranking of the hazard potential of the sites[42].

A study of all landfill sites in England, Scotland and Wales, investigated the risk of congenital malformations, and low and very low birth weight outcomes in populations living within 2 km of a landfill site, open or closed[43]. The study included over 9000 landfill sites. The study found that 80% of the population of Great Britain lived within 2 km of a landfill site. Statistically significant but small (<10%) increases in risk were reported around all sites combined for all congenital anomalies, neural tube defects, hypospadias, abdominal wall defects, and low birth weight. Findings for sites that were licensed to take special (hazardous) waste were generally similar to non-special sites. In this study, only 20% of the country was available as reference population and the comparability of the 'landfill' and 'reference' areas therefore raises questions. Also, if risks were associated with a particular group of 'high-risk' landfill sites such a finding would be lost in the overall comparison of over 9000 sites in this study. Excess risks of some specific anomalies were found in the period before the opening of the landfill sites in the subgroup of sites that opened during the study period.

There has been very little epidemiological study of the risk of congenital malformation or any other pregnancy outcome in populations living near incinerators. Incineration uses controlled combustion to dispose of a wide range of wastes. Airborne pollutants of concern for health impacts include a wide range of inorganic compounds (CO, NO_x, SO_x, HCl); heavy metals, specifically cadmium, lead, mercury, chromium and arsenic; and organic compounds specifically dioxins and furans, polychlorinated biphenyls (PCBs), and polycyclic aromatic hydrocarbons (PAHs)[44]. Incinerators with modern combustion design, practices and air pollution control equipment generally show much reduced emissions compared to old, uncontrolled incineration facilities. A multi-site study using births data from

1956 to 1993 in Cumbria, UK, found excesses in perinatal and infant mortality due to spina bifida and heart defects near incinerators, after controlling for social class[45].

Pesticides in agricultural areas

Non-occupational human exposure to pesticides can occur through domestic use in homes and gardens, consumption of treated foodstuffs, contaminated drinking water, or residence in agricultural areas associated with contaminated air, water or soil. The possible teratogenic effects of pesticides in humans have long been the subject of controversy. Reviews of pesticides and congenital malformations and other pregnancy outcomes have been published elsewhere[46–48]. Studies of residential exposure to pesticides have used proxy measures of exposure such as pesticide usage in the area of residence, residence in or near pesticide application areas, or residence near agricultural crops. A number of positive studies exist relating to residential exposure to agricultural usage, but it is difficult to arrive at any conclusions owing to the wide variety of pesticides and congenital anomalies studied, so that few studies can be compared or combined. An effect of environmental exposure is consistent with growing evidence that occupational exposure to some pesticides may be teratogenic.

Air pollution and industrial pollution sources

There are hardly any studies of the association between ambient air pollution (particulates, sulphur dioxide, nitrogen oxides, carbon monoxide) and the risk of congenital malformation despite growing evidence of a relationship with other pregnancy outcomes such as low birth weight and infant mortality. One recent study reports an increase in risk of cardiac defects, including ventricular septal defects, in relation to carbon monoxide exposure in California[49]. Odds ratios of around 3 were found comparing the lowest with the highest quartile of CO exposure during the second month of pregnancy. There were no relationships with NO_2, ozone, or PM_{10}. Orofacial clefts showed no relationship with any of the pollutants. Other birth defects were not studied.

The Brazilian town of Cubatao is reported to be one of the most polluted in the world. A study[50] compared the rate of congenital malformations in Cubatao with reference rates from a congenital anomaly registry network covering 102 hospitals in South America. A higher than expected prevalence rate was found for polydactyly only.

Five studies have investigated the risk of congenital malformation, in particular central nervous system defects, in areas where PVC polymerization plants were located in the USA and Canada because of prior concern about the potential teratogenicity of vinyl chloride[51–56]. Taken as a whole, these studies have not been able to establish a convincing relationship between residence near the plants and the congenital anomalies studied.

Since there is such a variety of industrial pollution, and so little existing evidence regarding its impact, one approach has been to first take a broad brush look at the impact of industrial pollution in general. Shaw *et al*[57] carried out a multi-site study on the risk of congenital malformations and low birth weight in areas with landfills, chemical dump sites, industrial sites, and hazardous treatment and storage facilities in the San Francisco Bay area, California. A 1.5-fold increase in risk was found for heart and circulatory malformations in the areas classified as having potential human exposure to such sites. Results were not adjusted for socio-economic status. A study of perinatal and infant mortality due to congenital anomalies in Cumbria, UK found no evidence of any increase near hazardous industrial sites, these sites not including landfills or incinerators[58].

A study following up suggested risks of residence near landfill sites found instead increased risks related to residence near solvent and metal emitting sites[37]. A number of specific types of factory or industry have been studied in a single community only. Some of these have found increased prevalence of congenital anomaly, but the evidence is insufficient to draw a causal interpretation.

Contamination of food

Contamination of food (as opposed to component nutrients or the use of preservatives) has been little studied in relation to congenital anomalies. One of the principal areas of current interest is organic mercury and persistent organochlorines in fish. Following disasters involving mercury contamination (see next section) it is known that damage to the fetal brain and microcephaly can result from high exposures[59]. Levels of mercury found in fish have been of concern and led to advice[60] to limit intake of tuna, shark, swordfish and marlin during pregnancy. Epidemiologic studies of adverse effects of *in utero* fish consumption have focused on neurodevelopmental effects rather than the morphological brain anomalies themselves, and brain anomalies would tend to be under-diagnosed and under-ascertained by congenital anomaly registers. A Swedish study found no difference in malformation and fetal death rates[61] between fishing communities on the east coast of Sweden where pollution with persistent organochlorine compounds including PCBs is high, and fishing communities on the west coast where such pollution is low.

Disasters involving environmental pollution

Up to now, we have been reviewing environmental pollution corresponding to normal licensed industrial and agricultural practice, although it is of course possible and even likely that substandard practice exists leading to more pollution than would ordinarily be predicted, either associated with chronic releases or undeclared accidents. We now briefly review some situations that can be classified as 'disasters' involving accidental, negligent or deliberate chemical releases of great magnitude, where there have been subsequent studies of congenital anomalies in the affected population. These studies are not only of interest in relation to the risks associated with such releases, but also shed some light on the risks associated with the chemicals involved for use in risk assessments relating to lower dose environmental exposures.

Methyl mercury is an established teratogen. Mercury poisoning during pregnancy in residents around the Minamata bay in Japan (1953–1971) caused central nervous system anomalies in new-borns[59,62]. Infants born to exposed mothers showed a complex of neurological symptoms, including cerebral palsy, ataxia, disturbed psychomotor development and mental retardation sometimes accompanied by microcephaly. The mothers did not show symptoms. In Iraq (1971–72), grain treated with methyl mercury was consumed by local populations and hundreds of people died from mercury poisoning. Brain damage and neurological effects were found among the infants exposed *in utero*[59,63].

The first indication that PCBs are teratogenic in man came in the consumption of rice oil contaminated with PCBs, causing the 'Yusho' (oil disease) epidemic in Japan in 1968. Apart from PCBs, other contaminants like furans were found in the rice oil. Among 13 exposed mothers, who all had the Yusho disease, two stillbirths were reported and babies were born with skin stains (cola-coloured babies), conjunctivitis and neonatal jaundice[64]. All live born babies were also below the mean weight for gestational age. After poisoning of cooking oil with PCBs (contaminated by dibenzofurans) in Taiwan in 1979, similar effects were noted: exposed children were shorter and lighter, and had skin, nail and teeth anomalies[65]. One study reported a very high rate of infant death in babies who had been born with hyperpigmentation (8/38, 20.5%)[66].

In 1976, an accident at a factory producing trichlorophenol released a cloud of toxic materials including 2,3,7,8-tetrachlorodibenzo-*p*-dioxin (TCDD) in the area of Seveso, Italy. The quantity of TCDD released into the environment was large but was never precisely determined. The prevalence of chloracne (an acute effect of exposure) was high in residents in the area immediately surrounding the factory, zone A, and lower but still raised in two areas with low and very low exposure, areas B and

R. Mastroiacovo et al[67] studied birth defects in the Seveso population up to 5 years after the accident. A total of 26 babies were born in zone A, none of which had major malformations, two of which had minor defects. Rates of major malformations in zone B and R were similar to those in a control area. The number of births in the highly exposed area was too small to draw firm conclusions about anything but the absence of very high absolute risks.

Agent Orange is a mixture of 2,4-D and 2,4,5-T sprayed by the US military in Vietnam as a defoliant to aid military manoeuvres, and thus exposing residents of those areas. Evidence for adverse reproductive effects of dioxins, Agent Orange and phenoxy herbicide spraying have been reviewed[68–71]. Two types of studies were carried out: studies of reproductive outcomes in couples who lived in areas that were sprayed with Agent Orange and studies of Vietnamese, American and Australian soldiers who served in South Vietnam, which in the context of this review of non-occupational exposures may serve as an indicator of community exposures of Vietnamese residents, though mainly limited to the potential for paternal preconceptional mutagenic effects. Rates of birth defects reported in the studies of Vietnamese residents were low compared to those found in Western countries, raising concerns about completeness of ascertainment. Although the quality of the studies is difficult to judge, they do report increased risks of miscarriages, stillbirths, molar pregnancy and birth defects in residents of the sprayed areas. Some increases in prevalence of specific congenital anomalies in offspring of fathers who served as soldiers were observed but no overall increase in anomalies and no clear dose–response relationship with dioxin exposure.

A study in north Cornwall examined outcomes of pregnancy after an incident where aluminium sulphate was added to the local water supply accidentally[72]. Ninety-two pregnancies in the affected area during the contamination incident were compared to pregnancies in the area before the incidents and pregnancies in an unexposed area. The study reports no excess of perinatal deaths, low birth weight, preterm birth, or congenital malformation in the affected area. There was however an increased rate of talipes among the exposed pregnancies.

In a Hungarian village, 11 out of 15 live births in a 2-year period were affected by congenital abnormalities and six were twins[73]. Four out of the 11 affected children had Down Syndrome. Trichlorofon (an organophosphate pesticide) was used in excess at local fish farms, and local people were known to have eaten the local fish. No congenital abnormalities occurred in the 2 years after the chemical treatment of fish was banned. The study did not manage to establish a full explanation for the cluster of congenital anomalies in terms of fish consumption, but the evidence was suggestive that Trichlorofon was at least a partial explanation.

Implications for clinical and public health practice

There are relatively few environmental pollution exposures for which we can draw strong conclusions about the potential to cause congenital anomalies, and, if so, the chemical constituents implicated, to provide an evidence base for public health and clinical practice. We lack a coherent surveillance strategy for licensed agricultural and industrial processes, and are necessarily in an even poorer position in relation to illegal practices. There is however enough evidence for the prevention of congenital anomalies to figure strongly among the health-related arguments reviewed in this volume for taking a precautionary approach and implementing measures to reduce community exposures to environmental pollution. High-risk groups with maximum exposure should in particular be identified.

Environmental pollution is almost by definition a by-product of agricultural and industrial processes in which we engage to increase our health and welfare, and there is therefore a balance between risks and benefits at both community and individual level. At community level, it is relevant to ask whose health and welfare is being improved, and who is suffering the negative consequences, and thus whether the benefits and risks are fairly distributed. At individual level, balancing decisions must also be made—for example, it might be counterproductive to limit intake of fresh fruit and vegetables in order to avoid pesticide exposure (although measures such as washing and peeling can be encouraged), or to limit recreation and exercise through swimming, walking or cycling in order to avoid exposure to contaminants of air and water.

In terms of clinical practice, one needs to tread a fine line between false reassurance and false alarm when counselling a future or current parent on the basis of current evidence. 'Absence of evidence is not evidence of absence'. It needs to be clear to parents that mechanisms are not necessarily in place to protect them against teratogenic exposures, especially weaker teratogenic exposures. A wise approach is to reduce personal exposure to environmental pollutants where possible and also to reduce personal exposure to chemical exposures in the home and garden. Periconceptional folic acid supplementation, one of the very few known preventive measures, can be advised both to reduce overall risk of congenital anomaly and also possibly to protect against risks associated with specific environmental exposures, although there is as yet little evidence to support this interaction. However it should be emphasized that the prevalence of major congenital anomalies is approximately 2% of births (or 2 in 1000 births for neural tube defects, and 2 in 10,000 for gastroschisis) and thus, even in relation to a doubling of risk, an exposed pregnant woman is still unlikely to have a child affected by a major congenital anomaly. Estimations of risks related to residence near landfill sites have for example been less than 1.5-fold.

Environmental concerns should be incorporated into both preconceptional and prenatal care. In order to prevent congenital anomalies, one must reduce exposure to potential teratogens before pregnancy is recognized (*i.e.* preconceptionally and in the first few weeks of pregnancy). This should be one part of the public health function in primary care. Targeting women preconceptionally has been found to be poor in another area of congenital anomaly prevention, periconceptional folic acid supplementation[74]. It is a challenge to develop effective strategies for preconceptional care within the primary care framework. Many women go on to have more than one child, and community midwives and health visitors are already ideally placed, as part of interagency working and with appropriate training, to give environmental health education and to refer at risk families for specialist interventions. Important opportunities are currently being missed even in regard to well-known environmental hazards, for example to verify that domestic water has low lead levels. Relevant prenatal services include prenatal screening and prenatal counselling. Prenatal service providers and counsellors need to be aware of the uncertainties regarding environmental pollution when addressing parental concerns.

Acknowledgements

We thank Maria Loane for literature review assistance and Mike Joffe for helpful editorial comments. The literature review was carried out with funding from the Department of Health; the views expressed in this publication are those of the authors and not necessarily those of the funding department.

References

1   Evrard P, Kadhim HJ, Saint-Goerges P, Gadisseux JF. Abnormal development and destructive processes of the human brain during the second half of gestation. In: Evrard P, Minkowski A (eds) *Developmental Neurobiology*. NY: Raven Press, 1989
2   Otake M, Schull WJ. In utero exposure to a bomb radiation and mental retardation; a reassessment. *Br J Radiol* 1984; **57**: 409–14
3   EUROCAT Working Group. EUROCAT Report 8: *Surveillance of Congenital Anomalies in Europe 1980–1999*. www.eurocat.ulster.ac.uk; University of Ulster, 2002
4   Rasmussen SA, Olney RS, Holmes LB *et al*. Guidelines for case classification for the National Birth Defects Prevention Study. *Birth Defects Res (Part A)* 2003; **67**: 193–201
5   Khoury JM, Erickson JD, James LM. Etiologic heterogeneity of NTD: clues from epidemiology. *Am J Epidemiol* 1982; **115**: 538–48
6   Dolk H *et al*. Heterogeneity of neural tube defects in Europe: the significance of site of defect and presence of other major anomalies in relation to geographic differences in prevalence. *Teratology* 1991; **44**: 547–59
7   Dolk H, Vrijheid M, Scott JES *et al*. Towards the effective surveillance of hypospadias. *Environ Health Perspect* 2003; doi 10.1289/ehp.6398 or http://ehp.niehs.nih.gov/docs/admin/newest.html

8 Wilson JG. Current status of teratology. In: Wilson JG, Fraser FC (eds) *Handbook of Teratology. I. General Principles and Etiology*. NY: Plenum Press, 1977; 47–74
9 WHO. Folic acid and the prevention of neural tube defects: from research to public health practice. Report of a WHO/EURO Meeting on the Regional Policy for Prevention of Congenital Anomalies, Rome, Italy, 11–12 November 2002. WHO, 2003; In press
10 Shaw GM, Ranatunga D, Quach T, Neri E, Correa A, Neutra RR. Trihalomethane exposures from municipal water supplies and selected congenital malformations. *Epidemiology* 2003; **14**: 191–9
11 Vrijheid M, Dolk H, Stone D, Abramsky L, Alberman E, Scott JES. Socioeconomic inequalities in risk of congenital anomaly. *Arch Dis Child* 2000; **82**: 349–52
12 Rothman KJ. A sobering start for the cluster busters' conference. *Am J Epidemiol* 1990; **132** (**Suppl 1**): S6–S13
13 Bradford Hill A. The environment and disease: association or causation? *Proc R Soc Med* 1965; **58**: 295–300
14 EUROCAT Working Group. Prevalence of neural tube defects in 20 regions of Europe and the impact of prenatal diagnosis, 1980–86. *J Epidemiol Community Health* 1991; **45**: 52–8
15 Di Tanna GL, Rosano A, Mastroiacovo P. Prevalence of gastroschisis at birth: a retrospective study. *BMJ* 2002; **325**: 1389–90
16 Toppari J, Larsen JC, Christiansen P *et al*. Male reproductive health and environmental xenoestrogens. *Environ Health Perspect* 1996; **104**: 741–803
17 Winder C. Lead, reproduction and development. *Neurotoxicology* 1993; **14**: 303–18
18 Bellinger D. Teratogen update: lead. *Teratology* 1994; **50**: 367–73
19 Bound JP, Harvey PW *et al*. Involvement of deprivation and environmental lead in neural tube defects: a matched case–control study. *Arch Dis Child* 1997; **76**: 107–12
20 Macdonell JE, Campbell H *et al*. Lead levels in domestic water supplies and neural tube defects in Glasgow. *Arch Dis Child* 2000; **82**: 50–3
21 Vinceti M, Rovesti S, Bergomi M *et al*. *Sci Total Environ* 2001; **278**: 23–30
22 Elwood JM, Little J, Elwood JH. *Epidemiology and Control of Neural Tube Defects*. Monographs in Epidemiology and Biostatistics no. 20. Oxford University Press, 1992
23 Aschengrau A, Zierler S *et al*. Quality of community drinking water and the occurrence of late adverse pregnancy outcomes. *Arch Environ Health* 1993; **48**: 105–13
24 Dorsch MM, Scragg RKR *et al*. Congenital malformations and maternal drinking water supply in rural south Australia: a case–control study. *Am J Epidemiol* 1984; **119**: 473–86
25 Arbuckle TE, Sherman GJ *et al*. Water nitrates and CNS birth defects: a population based case–control study. *Arch Environ Health* 1988; **43**: 162–7
26 Croen LA, Todoroff K, Shaw GM. Maternal exposure to nitrate from drinking water and diet and risk for neural tube defects. *Am J Epidemiol* 2001; **153**: 325–31
27 Cedergren MI, Selbing AJ, Lofman O, Kallen BA. Chlorination byproducts and nitrate in drinking water and risk for congenital cardiac defects. *Environ Res* 2002; **89**: 124–30
28 Needleman HL, Pueschel SM *et al*. Fluoridation and the occurrence of Down's syndrome. *N Engl J Med* 1974; **291**: 821–3
29 Erickson JD. Down syndrome, water fluoridation, and maternal age. *Teratology* 1980; **21**: 177–80
30 Lagakos SW, Wessen BJ *et al*. An analysis of contaminated well water and health effects in Woburn, Massachusetts. *J Am Stat Assoc* 1986; **81**: 583–96
31 Nieuwenhuijsen MJ, Toledano MB, Eaton NE *et al*. Chlorination disinfection byproducts in water and their association with adverse reproductive outcomes: a review. *Occup Environ Med* 2000; **57**: 73–85
32 Bove F, Shim Y, Zeitz P. Drinking water contaminants and adverse pregnancy outcomes: a review. *Environ Health Perspect* 2002; **110**: 61–74
33 Vrijheid M. Health effects of residence near hazardous waste landfill sites: a review of epidemiologic literature. *Environ Health Perspect* 2000; **108** (**Suppl 1**): 101–12
34 Goldman LR, Paigen B *et al*. Low birth weight, prematurity and birth defects in children living near the hazardous waste site, Love Canal. *Hazardous Waste Hazardous Mater* 1985; **2**: 209–23
35 Sosniak WA, Kaye WE *et al*. Data linkage to explore the risk of low birthweight associated with maternal proximity to hazardous waste sites from the National Priorities List. *Arch Environ Health* 1994; **49**: 251–5
36 Geschwind SA, Stolwijk JAJ *et al*. Risk of congenital malformations associated with proximity to hazardous waste sites. *Am J Epidemiol* 1992; **135**: 1197–207

37 Marshall E, Gensburg L *et al*. Maternal residential exposure to hazardous waste sites and risk of central nervous system and musculoskeletal birth defects. *Epidemiology* 1997; **6**: S63
38 Croen LA, Shaw GM *et al*. Maternal residential proximity to hazardous waste sites and risk of selected congenital malformations. *Epidemiology* 1997; **8**: 347–54
39 Orr M, Bove F, Kaye W, Stone M. Elevated birth defects in racial or ethnic minority children of women living near hazardous waste sites. *Int J Hyg Environ Health* 2002; **205**: 19–27
40 Dolk H, Vrijheid M *et al*. Risk of congenital anomalies near hazardous-waste landfill sites in Europe: the EUROHAZCON study. *Lancet* 1998; **352**: 423–7
41 Vrijheid M, Dolk H *et al*. Risk of chromosomal congenital anomalies in relation to residence near hazardous waste landfill sites in Europe. *Lancet* 2002; **359**: 320–2
42 Vrijheid M, Dolk H, Armstrong B *et al*. Hazard potential ranking of hazardous waste landfill sites and risk of congenital anomalies. *Occup Environ Med* 2002; **59**: 768–76
43 Elliott P, Briggs D *et al*. Risk of adverse birth outcomes in populations living near landfill sites. *BMJ* 2001; **829**: 363–8
44 National Research Council. *Waste Incineration and Public Health*. Washington, DC: National Academy Press, 2000
45 Dummer TJ, Dickinson HO, Parker L. Adverse pregnancy outcomes around incinerators and crematoriums in Cumbria, north west England, 1956–93. *J Epidemiol Community Health* 2003; **57**: 456–61
46 Garcia AM. Occupational exposure to pesticides and congenital malformations: A review of mechanisms, methods, and results. *Am J Ind Med* 1998; **33**: 232–40
47 Sever LE, Arbuckle TE *et al*. Reproductive and developmental effects of occupational pesticide exposure: the epidemiologic evidence. *Occup Med* 1997; **12**: 305–25
48 Nurminen T. The epidemiologic study of birth defects and pesticides. *Epidemiology* 2001; **12**: 145–6
49 Ritz B, Yu F *et al*. Ambient air pollution and risk of birth defect in Southern California. *Am J Epidemiol* 2002
50 Monteleone Neto R, Castilla EE. Apparently normal frequency of congenital anomalies in the highly polluted town of Cubatao, Brazil. *Am J Med Genet* 1994; **52**: 319–23
51 Theriault G, Iturra H *et al*. Evaluation of the association between birth defects and exposure to ambient vinyl chloride. *Teratology* 1983; **27**: 359–70
52 Rosenmann ND, Rizzo E *et al*. Central nervous system malformations in relation to two polyvinyl chloride production plants. *Arch Environ Health* 1989; **44**: 279–82
53 Edmonds LD, Anderson CE *et al*. Congenital central nervous system malformations and vinyl chloride monomer exposure: a community study. *Teratology* 1978; **17**: 137–42
54 Edmonds LD, Falk H *et al*. Congenital malformations and vinyl chloride. *Lancet* 1975; **2**: 1098
55 Infante PF, Wagoner JK *et al*. Carcinogenic, mutagenic and teratogenic risks associated with vinyl chloride. *Mutat Res* 1976; **41**: 131–42
56 Infante PF, Wagoner JK *et al*. Genetic risks of vinyl chloride. *Lancet* 1976; **1**: 734–5
57 Shaw GM, Schulman J, Frisch JD *et al*. Congenital malformations and birthweight in areas with potential environmental contamination. *Arch Environ Health* 1992; **47**: 147–54
58 Dummer TJ, Dickinson HO, Parker L. Prevalence of adverse pregnancy outcomes around hazardous industrial sites in Cumbria, north-west England, 1950–93. *Paediatr Perinat Epidemiol* 2003; **17**: 250–5
59 Clarkson TW. The three modern faces of mercury. *Environ Health Perspect* 2002; **110 (Suppl 1)**: 11–23
60 Committee on Toxicity of Chemicals in Food, Consumer Products and the Environment. *Statement on a Survey of Mercury in Fish and Shellfish*. Food Standards Agency, UK, 2002
61 Rylander L, Hagmar L. No evidence for congenital malformations or prenatal death in infants born to women with a high dietary intake of fish contaminated with persistent organochlorines. *Int Arch Occup Environ Health* 1999; **72**: 121–4
62 Harada M. Congenital Minamata disease: intrauterine methylmercury poisoning. *Teratology* 1978; **18**: 285–8
63 Marsh DO, Myers GJ *et al*. Fetal methylmercury poisoning: clinical and toxicological data on 29 cases. *Ann Neurol* 1980; **7**: 348–53
64 Kuratsune M, Yoshimura T *et al*. Epidemiologic study on Yusho, a poisoning caused by ingestion of rice oil contaminated with a commercial brand of polychlorinated biphenyls. *Environ Health Perspect* 1972; **1**: 119–28

65 Rogan WJ, Gladen BC et al. Congenital poisoning by polychlorinated biphenyls and their contaminants in Taiwan. *Science* 1988; **241**: 334–6
66 Hsu ST, Ma CI et al. Discovery and epidemiology of PCB poisoning in Taiwan: a four-year follow-up. *Environ Health Perspect* 1985; **59**: 5–10
67 Mastroiacovo P, Spagnolo A et al. Birth defects in the Seveso area after TCDD contamination. *JAMA* 1988; **259**: 1668–72
68 Lilienfeld DE, Gallo MA. 2,4-D, 2,4,5-T, and 2,3,7,8-TCDD: an overview. *Epidemiol Rev* 1989; **11**: 28–58
69 Skene SA, Dewhurst IC et al. Polychlorinated dibenzo-*p*-dioxins and polychlorinated dibenzofurans: the risks to human health: A review. *Hum Toxicol* 1989; **8**: 173–203
70 Sterling TD, Arundel A. Review of recent Vietnamese studies on the carcinogenic and teratogenic effects of phenoxy herbicide exposure. *Int J Health Serv* 1986; **16**: 265–78
71 Constable JD, Hatch MC. Reproductive effects of herbicide exposure in Vietnam: recent studies by the Vietnamese and others. *Teratog Carcinog Mutagen* 1985; **5**: 231–50
72 Golding J, Rowland A et al. Aluminium sulphate in water in north Cornwall and outcome of pregnancy. *BMJ* 1991; **302**: 1175–7
73 Czeizel A, Elek C, Gundy S et al. Environmental trichloroform and cluster of congenital abnormalities. *Lancet* 1993; **341**: 539–42
74 EUROCAT Special Report. *Periconceptional Folic Acid Supplementation and Prevention of Neural Tube Defects*. www.eurocat.ulster.ac.uk; University of Ulster, 2003

Infertility and environmental pollutants

Michael Joffe

Department of Epidemiology and Public Health, Imperial College London, London, UK

While it has long been known that female fertility is impaired by oestrogen exposure, it is unclear whether environmental pollutants with weak oestrogenic effects are sufficiently potent and prevalent to have biological effects in humans. Male fertility, or sperm concentration at least, appears to have deteriorated, and there is substantial spatial variation at both national and global level, as well as a genetic component. Sperm morphology and motility are implicated too. There is good evidence for an increase in testicular cancer, and possibly in other conditions that certain spatial characteristics plus evidence on heritability suggest are linked to impaired spermatogenesis. A candidate agent would need to have started increasing in the early 20th century. Weak environmental oestrogens are not responsible. Candidates include agents affecting endogenous maternal oestrogen levels, environmental anti-androgens (although these cannot explain the epidemiological findings), and dioxin and related compounds. Genetic damage should be considered as a unifying hypothesis, possibly focused on the Y-chromosome.

Introduction

Correspondence to:
Michael Joffe,
Department of
Epidemiology and Public
Health, Imperial College
London, St Mary's
Campus, Norfolk Place,
London W2 1PG, UK.
E-mail: m.joffe@
imperial.ac.uk

A review of the literature on the determinants of fertility raises more questions than can be answered in the current state of knowledge. Even so, at least this is an advance on the situation that previously existed, when the topic was largely ignored. This state of affairs is mainly the result of the lack of priority that has been given to research on reproduction in basic biology, as well as in epidemiology and toxicology.

Reproduction and/or development can be affected by exposure to a wide variety of agents, including dioxins, poly-chlorinated biphenyls (PCBs), phytoestrogens such as isoflavones, heavy metals (*e.g.* organic mercury, lead); chlorination disinfection by-products in water, organic solvents, poly-aromatic hydrocarbons, particulate air pollution, substances emitted from landfill sites and caffeine. Often, effects on the reproductive system and/or infant development have been detected at lower doses of the substance in question than is the case for other endpoints. Most of these agents have featured in recent discussions of official regulatory bodies, which have highlighted the paucity of good quality epidemiological evidence.

In principle, fertility can be studied in humans or in laboratory animals by using biomarkers such as measures of semen quality, and/or using a functional measure of the ability to conceive. In practice, for many exposures, little or no epidemiological research has been carried out with fertility as an endpoint, and the toxicology database is typically seriously incomplete in this respect as well.

The much-publicised concern over the possibility of 'falling sperm counts' has altered the position in the past 10 years, but during that time the research initiative has been dominated by just one hypothesis, that endocrine disrupting agents are responsible. As a result, we are in the situation of having the following rather fragmentary information:

(i) the effects of a few known agents on fertility in either sex;

(ii) descriptive epidemiology, which suggests that there may be a problem with male fertility, and that this is likely to be linked to deterioration in related conditions such as testicular cancer; it also indicates when and where such a problem may be at its greatest, so that one can begin to outline the epidemiological characteristics of the responsible agent(s);

(iii) some additional observations that allow preliminary assessment of the type of biological process underlying any such effect, in particular whether endocrine disruption or genetic damage are responsible.

This paper reviews these three areas in turn, concentrating on epidemiology. Evidence from toxicology, endocrinology, genetics and research on wildlife is not presented, but informs the discussion. Priority is given to agents or processes that could affect whole populations, or substantial proportions of whole populations, rather than relatively small groups such as those with occupational exposure. The next section focuses on effects of known agents. As the main evidence for widespread impairment comes from epidemiological observations on males, the following section discusses this in detail. In view of the limited data available on male fertility itself, epidemiological information is also presented on possibly parallel variations in other disorders of the male reproductive system that arise early in life, especially on testicular cancer. Genetics is covered, as well as temporal and spatial variation. The available evidence is that the epidemiological observations are not explicable in terms either of the known agents or of agents that are currently known to have endocrine disrupting effects.

Effects of known agents

Infertility in females

In the 1940s, reduced fertility was noticed in Australian ewes, and it was established that this was due to the clover that they were grazing.

This condition, which came to be known as 'Clover Disease', was traced to phytoestrogens—plant compounds that have oestrogenic properties—in the clover. Ewes feeding on Australian clover developed abnormal plasma concentrations of endogenous hormones, with subsequent reduced fertility[1].

In humans, dietary phytoestrogen consumption is considerable in some populations, and constitutes by far the major route of exposure to exogenous 'endocrine disruptors' (Table 1). Probably the most important of these are isoflavones such as genistein, which occur in legumes and are particularly abundant in soybeans and soy-based foodstuffs. The isoflavone

Table 1 Principal exogenous substances that may affect sex hormone function

A. Oestrogenic effects
1. *High potency—pharmacological agents*
- DES (diethylstilboestrol)
- ethinyl oestradiol (component of contraceptive pill)

2. *Medium potency—dietary phytoestrogens*
- isoflavones, *e.g.* genistein, daidzein
- coumestans, *e.g.* coumestrol
- lignans

3. *Low potency—environmental or occupational agents*
- bisphenol A
- octylphenol and nonylphenol

  pesticides, including chlordecone, DDT, dieldrin, endosulphan, p,p'-methoxychlor, toxaphene

B. Anti-androgenic effects
- p,p'-DDE (the major breakdown product of DDT)
- certain phthalates, *e.g.* DBP, DEHP
- pesticides, including linuran, procymidone, metabolites of vinclozolin
- hydroxyflutamide

C. Others
- dioxins, furans and 'dioxin-like' PCBs

Adapted with permission from Joffe[20].
This classification is an over-simplification: it conflates receptor-mediated effects with those due to other mechanisms, *e.g.* interference with hormone synthesis. Moreover, several of the 'oestrogens' show considerable affinity for the androgen receptor. In addition, many of these compounds have important biological actions that are not endocrine in mechanism.
For the reasons given in the text, compounds in the 'Low potency' category cannot plausibly be considered responsible for the types of impairment of the male reproductive system considered in this paper, and the exposures may also be too low to affect females.

content of soy varies in relation to many factors, including plant species, strain, crop year and geographical location. The concentrations are sufficient to cause biological effects in humans, even after cooking or other processing. Populations that traditionally consume large quantities of soy, notably Chinese and Japanese people, tend to have relatively high phytoestrogen exposure. Less is known about other types of phytoestrogen; exposure to lignans is probably widespread, but the potency is lower and may not have biological significance[1].

In rodents, exposure to isoflavones and other phytoestrogens has been shown to alter a number of functions of the female reproductive system, including advancement of puberty, subfertility and irregular oestrus cycling. Perinatal, neonatal or prepubertal exposure appears to produce the most marked effects. It is unclear to what extent these findings are relevant to humans, owing to species differences in sexual development, experimental considerations such as route of administration, and uncertainty over comparability of plasma concentrations[1].

A different source of oestrogen exposure was the synthetic compound, diethylstilboestrol (DES), which has oestrogenic potency comparable to that of oestradiol. From the late 1940s onwards, it was widely used during pregnancy, especially in the USA, in the belief that it could prevent miscarriage and a range of pregnancy complications. It is estimated that more than two million women were exposed to this drug. Pharmacological doses were given, often at the stage of pregnancy during which the sexual organs develop.

A randomized controlled trial published in 1953[2] showed that DES was ineffective for the conditions for which it was being prescribed. However, clinical use of the drug continued until it was banned in 1971, after the discovery that *in utero* exposure of female fetuses led to a risk of developing clear cell adenocarcinoma some 15 years later[3]. While this particular risk is fortunately rare, DES-exposed girls have reproductive tract anomalies, and they subsequently have reduced fertility and increased rates of ectopic pregnancy, spontaneous abortion and preterm delivery[3].

The sensitivity of the developing female reproductive tract to oestrogens raises the question of whether exposure to environmental chemicals having oestrogenic activity (Table 1) might affect fertility through the female route. There is insufficient evidence to answer this question definitively, as research on these exposures has tended to focus on the male, even though toxicological experiments consistently find stronger effects of oestrogens on females than males[1]. As their potency combined with exposure concentrations are many orders of magnitude lower than endogenous hormones, or even than phytoestrogen intake in oriental populations, a strong effect seems unlikely.

Aside from the specific question of oestrogens, high maternal but not paternal consumption of sport fish from the heavily polluted Lake Ontario has been associated with reduced fertility as measured using TTP (see below), but the findings are inconsistent. This could be due to PCBs, or to other pollutants (including oestrogens)[4].

Other specific populations who have high exposures, generally occupational, to particular agents have been identified as having an increased risk of subfertility. Such agents include solvents[5,6], inorganic mercury[7], nitrous oxide[8,9] and antineoplastic agents[10], but these do not have implications for environmental exposures to the same substances, which are too low.

Infertility in males

The nematocide dibromochloropropane (DBCP), used for soil fumigation on fruit plantations, is a potent testicular toxin. This was discovered when the wives of occupationally exposed male workers were discussing the problems they had had in becoming pregnant, and was subsequently confirmed by epidemiological studies[11]. High exposure causes permanent azoospermia. DBCP was banned in the late 1970s in the USA, although it is still used elsewhere.

The other exposures that are known to affect fertility in human males are also predominantly occupational, and include other pesticides, and heavy metals such as lead and cadmium[12]. That these risks are not generalizable to the general population (*e.g. via* pesticide residues) is illustrated by evidence for a threshold for inorganic lead and sperm concentration: no effect was seen below a blood level of about 44 µg/dl[13]—and even in the occupational context few men have higher exposures than this, in the economically developed world.

Descriptive epidemiology: conditions affecting the male reproductive system

Fertility and semen quality

Two types of endpoint can be studied: semen quality, and fertility as measured by the time taken to conceive (Time To Pregnancy, or TTP). TTP reflects the probability of conception for couples having unprotected intercourse. It is a functional measure of biological fertility at the level of the couple, and validity studies have shown that it can be studied retrospectively as well as prospectively[14]. Care in the design and analysis of TTP studies is important, to avoid potential pitfalls.

Methodological issues also affect the interpretation of studies of semen quality, which is usually taken to include sperm concentration, motility and morphology. All are subject to large degrees of within-person biological variation and/or measurement error that varies between centres and very likely over time. In addition, representative samples of the general population, which are so important for descriptive epidemiology, are unachievable as participation rates are too low. The best evidence is from candidates for semen donation and for vasectomy; data from men in contact with medical services for a fertility-related problem, or from those accepted for semen donation, are too unreliable to use.

Trends

For various reasons, long-term trends in fertility and semen quality are difficult to confirm. One study has found that fertility as measured by TTP increased (not decreased, as was hypothesized) in the period 1961–93, based on a representative sample of the British population[15]. However, firm conclusions cannot be drawn from a single report. Another study, from Sweden, that reported a decline in clinical subfertility in 1983–93[16] was shown to be likely to have resulted from truncation bias that had not been allowed for[17].

The 'time trend' debate regarding semen quality has focused mainly on sperm concentration. A much-cited paper published in 1992 reviewed the world literature, relating this variable to the date of publication[18]—a crude exercise in terms both of the methodology and of the hypothesis, which treated location as irrelevant. Its claim of a 50% decline in mean concentration over 50 years, from 113 to 66 million/ml, should be treated with great caution. An attempt at a more rigorous analysis along the same lines, but dividing the world into three, found the decline in sperm density to be much steeper in Europe than in America; studies from elsewhere were too sparse and diverse to draw confident conclusions[19].

However, the 1992 paper did stimulate several centres to analyse their semen quality data, which had been continuously collected for some two decades. Those data are less likely to have been distorted by possible changes in the method of semen examination and/or in selection processes affecting the populations studied. The principal conclusions that emerge are that: (i) declines in semen quality have occurred in some places (*e.g.* Paris, Edinburgh, Gent) but not in others (Toulouse, Finland and the five US cities with published data)[20]; (ii) at most, the available data go back to the early 1970s; and (iii) where concentration has deteriorated, so usually have sperm motility and morphology.

Where a decline has occurred, the findings are compatible not only with a period effect but also with a birth cohort effect, men born in the 1940s having better quality semen than those born in the 1960s. As the observed decline, with either method of analysis, is already visible in the earliest available data in all affected centres, it is impossible to locate the year when the decline started or what the pre-decline values were. As semen quality is inferior in humans compared with other mammalian species, it is possible that deterioration from a 'natural' level has a much longer history than we have the data to substantiate[20].

While it is difficult to be confident about drawing conclusions from this literature, it is likely that semen quality deteriorated in some parts of Europe for two decades after the early 1970s as a period effect, or the mid-1940s as a birth cohort effect. This deterioration involved not only sperm concentration, but also morphology and motility. No evidence is available on earlier periods, so that a decline may possibly have begun earlier. The evidence for a similar trend in America is unconvincing[20].

Spatial variation
Substantial spatial variation in sperm concentration has been demonstrated, within both Europe and America[20]. Based on the available evidence, concentrations appear to be relatively high in New York and Finland and low in California and north-western Europe including Denmark and Britain. Couple fertility assessed by TTP is high in parts of southern Europe compared with the north[21], with the exception that it is also high in Finland[22]. The congruence of the findings for Finland suggests that the higher levels of sperm concentration observed there are not the result of differences in methodology or to longer abstinence (less frequent intercourse).

Genetic factors
Male fertility problems tend to aggregate in families[23,24], infertile men have relatively few siblings[24], and their brothers have inferior semen quality[25]. However, most fertility-affecting genetic aberrations cannot be detected using current clinical laboratory methods[24]. A recent report based on a twin study reported the heritability of sperm concentration, uncorrected for biological variation and measurement error (and therefore an underestimate), as 20%. The heritability of sperm morphology was 41%, and that of chromatin stability was 68%[26]. Certain paternal lineages, identified through their Y chromosomes, are predisposed to low sperm counts[27,28]. There is also evidence for heritability of TTP, probably by nonadditive polygenic inheritance[29].

Testicular cancer

Epidemiological information on cancer of the testis is very reliable. As a disease of relatively young men that has unmistakable features, it is likely to be rarely missed or misdiagnosed, so that only an efficient collating system is required to produce high quality ascertainment. Good incidence data have been available from cancer registries in developed countries for some decades. Mortality data are also available for certain countries going back a hundred years, and since the disease is invariably fatal if untreated, these are reliable for the early 20th century, although not more recently as cure rates are now high.

Testicular cancer is strongly and consistently associated with subfertility, and this has been shown to be present before the cancer appears[30].

Trends

This disease has shown an increasing trend in recent decades throughout the developed world, typically with rates being trebled or more. An important and often overlooked question is when this began. In England and Wales, mortality started rising around 1920, having been stable before World War I[31]. In Denmark, a continuous rise in age-standardized incidence is observable since cancer registration began in 1943[32].

Clinical research strongly suggests that the predisposition to testicular cancer is present from an early age, probably *in utero*[33], so that the possible influence of environmental agents needs to be evaluated in relation to time of birth rather than of diagnosis or death. Accordingly, if these trends are examined in terms of birth cohorts, mortality started rising among men born before 1900 in England and Wales[31], and incidence in Denmark, Norway and Sweden started rising among men born around 1905[34]. In these latter three countries, rates stabilized or fell for men born during 1935–45, whereas the rise was rapid and inexorable among men born from 1920 until at least 1960 in East Germany, Finland and Poland[34]. Recent data indicate that the rates may be stabilizing for Danish men born since about 1960[35], but the 1965 birth cohort shows a continuing rise in other countries[34].

Spatial and ethnic variation

There is considerable spatial and ethnic variation. Denmark has the highest incidence in the world, the lifetime risk now being almost 1%. However, the Nordic countries do not have a uniformly high risk, as Finnish men have comparatively low rates, with Norway and Sweden in intermediate positions[32,36]. The spatial pattern for testicular cancer in the Nordic countries does not resemble that of other hormone-sensitive carcinomas such as those of the prostate or female breast, but is similar to that of colo-rectal cancer in both sexes[36].

Other high-risk populations include Switzerland and New Zealand (including Maoris), whereas the Baltic states and African-Americans have comparatively low rates[20]. The tumour is rare among Chinese and Japanese men[20].

Genetic factors

Whereas the rapid trends in testicular cancer indicate the importance of environment in the broadest sense, migrant studies suggest a genetic component as well: for example, a high risk among European immigrants to Israel was still present, albeit reduced, in the next generation[37]. This is confirmed by family[38] and case-control studies[39]. However, brothers have a far higher risk than father–son pairs, suggesting the importance of shared maternal characteristics as well as shared genes[40,41]; dizygous twins have a particularly high risk, which may indicate that endogenous maternal oestrogen levels play a part[42,43], although twins also share a time-specific maternal (and paternal) environment.

Anomalies of the male genitalia

Hypospadias and cryptorchidism have been grouped with male infertility and cancer of the testis into the 'testicular dysgenesis syndrome'[44], on the grounds that they occur together more often than expected by chance, and that they all probably originate early in life. Hence, it is argued, they probably share at least some risk factors.

Both anomalies are likely to be unreliably ascertained at birth, particularly in mild cases, and the study of cryptorchidism is further complicated by the difficulty of distinguishing testes that have not descended from those that readily but reversibly retract back into the abdominal cavity in early infancy. The consequence is that published data from congenital malformation registries cannot be relied on to reflect real variations: reported time trends and differences between registries may both merely reflect differences in ascertainment and reporting[45]. Self-reported data (by mothers) are similarly unreliable.

Trends and spatial/ethnic variation

For hypospadias, the apparent increase in many countries may well be because of variations in the registry system rather than a real change[45], apart from a step increase between 1982 and 1985 in the severe form in Atlanta, Georgia[46]. Recent studies in Denmark and Finland using strict criteria have shown a higher rate in Denmark[45].

In the case of cryptorchidism, a study was carried out in Oxford during the 1950s, using strict diagnostic criteria and examination of the baby boys at 3 months when the diagnosis is more reliable. A subsequent study in southern England using the same criteria found almost double the proportion of boys having cryptorchidism[47]. Recent studies in New York[48] and in Finland[45] using the same criteria found a similar proportion to the original (lower) Oxford estimate, whereas in Denmark it was close to the later English value[45]. Unlike for testicular cancer, African-Americans do not appear to have a lower risk[48,49].

Genetic factors

Both hypospadias and cryptorchidism show familial association[50,51], with different modes of inheritance having been suggested, probably reflecting the importance of several different genes[51]. They tend to occur together more often than would be expected by chance, in individuals though not in families[50]. Hypospadias is also associated with parental subfertility[52] and with impaired paternal sperm motility and morphology[53].

Sex ratio

Although the proportion of births that are male, conventionally but inaccurately called the sex ratio, is not included in the definition of the testicular dysgenesis syndrome, it is prudent to consider what is known about its temporal and spatial variation in this discussion.

Trends

A reduction in the sex ratio has been observed in several countries, including Denmark and other Nordic countries, the Netherlands, the USA and Canada (but not Australia)[54]. For those countries with data going back that far, this began around 1950 and continued at least into the 1990s. The changes are small—typically the decline is in the order of 2 per 1000 births—but they may nevertheless indicate a biological process that is important for other reasons. Before 1950, an increasing trend was observed, which has been attributed to a fall in the proportion of stillbirths (a disproportionate number of whom are male) and therefore discounted as the mirror image of the more recent trend. However, an alternative possibility is that long-term cyclical fluctuations occur[55], possibly as a result of an adjustment process[56].

Spatial and other variation

While a higher proportion of male births has been observed within southern Europe compared to the north[57], spatial variation is not a major feature of the literature on sex ratio. However, certain chemical exposures have been associated with altered sex ratio, always in the direction of reducing the proportion of males, which may reflect the greater vulnerability of males at all ages from conception onwards. The best documented is of men occupationally exposed to DBCP (see above): those who retained or regained some degree of fertility fathered predominantly daughters, and the impairment of sex ratio was correlated with the degree of oligospermia[11]. Dioxin exposure following the Seveso incident was also followed by a reduction in male births[58].

The sex ratio can be affected either by an endocrine process, which may be the case for dioxin, or by genetic damage, which is more likely for DBCP as it is mutagenic. These are discussed further below.

Epidemiological characteristics of the responsible agent(s)

The question of linkage

A key question is, to what extent (if at all) these five conditions are linked by shared aetiological factors, as indicated by parallelism in their epidemiological characteristics. Clearly, for the congenital anomalies or sex ratio to be included, the exposure would need to be acting prenatally. As testicular cancer epidemiology is by far the most robust, a useful starting point is to consider its main features:

(i) The incidence has increased at least three-fold in all developed countries, suggesting the importance of one or more environmental risk factors that have changed over the same period.

(ii) The increasing trend started around 1900 in at least some countries, assessed as a birth cohort effect, and continued until at least the 1960s.

(iii) The rise was interrupted for some years around 1940 in Scandinavia, but not in nearby countries that were more severely affected by World War II.

(iv) Marked variations exist between different nationalities (*e.g.* Denmark *versus* Finland), and between ethnic groups.

(v) Migrant, family and case-control studies suggest some degree of heritability. This may be related to international and inter-ethnic differences, but cannot explain the rapid trends.

The first three of these refer to trends, so the question is, to what extent the other endpoints have parallel trends, making due allowance for latent periods, *etc.*: linked trends in congenital anomalies or sex ratio and the adult-life endpoints will tend to be a few decades apart, or to put it another way, comparison would have to be by the year of birth. Even if parallel variations were found, this would be unreliable evidence for linkage, as so many factors vary with time. Whereas recognition of *discrepancies* would, in principle, be good evidence *against* the hypothesis of linkage, so that the latter could be tested in a negative way (*i.e.* how well it survives attempts at refutation), the paucity and low quality of the available data place severe limits on this course of action.

The timing for each observed trend does not appear to be exactly parallel: in particular, the beginning of the fall in the sex ratio (for which the evidence is robust) was later than that of the rise in testicular cancer incidence. Also, the trends in sex ratio and possibly in semen quality have continued later than the early 1960s, which may mark the end of the trend in testicular cancer. However, the latter observation is currently tentative, and even if true may not represent a permanent interruption; also, it applies just to Denmark. More generally, if one allows for the possibility that trends can be interrupted or even temporarily reversed, and that all these features are likely to vary across populations and sub-populations, the available data on the timing of trends cannot be confidently regarded as evidence for or against linkage.

The spatial location of trends also does not correspond precisely for the different endpoints: whereas those for testicular cancer and the sex ratio are apparently widespread, at least in Europe and North America, semen quality does not seem to have deteriorated in America.

The pattern of spatial variation for the four endpoints of the testicular dysgenesis syndrome, notably the sharp and consistent contrast between Finland and Denmark, suggests possible linkage. The observations on cryptorchidism, while sparse, also show some resonance with data on other endpoints.

Overall, there is enough parallelism to suggest that at least some of the endpoints share one or more risk factors, but that there must also be some additional harmful and/or protective factors, which is perhaps unsurprising. A 'linked' risk factor would need to have started increasing (or decreasing, in the case of a protective factor) in its exposure level during the early 20th century in developed countries, which rules out chemicals introduced since the mid-20th century. Possible explanations include environmental pollution, and dietary changes involving macro- or micro-nutrients or contaminants, both natural and man-made. While the epidemiological pattern is compatible with something that increased with rising prosperity, such as increasing meat consumption, there is no direct evidence for this.

In the case of semen quality, a possible 'unlinked' factor could include an increasingly sedentary way of life, possibly together with tight clothing, since raising the intra-testicular temperature strongly affects the quality as well as the quantity of sperm—sufficient to cause delayed conception in men with sedentary occupations such as driving—and on the viability of the offspring[59]. This would be a period effect not a birth cohort effect.

Heritability

All four conditions of the testicular dysgenesis syndrome show some degree of heritability. This is compatible with the idea that they are linked.

Heritability of subfertility may at first sight appear to be impossible: a 'gene for infertility' is surely impossible, on evolutionary grounds, as it would quickly be eliminated. (Theoretically, this would not apply if the heterozygous state of a recessive gene carried some advantage, but evidence from a Hutterite population is that inbreeding is unrelated to male fertility[60].) An alternative explanation is, however, possible. In a steady state, a balance would exist between selection against polymorphisms that impair fertility and their *de novo* creation as a result of genetic damage. As there is nothing to constrain these two processes to be equal, new damage could occur at a rate greater than elimination, leading to an increase in incidence. The same argument applies, but with less strong elimination, to testicular cancer, hypospadias and cryptorchidism, which are associated with subfertility, and which show evidence of heritability.

The selection process can also vary in its intensity. Czeizel has pointed out that as family size decreased markedly during the 20th century, births to the biologically most fertile couples became a less dominant proportion of births at the population level[61]. Secondly, towards the end of the century, assisted reproduction meant that the proportion of births to clinically subfertile couples increased. These two tendencies would have the effect of decreasing the rate of removal from the population of polymorphisms that reduce fertility and that might also predispose to testicular cancer or one of the other endpoints.

The first of these at least may well be important to consider. However, it is unlikely that on its own it would be strong enough to bring about, for example, a three-fold rise in testicular cancer incidence. Furthermore, this hypothesis depends on the existence of polymorphisms that decrease fertility, raising the question, why had they not already been eliminated from the population, even before 1900. New generation of such polymorphisms is required to complete the picture, and once this is accepted it becomes important to consider:

(i) the nature of the defects and their location in the genome,

(ii) the rate of their appearance,

(iii) the identification of agents that could affect their generation, and

(iv) the possible other effects that this process could have.

Before considering genetic mechanisms, it is necessary to review other possibilities.

The type of underlying biological process

Endocrine disruption

Oestrogens

The original version of the endocrine disruption hypothesis was concerned with exposure to oestrogens. As we have seen, effects on female fertility are well established, although it is unlikely that current environmental exposure levels are sufficient, except possibly in the case of dietary phytoestrogens.

In 1979, Henderson *et al*[62] observed that factors such as high maternal weight and excessive vomiting during pregnancy, which are associated with high levels of endogenous maternal oestrogens such as oestradiol, increased the risk of testicular cancer. Cryptorchidism is also implicated[49]. This invites the question, whether trends or spatial variations in these or related conditions could be due to variations in maternal oestrogens, as a result of exposure to environmental, nutritional or other factors, but there has been insufficient research in this area[20].

The idea was subsequently extended to exogenous substances with oestrogenic activity[63]. The criticism that the exposure levels and the potency of such substances are too low by several orders of magnitude, compared with oestradiol[64], was refuted by the argument that endogenous oestrogen does not reach the fetus (*e.g.* because of protein binding)—although this postulates a near-absolute barrier that seems implausible, and is directly contradicted by the evidence just cited for Henderson's hypothesis[20].

Nevertheless, the 'oestrogen hypothesis' became influential, both scientifically and in society at large, where public concern about 'gender benders' ensued. While it is superficially plausible that oestrogens 'demasculinize' the developing male, this is biologically naïve because mammals are adapted to starting life inside their mothers, whose internal environment is oestrogen-rich (even before the early pregnancy surge). In contrast to other vertebrates, the mammalian default sex is female, and masculinization of the gonads and central nervous system depends on the presence of androgens.

In contrast to the marked impact on girls, boys exposed *in utero* to DES show relatively minor effects. They tend to have genital abnormalities such as cysts and urethral stenosis, but among the features of the testicular dysgenesis syndrome only cryptorchidism is clearly and strongly affected, despite the high exposures. The risk of testicular cancer may be raised, but by less than the trend observed throughout the developed world, and the position for sperm concentration is similar; hypospadias has only been implicated because of a propagated error in the literature[65].

Although the DES disaster has often been cited in favour of the hypothesis that exogenous oestrogens are responsible for the observed deterioration in the health of the male reproductive system, it is rather strong evidence against. This is reinforced by the low incidence of testicular cancer among Chinese and Japanese men who are exposed *in utero* to high levels of phytoestrogens in soy. In relation to the 'environmental oestrogens' listed in Table 1, which are orders of magnitude less potent, it is now accepted that their uniformly weak oestrogenicity excludes the possibility that they could induce these disorders[66].

Anti-androgens and other types of endocrine disruption

Interference with either the synthesis or the action of androgens could prevent the normal masculinization of the male fetus, and could also affect male infants postnatally. There is toxicological evidence that p,p′-DDE, the stable breakdown product of DDT, can block the androgen receptor, as can certain other pesticides, and that some phthalates inhibit testosterone synthesis (see Table 1)[66]. More nuanced hypotheses that relate, for example, to the balance between oestrogens and androgens, or to their interconversion *via* aromatase[66], are interesting, but lack candidate substances that could explain the epidemiological findings.

It is therefore plausible that exposure to anti-androgens can affect male fertility, as well as other related endpoints, and this is supported by toxicological evidence[66]. Could pollution with DDE, phthalates or other anti-androgens explain any of the epidemiological findings presented above?

One obvious objection is that the rising trend, at least in testicular cancer, started before any of the known anti-androgens were introduced. Secondly, the striking contrast between Denmark and Finland cannot be explained by exposure to DDE, which has been monitored in human breast milk, and the concentrations were similar in all the Nordic countries[67]. Thirdly, high levels of exposure to DDE in developing countries, in the course of attempts at malaria control, have not resulted in an epidemic of testicular cancer[20].

The answer, then, is no; the idea that disorders of the male reproductive system are due to chemicals that interfere with the sex hormone system, in any of its variants, cannot explain any of the main features of the epidemiological evidence.

The expected spectrum of effects

It is far from clear that endocrine disruption would affect non-quantitative aspects of semen quality, especially morphology. However, it is necessary to go further. In addition to focusing on the various endpoints and asking 'could this be due to endocrine disruption?', it is important also to turn the question around and ask: 'if an endocrine-disrupting substance were responsible, what spectrum of effects would be predicted?' One plausible expectation is of a coherent pattern in hormone-sensitive cancers, but this is not observed[20].

A second is that endocrine agents would be expected to influence growth and development, secondary sexual characters and the timing of puberty. No such change has been reported among boys, either in Europe or America[20].

As mentioned earlier in the section 'Infertility in females', such effects have, however, been found in female rodents fed with phytoestrogens. Precocious puberty has been reported to be widespread among girls in the USA, especially African-American girls[68], and this raises the question of possible excessive oestrogenic stimulation. The potential effect on their subsequent fertility is unknown.

Dioxin

'Dioxin' (2,3,7,8-tetrachlorodibenzo-p-dioxin) is the most potent of a group of chemicals, dioxins and furans; dioxin-like activity is also displayed by some PCBs. While it is sometimes called an endocrine disruptor, as it has some effects on the endocrine system such as anti-oestrogenicity, it mainly acts through a distinct receptor, the aryl hydrocarbon receptor (AhR)[69]. Toxicological evidence, albeit with some inconsistencies, shows that sperm production and morphology are adversely affected even by very low doses given during pregnancy and lactation[69,70].

There is little information on infertility and related conditions based on human exposure. In men, a case–control study has suggested lowered serum testosterone and raised follicle stimulating hormone and luteinizing hormone levels with occupational dioxin exposure[71]. Vietnam veterans tended to have lower sperm concentrations and fewer morphologically 'normal' cells than non-Vietnam veterans, but few of the former group were greatly exposed to Agent Orange, a pesticide that was heavily contaminated with dioxin[72]. In US military working dogs, an excess of testicular cancer (seminoma) was found among those that had worked in Vietnam, and possibly exposed to Agent Orange[73]. All these findings would relate to a period, not a birth cohort, effect, following exposure of adult males.

For exposure of women, we have already noted the suggestive findings of longer TTP associated with high consumption of sport fish from Lake Ontario[4], where the pollution includes PCBs, and the low sex ratio in births to women with relatively high dioxin exposure following the Seveso incident[58].

Dioxin is persistent, in the environment and in the body: in humans, it has a half-life of 6–11 years. The intake estimates for the UK in 1997 are below the levels thought necessary to affect the reproductive system, but this may not be true for earlier periods; for example, in 1982, intakes were four times as high as in 1997[69]. Interpretation of the possible role of dioxin and related compounds in reproductive health requires information on the spatial and temporal variation of exposure to be considered alongside the descriptive epidemiological findings discussed above.

Genetic damage

The hypothesis

A possible unifying hypothesis for impairment of the male reproductive system is that genetic damage in the germ line is responsible. If so, the health implications could extend beyond the conditions discussed in this paper, and could include chromosomal abnormalities and other genetic anomalies, malformations and cancer in the offspring and in future

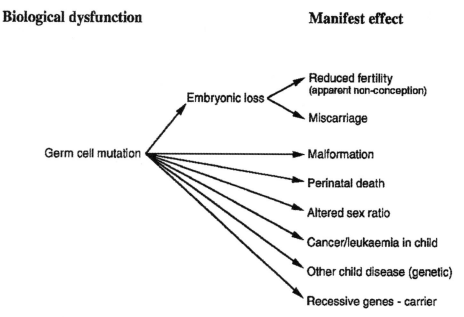

Fig. 1 Possible consequences of germ line genetic damage (reproduced with permission from Joffe[14]).

generations (Fig. 1)[14,74]. With such heterogeneous outcomes, each being uncommon, it is possible that increasing trends have escaped detection.

The evidence that these endpoints have both genetic and environmental determinants could lead discussion in the direction of trying to apportion causation between the two, and/or to consideration of gene–environment interaction[75]. But another possibility is more interesting: the routine distinction between environmental and genetic factors breaks down when we consider germ-line genetic damage. Unless such damage fails to be passed on, for example due to lethality or sterility, there is a heritable element—but the origin of the defect lies in an environmental cause (Fig. 2). If this is true, then it would no longer make sense to refer to 'environmental *or* genetic influences', nor to equate 'genetic' with 'inherited', as is commonly done. (Obviously the three possibilities are not mutually exclusive.)

This hypothesis would accord with the observations outlined previously, on inheritance of the elements of the testicular dysgenesis syndrome. Exposure to a genotoxic agent would lead to some form of mutation; its survival in subsequent generations would depend on:

(i) the degree to which it affects health (including lethality at one extreme), at all stages of life from conception to the end of reproductive life;

(ii) the extent to which it affects biological fertility, the probability of achieving a fertilized ovum, given unprotected intercourse;

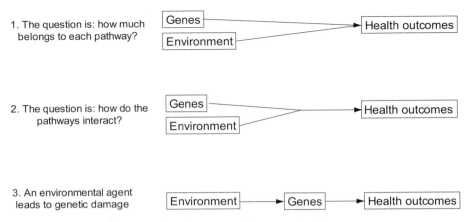

Fig. 2 Ways in which genes and environment relate to each other.

(iii) additional factors that involve volition and control, as well as biology: contraception and achieved family size, and the use of artificial reproductive technology.

The epidemiological implication would be to introduce a degree of inertia into the time trend: whereas an increase in the health outcome would still directly follow an increase in the causal agent (allowing for latency), there would be a gap of some generations between their respective disappearance. At the individual level (*e.g.* in a cohort study), the relevant exposure would not necessarily be to a parent, as it could well be to a grandparent or earlier ancestors.

Mechanism and genomic localization

Is genetic damage a plausible mechanism? Cytogenetic abnormalities are more frequent among infertile men than in the general population[76]. More specifically, severe infertility is a consequence of micro-deletions in three non-overlapping regions of the Y chromosome AZF a-b-c[77]; they may arise by a common *de novo* event[78]. The generally poor semen quality in Denmark is not attributable to micro-deletions[79], but more minor impairments, for example single-nucleotide polymorphisms (SNPs), could well lead to lesser degrees of impairment of semen quality, not only sperm concentration but also *e.g.* morphology[74].

The Y chromosome is a likely target for genetic damage. The probability of mutation is increased by the rapid division of the germ cells, both in fetal and in adult life. In the former case the exposure would be to the pregnant mother, in the latter to the father before conception; either would result in a birth cohort effect. The Y chromosome is not shielded from a mutagenic environment, as are the other chromosomes, by long inert periods in the ovum[77]. The far higher number of cell divisions in spermatogenesis compared with oogenesis has led to the hypothesis that evolution is 'male driven'[80], and these provide extra opportunities for error. Furthermore, the Y chromosome may be unable to undergo DNA repair, as this depends on having an opposite number to pair with during cell division. Finally, as all its genes are haploid, defects in a single gene are likely to have effects[77]—although this is complicated by the presence of multi-copy genes on the Y chromosome.

Genotoxicity can readily explain carcinogenesis, but more work is required to identify the particular gene(s) involved; a possible candidate on the Y chromosome is TSPY[81], or another nearby locus in the gonadoblastoma region[82]. A deficit in male births could result from a selective effect on Y-chromosome-carrying spermatozoa, selective loss of male embryos or fetuses, or mutation of the sex-determining gene SRY on the Y chromosome[54].

Most cases of cryptorchidism and hypospadias are likely to have an endocrine rather than a genetic mechanism. However, this does not contradict the suggestion of a genetic aetiology, as the impairment may originate upstream, in the gene(s) that control(s) the more distal endocrine processes. The same two-stage (genetic-endocrine) principle could also play a part in the other manifestations of the testicular dysgenesis syndrome, including the hormone-sensitive processes underlying fertility.

The cluster of endpoints, all of them concerned with the male reproductive system, is not a coincidence. The gene determining male sex, and those controlling spermatogenesis and other male-reproduction-related functions, have migrated to the human Y chromosome in the course of evolution[83].

A possible causal agent would need to have exposure characteristics that correspond to the epidemiological observations—at least in part, as multi-factorial causation is almost certain to apply. It would need to be genotoxic, to be absorbed, and to localize in the testis. If female (*in utero*) exposure were responsible, it would also have to cross the placenta. Male exposure could in principle affect the stem cells (a permanent effect), or the gamete during spermato- or spermio-genesis (a transient effect) in the weeks before fertilization; maximal damage would be seen after post-meiotic exposure, as elongating spermatids and spermatozoa no longer have the ability to undergo DNA repair or to initiate apoptosis[84].

The sperm chromatin stability assay (SCSA) has demonstrated DNA damage following chemical exposures, as well as heat stress[85], and this test is associated with fertility[86]. Oxidative damage has been suggested as a mechanism[87], but it is unclear what agents could be responsible. Paternal smoking is one possibility, and has been linked with childhood cancer[88], but the epidemiological evidence does not suggest that it is implicated in impaired fertility[89].

One candidate agent is the heterocyclic amine PhIP, which occurs in meat or fish cooked at high temperatures, a powerful mutagen that has been implicated in colorectal[90] and other cancers[91], and fulfils these criteria. Localization in the testis would occur because it binds to the oestrogen receptor[92]. As meat consumption increased in the early decades of the 20th century with rising prosperity, it could be that in cultures whose cooking methods generate heterocyclic amines, and/or who lack protective exposures, PhIP-induced damage has increased.

Conclusions

There is little epidemiological information on trends or spatial variation in female infertility. Certain occupational exposures have been shown to impair female fertility, but the agents are not sufficiently widespread in the general environment to have any effect on the general population. As females are relatively sensitive to oestrogens, agents with oestrogenic activity should be considered in relation to disturbances in female reproductive function, for example precocious puberty. One reassuring finding is that couple fertility has increased in recent decades, but so far this is based on only one report.

While the observed deterioration in semen quality, and in other possibly-linked conditions affecting the male reproductive tract, have been widely discussed in relation to the 'oestrogen hypothesis', pollution with weak environmental oestrogens cannot plausibly be responsible. The anti-androgen variant of the endocrine disruption hypothesis, or androgen/oestrogen balance, may be important, but cannot explain the existing epidemiological findings. A hypothesis that deserves more detailed consideration is the role of dioxin and dioxin-like effects.

Several lines of evidence point towards genetic damage as an explanation of various types of impairment of the male reproductive system. In principle, this could arise through male or female exposure. Possibly the Y chromosome is especially important as a target for mutation. A genetic aetiology raises the possibility that additional health endpoints are also affected.

References

1 Committee on Toxicity of Chemicals in Food, Consumer Products and the Environment. *Phyto-estrogens and Health*. London: Food Standards Agency, 2003. http://www.foodstandards.gov.uk/multimedia/pdfs/phytoreport0503, last accessed November 7, 2003
2 Dieckmann WJ, Davis ME, Rynkiewicz LM, Pottinger RE, Gabbe SG. Does the administration of diethylstilbestrol during pregnancy have therapeutic value? 1953. *Am J Obstet Gynecol* 1999; **181**: 1572–3

3   Goldberg JM, Falcone TF. Effect of diethylstilbestrol on reproductive function. *Fertil Steril* 1999; **72**: 1–7
4   Buck G, Vena J, Schisterman E *et al*. Parental consumption of contaminated sport fish from Lake Ontario and predicted fecundability. *Epidemiology* 2000; **11**: 388–93
5   Sallmén M, Lindbohm M-L, Kyyrönen P *et al*. Reduced fertility among women exposed to organic solvents. *Am J Ind Med* 1995; **27**: 699–713
6   Wennborg H, Bodin L, Vainio H, Axelsson G. Solvent use and time to pregnancy among female personnel in biomedical laboratories in Sweden. *Occup Environ Med* 2001; **58**: 225–31
7   Rowland AS, Baird DD, Weinberg CR, Shore DL, Shy CM, Wilcox AJ. The effect of occupational exposure to mercury vapor on the fertility of female dental assistants. *Occup Environ Med* 1994; **51**: 28–34
8   Rowland AS, Baird DD, Weinberg CR, Shore DL, Shy CM, Wilcox AJ. Reduced fertility among women employed as dental assistants exposed to high levels of nitrous oxide. *N Engl J Med* 1992; **327**: 993–7
9   Ahlborg Jr G, Axelsson G, Bodin L. Shift work, nitrous oxide exposure and subfertility among Swedish midwives. *Int J Epidemiol* 1996; **25**: 783–90
10 Valanis B, Vollmer W, Labuhn K, Glass A. Occupational exposure to antineoplastic agents and self-reported infertility among nurses and pharmacists. *J Occup Environ Med* 1997; **39**: 574–80
11 Goldsmith JR. Dibromochloropropane: epidemiological findings and current questions. *Ann N Y Acad Sci* 1997; **837**: 300–6
12 Bonde JP, Giwercman A. Occupational hazards to male fecundity. *Reprod Med Rev* 1995; **4**: 59–73
13 Bonde JP, Joffe M, Apostoli P *et al*. Sperm count and chromatin structure in men exposed to inorganic lead: lowest adverse effect levels. *Occup Environ Med* 2002; **59**: 234–42
14 Joffe M. Asclepios. Time to pregnancy: a measure of reproductive function in either sex. *Occup Environ Med* 1997; **54**: 289–95
15 Joffe M. Time trends in biological fertility in Britain. *Lancet* 2000; **355**: 1961–5
16 Akre O, Cnattingius S, Bergström R, Kvist U, Trichopoulous D, Ekbom A. Human fertility does not decline: evidence from Sweden. *Fertil Steril* 1999; **71**: 1066–9
17 Jensen TK, Keiding N, Scheike T, Slama R, Spira A. Declining human fertility? *Fertil Steril* 2000; **73**: 421–2
18 Carlsen E, Giwercman A, Keiding N, Skakkebaek NE. Evidence for decreasing quality of semen during past 50 years. *BMJ* 1992; **305**: 609–13
19 Swan SH, Elkin EP, Fenster L. The question of declining sperm density revisited: an analysis of 101 studies published 1934–1996. *Environ Health Perspect* 2000; **108**: 961–6
20 Joffe M. Are problems with male reproductive health caused by endocrine disruption? *Occup Environ Med* 2001; **58**: 281–8
21 Karmaus W, Juul S, European Infertility and Subfecundity Group. Infertility and subfecundity in population-based samples from Denmark, Germany, Italy, Poland and Spain. *Eur J Public Health* 1999; **9**: 229–35
22 Joffe M. Lower fertility in Britain compared with Finland. *Lancet* 1996; **347**: 1519–20
23 Lilford R, Jones AM, Bishop DT *et al*. Case control study of whether subfertility in men is familial. *BMJ* 1994; **309**: 570–3
24 Meschede D, Lemke B, Behre HM, De Geyter C, Nieschlag E, Horst J. Clustering of male infertility in the families of couples treated with intracytoplasmic sperm injection. *Hum Reprod* 2000; **15**: 1604–8
25 Auger J, Kunstmann JM, Czyglik F, Jouannet P *et al*. Decline in semen quality among fertile men in Paris during the past 20 years. *N Engl J Med* 1995; **332**: 281–5
26 Storgaard L, Bonde JP, Ernst E *et al*. The impact of genes and environment on semen quality: an epidemiological twin study. In: Storgaard L, *Genetical and Prenatal Determinants for Semen Quality: An Epidemiological Twin Study*. PhD thesis. Aarhus: University of Aarhus, 2003
27 Kuroki Y, Iwamoto T, Lee J *et al*. Spermatogenic ability is different among males in different Y chromosome lineage. *J Hum Genet* 1999; **44**: 289–92
28 Krause C, Quintana-Murci L, Rajpert-De-Meyts E *et al*. Identification of a Y chromosome haplogroup associated with reduced sperm counts. *Hum Mol Genet* 2001; **10**: 1873–7

29 Christensen K, Kohler H-P, Basso O, Olsen J, Vaupel JW, Rodgers JL. The correlation of fecundability among twins: evidence of a genetic effect on fertility? *Epidemiology* 2003; **14**: 60–4

30 Møller H, Skakkebaek NE. Risk of testicular cancer in subfertile men: case-control study. *BMJ* 1999; **318**: 559–62

31 Davies JM. Testicular cancer in England and Wales: some epidemiological aspects. *Lancet* 1981; **i**: 928–32

32 Adami H-O, Bergstrom R, Mohner M *et al*. Testicular cancer in nine northern European countries. *Int J Cancer* 1994; **59**: 33–8

33 Skakkebaek NE, Berthelsen JG, Giwercman A, Muller J. Carcinoma-in-situ of the testis: possible origin from gonocytes and precursors of all types of germ cell tumours except spermatocytoma. *Int J Androl* 1987; **10**: 19–28

34 Bergström R, Adami H-O, Möhner M *et al*. Increase in testicular cancer incidence in six European countries: a birth cohort phenomenon. *J Natl Cancer Inst* 1996; **88**: 727–33

35 Møller H. Trends in incidence of testicular cancer and prostate cancer in Denmark. *Hum Reprod* 2001; **16**: 1007–11

36 Møller Jensen O, Carstensen B, Glattre E *et al*. *Atlas of Cancer Incidence in the Nordic Countries*. Nordic Cancer Union, 1988

37 Parkin DM, Iscovich J. Risk of cancer in migrants and their descendants in Israel: II carcinomas and germ-cell tumours. *Int J Cancer* 1997; **70**: 654–60

38 Forman D, Oliver RT, Brett AR *et al*. Familial testicular cancer: a report of the UK family register, estimation of risk and an HLA class 1 sib-pair analysis. *Br J Cancer* 1992; **65**: 255–62

39 Swerdlow AJ, De Stavola BL, Swanwick MA, Mangtani P, Maconochie NE. Risk factors for testicular cancer: a case-control study in twins. *Br J Cancer* 1999; **80**: 1098–102

40 Westergaard T, Olsen JH, Frisch M, Kroman N, Nielsen JW, Melbye M. Cancer risk in fathers and brothers of testicular cancer patients in Denmark. A population-based study. *Int J Cancer* 1996; **66**: 627–31

41 Sonneveld DJA, Sleijfer DTh, Schraffordt Koops H *et al*. Familial testicular cancer in a single-centre population. *Eur J Cancer* 1999; **35**: 1368–73

42 Braun MM, Alhbom A, Floderus B, Brinton LA, Hoover RN. Effect of twinship on incidence of cancer of the testis, breast and other sites. *Cancer Causes Control* 1995; **6**: 519–24

43 Swerdlow AJ, De Stavola BL, Swanwick MA, Maconochie NES. Risks of breast and testicular cancers in young adult twins in England and Wales: evidence on prenatal and genetic aetiology. *Lancet* 1997; **350**: 1723–8

44 Skakkebaek NE, Rajperts-de Meyts E, Main KM. Testicular dysgenesis syndrome: an increasingly common developmental disorder with environmental aspects. *Hum Reprod* 2001; **16**: 972–8

45 Toppari J, Kaleva M, Virtanen HE. Trends in the incidence of cryptorchidism and hypospadias, and methodological limitations of registry-based data. *Hum Reprod Update* 2001; **7**: 282–6

46 Paulozzi LJ. International trends in rates of hypospadias and cryptorchidism. *Environ Health Perspect* 1999; **107**: 297–302

47 John Radcliffe Hospital Cryptorchidism Study Group. Cryptorchidism: a prospective study of 7500 consecutive male births, 1984–8. *Arch Dis Child* 1992; **67**: 892–9

48 Berkowitz GS, Lapinski RH, Dolgin SE, Gazella JG, Bodian CA, Holzman IR. Prevalence and natural history of cryptorchidism. *Pediatrics* 1993; **92**: 44–9

49 Depue RH. Maternal and gestational factors affecting the risk of cryptorchidism and inguinal hernia. *Int J Epidemiol* 1984; **13**: 311–8

50 Weidner IS, Møller H, Jensen TK, Skakkebaek NE. Risk factors for cryptorchidism and hypospadias. *J Urol* 1999; **161**: 1606–9

51 Fredell L, Iselius L, Collins A *et al*. Complex segregation analysis of hypospadias. *Hum Genet* 2002; **111**: 231–4

52 Czeizel A, Toth J. Correlation between the birth prevalence of isolated hypospadias and parental subfertility. *Teratology* 1990; **41**: 167–72

53 Fritz G, Czeizel AE. Abnormal sperm morphology and function in the fathers of hypospadiacs. *J Reprod Fertil* 1996; **106**: 63–6

54 Davis DL, Gottleib MB, Stampnitzky JR. Reduced ratio of male to female births in several industrial countries: a sentinel health indicator? *JAMA* 1998; **279**: 1018–23

55 James WH. Declines in population sex ratios at birth. *JAMA* 1998; **280**: 1139
56 Lummaa V, Merila J, Kause A. Adaptive sex ratio variation in preindustrial human (Homo sapiens) populations? *Proc R Soc Lond B Biol Sci* 1998; **265**: 563–8
57 Grech V, Vassallo-Agius P, Savona-Ventura C. Declining male births with increasing geographical latitude in Europe. *J Epidemiol Community Health* 2000; **54**: 244–6
58 Mocarelli P, Gerthoux PM, Ferrari E *et al.* Paternal concentrations of dioxin and sex ratio of offspring. *Lancet* 2000; **355**: 1858–63
59 Setchell BP. Heat and the testis. *J Reprod Fertil* 1998; **114**: 179–84
60 Ober C, Hyslop T, Hauck WW. Inbreeding effects on fertility in humans: evidence for reproductive compensation. *Am J Hum Genet* 1999; **64**: 225–31
61 Czeizel AE, Rothman KJ. Does relaxed reproductive selection explain the decline in male reproductive health? A new hypothesis. *Epidemiology* 2002; **13**: 113–4
62 Henderson BE, Benton B, Jing J, Yu MC, Pike MC. Risk factors for cancer of the testis in young men. *Int J Cancer* 1979; **23**: 598–602
63 Sharpe RM, Skakkebaek NE. Are oestrogens involved in falling sperm counts and disorders of the male reproductive tract? *Lancet* 1993; **341**: 1392–5
64 Safe SH. Environmental and dietary estrogens and human health: is there a problem? *Environ Health Perspect* 1995; **103**: 346–51
65 Joffe M. Myths about endocrine disruption and the male reproductive system should not be propagated. *Hum Reprod* 2002; **17**: 101–2
66 Sharpe RM. The 'oestrogen hypothesis'—where do we stand now? *Int J Androl* 2003; **26**: 2–15
67 Ekbom A, Wicklund-Glynn A, Adami H-O. DDT and testicular cancer. *Lancet* 1996; **347**: 553–4
68 Herman-Giddens ME, Slora EJ, Wasserman RC *et al.* Secondary sexual characteristics and menses in young girls seen in office practice: a study from the Pediatric Research in Office Settings network. *Pediatrics* 1997; **99**: 505–12
69 Committee on Toxicity of Chemicals in Food, Consumer Products and the Environment. *Statement on the Tolerable Daily Intake for Dioxins and Dioxin-like Polychlorinated Biphenyls.* London: Food Standards Agency, 2001. http://www.food.gov.uk/science/ouradvisors/toxicity/statements/, last accessed November 7, 2003
70 Faqi AS, Dalsenter PR, Merker HJ, Chahoud I. Reproductive toxicity and tissue concentrations of low doses of 2,3,7,8-tetrachlorodibenzo-p-dioxin in male offspring of rats exposed throughout pregnancy and lactation. *Toxicol Appl Pharmacol* 1998; **150**: 383–92
71 Egeland GM, Sweeney MH, Fingerhut MA, Wille KK, Schnorr TM, Halperin WE. Total serum testosterone and gonadotropins in workers exposed to dioxin. *Am J Epidemiol* 1994; **139**: 272–81
72 The Centers for Disease Control Vietnam Experience Study. Health status of Vietnam veterans. II Physical health. *JAMA* 1988; **259**: 2708–14
73 Hayes HM, Tarone RE, Casey HW, Huxsoll DL. Excess of seminomas observed in Vietnam service U.S. military working dogs. *J Natl Cancer Inst* 1990; **82**: 1042–6
74 Wyrobek AJ. Methods and concepts in detecting abnormal reproductive outcomes of paternal origin. *Reprod Toxicol* 1993; **7**: 3–16
75 Harland SJ. Conundrum of the hereditary component of testicular cancer. *Lancet* 2000; **356**: 1455–6
76 Shi Q, Martin RH. Aneuploidy in human spermatozoa: FISH analysis in men with constitutional chromosomal abnormalities, and in infertile men. *Reproduction* 2001; **121**: 655–66
77 Hargreave TB. Genetic basis of male fertility. *Br Med Bull* 2000; **56**: 650–71
78 Katz MG, Chu B, McLachlan R, Alexopoulos NI, de Kretser DM, Cram DS. Genetic follow-up of male offspring born by ICSI, using a multiplex fluorescent PCR-based test for Yq deletions. *Mol Hum Reprod* 2002; **8**: 589–95
79 Krausz C, Rajpert-de-Meyts E, Frydelund-Larsen L, Quintana-Murci L, McElreavey K, Skakkebaek NE. Double-blind Y chromosome microdeletion analysis in men with known sperm parameters and reproductive hormone profiles: microdeletions are specific for spermatogenetic failure. *J Clin Endocrinol Metab* 2001; **86**: 2638–42
80 Ellegren H, Fridolfsson A-K. Male-driven evolution of DNA sequences in birds. *Nature Genet* 1997; **17**: 182–5
81 Lau Y-FC. Gonadoblastoma, testicular and prostate cancers, and the *TSPY* gene. *Am J Hum Genet* 1999; **64**: 921–7

82 Lahn BT, Page DC. Functional coherence of the human Y chromosome. *Science* 1997; **278**: 675–80
83 Marshall Graves JA. Human Y chromosome, sex determination, and spermatogenesis—a feminist view. *Biol Reprod* 2000; **63**: 667–76
84 Robaire B, Hales BF. The male germ cell as a target for drug and toxicant action. In: Gagnon C (ed.) *The Male Gamete: From Basic Science to Clinical Applications*. Vienna, IL: Cache River, 1999
85 Sakkas D. The need to detect DNA damage in human spermatozoa: possible consequences on embryo development. In: Gagnon C (ed.) *The Male Gamete: From Basic Science to Clinical Applications*. Vienna, IL: Cache River, 1999
86 Spanò M, Bonde JP, Hjøllund HI *et al*. Sperm chromatin damage impairs human fertility. *Fertil Steril* 2000; **73**: 43–50
87 Aitken RJ. The Amoroso lecture: The human spermatozoon—a cell in crisis? *J Reprod Fertil* 1999; **115**: 1–7
88 Sorahan T, Lancashire RJ, Hulten MA, Peck I, Stewart AM. Childhood cancer and parental use of tobacco: deaths from 1953 to 1955. *Br J Cancer* 1997; **75**: 134–8
89 Joffe M, Li Z. Male and female factors in fertility. *Am J Epidemiol* 1994; **140**: 921–9
90 Butler LM, Sinha R, Millikan RC *et al*. Heterocyclic amines, meat intake, and association with colon cancer in a population-based study. *Am J Epidemiol* 2003; **157**: 434–45
91 Sinha R. An epidemiological approach to studying heterocyclic amines. *Mutat Res* 2002; **506–507**: 197–204
92 Gooderham NJ, Zhu H, Lauber S, Boyce A, Creton S. Molecular and genetic toxicology of 2-amino-1-methyl-6-phenylimidazo[4,5-b]pyridine (PhIP). *Mutat Res* 2002; **506–507**: 91–9

Contribution of environmental factors to cancer risk

Paolo Boffetta[*,†] and **Fredrik Nyberg**[‡]

International Agency for Research on Cancer, Lyon, France, †Department of Medical Epidemiology and Biostatistics, Karolinska Institute, Stockholm and ‡Institute of Environmental Medicine, Karolinska Institute, Stockholm, Sweden

Environmental carcinogens, in a strict sense, include outdoor and indoor air pollutants, as well as soil and drinking water contaminants. An increased risk of mesothelioma has consistently been detected among individuals experiencing residential exposure to asbestos, whereas results for lung cancer are less consistent. At least 14 good-quality studies have investigated lung cancer risk from outdoor air pollution based on measurement of specific agents. Their results tend to show an increased risk in the categories at highest exposure, with relative risks in the range 1.5–2.0, which is not attributable to confounders. Results for other cancers are sparse. A causal association has been established between exposure to environmental tobacco smoke and lung cancer, with a relative risk in the order of 1.2. Radon is another carcinogen present in indoor air which may be responsible for 1% of all lung cancers. In several Asian populations, an increased risk of lung cancer is present in women from indoor pollution from cooking and heating. There is strong evidence of an increased risk of bladder, skin and lung cancers following consumption of water with high arsenic contamination; results for other drinking water contaminants, including chlorination by-products, are inconclusive. A precise quantification of the burden of human cancer attributable to environmental exposure is problematic. However, despite the relatively small relative risks of cancer following exposure to environmental carcinogens, the number of cases that might be caused, assuming a causal relationship, is relatively large, as a result of the high prevalence of exposure.

Correspondence to:
Dr Paolo Boffetta, Chief, Unit of Environmental Cancer Epidemiology, International Agency for Research on Cancer, 150 cours Albert-Thomas, 69008 Lyon, France.
E-mail: boffetta@iarc.fr

Introduction

The concept of environment is often used with a broad scope in the medical literature, including all non-genetic factors such as diet, lifestyle and infectious agents. In this broad sense, the environment is implicated in the causation of the majority of human cancers[1]. In a more specific sense, however, environmental factors include only the (natural or man-made) agents encountered by humans in their daily life, upon which they have no or limited personal control. The most important 'environmental' exposures, defined in this strict sense, include outdoor and indoor air pollution and soil and drinking water contamination.

In this review of the evidence linking exposure to selected (narrowly defined) environmental factors and risk of cancer, we consider the following sources of environmental exposure to possible carcinogens: asbestos, outdoor air pollution including residence near major industrial emission sources, environmental tobacco smoke (ETS), indoor radon, other sources of indoor air pollution, arsenic in drinking water, chlorination by-products in drinking water, and other drinking water pollutants. We do not consider agents whose exposure depends on lifestyle, such as solar radiation and food additives, nor agents occurring in the environment as a consequence of accidents or warfare. Whenever possible, we attempt a quantification of the burden of environmental cancer in the European Union, comprising 15 countries, as of 2003.

Cancer risk from environmental exposure to asbestos

Asbestos and asbestiform fibres are naturally occurring fibrous silicates with an important commercial use, mainly in acoustical and thermal insulation. They can be divided into two groups: chrysotile and the group of amphiboles, including amosite, crocidolite, anthophyllite, actinolite and tremolite fibres. Chrysotile is the most widely used type of asbestos. Although all types are carcinogenic to the lung and mesothelioma, the biological effects of amphiboles on the pleura and peritoneum seem to be stronger than those of chrysotile[2]. The use of asbestos has been restricted or banned in many countries.

In contrast to the many epidemiological studies available on asbestos-exposed workers, there are few studies on the health effects of non-occupational (household and residential) exposure to asbestos. One type of household exposure concerns cohabitants of asbestos workers and arises from dust brought home on clothes. Other household sources of asbestos exposure are represented by the installation, degradation, removal and repair of asbestos-containing products. Residential exposure mainly results from outdoor pollution related to asbestos mining or manufacturing, in addition to natural exposure from the erosion of asbestos or asbestiform rocks. The assessment of non-occupational exposure to asbestos presents difficulties, since levels are generally low, and the duration and frequency of exposure and the type of fibre are seldom known with precision.

Table 1 summarizes the results of studies on risk of pleural mesothelioma and lung cancer from environmental (residential) exposure to asbestos. Studies were available from various countries and, in most cases, exposure was defined as residence near a mine or another major source of asbestos

Table 1 Studies of risk of mesothelioma and lung cancer from environmental exposure to asbestos

Country	SD	TF	Source of exposure	Mesothelioma			Lung cancer			Reference
				Ca	RR	95% CI	Ca	RR	95% CI	
South Africa	Ec	A	Res. in mining area	61	8.7	6.7–11.4	86	1.7	1.2–2.5	Botha et al[3a]
South Africa	CC	A	Res. in mining areas				16	3.6	1.4–9.3	Mzileni et al[4a]
Canada	Co	C	Res. in mining area	7	1.3	0.5–3.0				Theriault and Grand-Bois[5a]
Canada	Ec	C	Res. in mining area	7	7.6	3.4–14.9	71	1.1	0.9–1.4	Camus et al[6b]
USA	CC	A	Res. near to asbestos plant				41	0.9	0.6–1.3	Hammond et al[7]
Austria	Ec	A	Res. in polluted town				36	0.8	0.4–1.6	Neuberger et al[8a]
Italy	Co	UM	Res. <1 km from asbestos cement plant	36	6.6	4.1–11				Magnani et al[9a]
Italy	CC	UM	Res. in polluted city	32	14.6	4.9–43.1				Magnani et al[10a]
Italy, Spain, Switzerland	CC	UM	Res. <2 km from potential source	17	11.5	3.5–38.2				Magnani et al[11]
UK	CC	A	Res. <0.5 miles from asbestos factory	11	5.4	1.8–17				Newhouse and Thompson[12]
UK	CC	A	Res. <0.5 km from potential source	5	6.6	0.9–50				Howel et al[13]
China	CC	C	Res. >20 years <0.2 km from asbestos plant	NA	182	NA	47	1.9	0.5–6.4	Xu et al[14a]
China	Co	A	Res. in polluted area				NA	5.7	NA	Luo et al[15]
New Caledonia	CC	A	Use of contaminated building materials	14	40.9	5.1–325	56	0.9	0.6–1.3	Luce et al[16a]
Australia	CC	A	Res. >5 years in mining area[c]	14	6.7	2.0–22.2				Hansen et al[17]

SD, study design: CC, case-control study; Co, cohort study; Ec, ecological study; TF, predominant type of fibre: A, amphiboles; C, chrysotile; UM, unspecified and mixed; Res., residence; Ca, number of exposed cases; RR, relative risk; CI, confidence interval.
[a]Results derived from raw data reported in the publication.
[b]Women only.
[c]Compared to residence <1 year.

exposure. A potential limitation of these studies, in particular those without assessments of exposure at the individual level ('ecological' studies), is possible concomitant occupational or household exposure to asbestos. The risk of mesothelioma was greatly increased in all but one study among individuals with environmental exposure to asbestos. Results for lung cancer, however, are less consistent, with an increased risk detected in studies from South Africa and China, but not in studies from Europe and North America. Imperfect control of confounding by smoking and other lung carcinogens may explain the lack of consistency.

According to a model used by WHO[18], 5% of the European population experience residential exposure to asbestos. A meta-analysis estimated the relative risk (RR) of mesothelioma from environmental exposure to asbestos at 3.5 (95% confidence interval [CI] 1.8–7.0). The corresponding RR of lung cancer was 1.1 (95% CI 0.9–1.5)[19]. Combining these results with a prevalence of exposure of 5%, leads to estimated annual numbers of 425 mesotheliomas in men and 56 in women, and (if one assumes a causal association between environmental exposure to asbestos and lung cancer) of 771 lung cancers in men and 206 in women in the European Union. The figure of 5% might however over-estimate the prevalence of exposure to circumstances comparable to those investigated in the studies listed in Table 1. A conservative estimate of 1% of the exposed population leads to estimates of 92 mesotheliomas in men and 12 in women, as well as 153 lung cancers in men and 41 in women. It should be stressed that in specific areas, such as Casale Monferrato in Italy[9] and Metsovo in Greece[20], the prevalence of heavy environmental exposure is relatively high, leading to a substantial burden of cancer.

Cancer risk from outdoor air pollution

Ambient air pollution has been implicated as a cause of various health effects, including cancer. Air pollution is a complex mixture of different gaseous and particulate components, and it is difficult to define an exposure measure of relevance when the biological mechanisms are largely unknown. The air pollution mix varies greatly by locality and time. In recent decades, emissions and air concentrations of traditional industrial air pollutants, such as SO_2 and smoke particles, have decreased, whereas there is an increasing or continued problem with air pollution from vehicles, with emissions of engine combustion products including volatile organic compounds, nitrogen oxides and fine particulates, as well as with secondarily increased ozone levels. There is biological rationale for a carcinogenic potential of numerous components of the air pollution mix, including benzo[a]pyrene, benzene, some metals, particles (especially fine particles) and possibly ozone.

Many definitions of outdoor air pollution exposure have been used in epidemiological studies. Earlier analytical studies generally compared residence in urban areas, where the air is considered more polluted, to residence in rural areas (for a review, see Katsouyanni and Pershagen[21]), sometimes providing limited data on the typical levels of some pollutants in the areas studied. Other studies have attempted to address exposure to specific components of outdoor air, providing risk estimates in relation to quantitative or semi-quantitative air pollution exposure assessments[22–35] or, in some cases, to more qualitative exposure assessments[14,36]. Another type of study has addressed residence in the proximity of specific sources of pollution, such as major industrial emission sources or heavy road traffic.

The evidence regarding outdoor air pollution and lung cancer has been the subject of several reviews[21,37–40]. We do not further review the evidence from ecological studies, given the abundance of analytical studies. Eleven cohort studies of outdoor air pollution have been reported[27,29,31,35,41–56], as well as a number of case–control studies[14,22–26,28,30,32–34,36,57–65].

In these studies, most reported RRs were adjusted for age and active smoking, but generally information on other potential confounders, such as occupational exposure, radon, passive smoking and dietary habits, was lacking. Overall, the studies suggest RRs of up to about 1.5 for urban *versus* rural residence or high *versus* low estimated air pollution exposure. There is no clear indication if early or late exposure is more important, and data on possible interaction with smoking or occupational exposures are inadequate.

Among these studies, four cohort[27,29,31,35,54–56] and 10 case–control[22–26,28,30,32,34,65] studies were based on measurements of specific air components. Selected RR estimates from these studies, with the corresponding air pollution differentials, are presented in Table 2.

Although the results reported in Table 2 are not directly comparable, mainly because of differences in exposure assessment, they tend to show an increased risk of lung cancer in the categories at highest exposure, which does not seem to be attributable to confounding factors. The studies of lung cancer and air pollution, however, suffer from several weaknesses. Exposure measurements are often crude, and sometimes only represent urban/rural contrasts. Population exposure estimates suffer from measurement errors due to mobility, not only long-term residential but also short-term around the area of residence. Even when quantitative exposure assessments were attempted, limited air monitoring data were a problem. Another limitation of many studies is that the sufficiently exposed population is diluted by a considerable number of minimally exposed persons.

An apparent inconsistency is that cancer mortality rates are often highest in medium-sized cities and lower in larger agglomerations. An ecological analysis of environmental correlates and total cancer mortality in 98 US cities found that vehicle density was an excellent predictor[66].

Table 2 Relative risk of lung cancer and outdoor air pollution measurements in some studies with quantitative or semi-quantitative exposure assessment

Location, study period	Reference	Sex	RR	95% CI	Exposure contrast[a]	Basis for exposure assessment for individuals and/or areas	Comments
Cohort studies							
USA, 1975–91	Dockery et al[29]	M + F	1.37	0.81–2.31	per 18.9 µg/m³ PM$_{2.5}$	City of residence in 1975. Pollutant average 1979–85	Study range 11–29.6 µg/m³ PM$_{2.5}$ across six studied cities
USA, 1982–98	Pope et al[56]	M + F	1.08	1.01–1.16	per 10 µg/m³ PM$_{2.5}$ (1979–1983)	City of residence in 1982. Pollutant averages 1979–83, 1999–2000 or average of 1979–1983 and 1999–2000	Mean (s.d.): 21.1 (4.6), 14.0 (3.0), 17.7 (3.7) for pollutant averages 1979–83, 1999–2000 or average of 1979–1983 and 1999–2000, respectively
			1.13	1.04–1.22	per 10 µg/m³ PM$_{2.5}$ (1999–2000)		Study range roughly 10–30 or 5–20 µg/m³ PM$_{2.5}$ (1979–1983 and 1999–2000, respectively)
			1.14	1.04–1.23	per 10 µg/m³ PM$_{2.5}$ (average of 1979–1983 & 1999–2000)		
	Pope et al[31]	M + F	1.36	1.1–1.7	per 19.9 µg/m³ SO$_4$	1982–1989. City of residence in 1982. Pollutant average 1980	Study range 3.6–23.5 µg/m³ SO$_4$ across 151 studied cities
USA, California, 1977–92	Beeson et al[53]	M	3.56	1.4–9.4	per 556 h/year above 200 µg/m³ O$_3$	Residential history 1973–92 and local monthly pollutant levels 1973–92 used to calculate individual subject averages over 1973–92	Range 988 h over study subjects
		M	5.21	1.9–14.0	per 24 µg/m³ PM$_{10}$		
		M	2.66	1.6–4.4	per 11 µg/m³ SO$_2$ (3.72 ppb)		
		F	2.14	1.4–3.4	per 11 µg/m³ SO$_2$ (3.72 ppb)		
	McDonnell et al[54]	M	2.23	0.56–8.94	per 24.3 µg/m³ PM$_{2.5}$	Residential history 1966–92 or 1973–92 (for PM$_{2.5}$ and PM$_{10}$, respectively) and local monthly pollutant estimates based on airport visibility data 1966–92 (for PM$_{2.5}$) and TSP 1973–86 and measured PM$_{10}$ 1987–1992 (for PM$_{10}$)	Results for smaller sub-cohort of subjects living near an airport

(Continued on next page)

Table 2 (continued from opposite page)							
Netherlands, 1986–94	Hoek et al[55]	M	1.84	0.59–5.67	per 29.5 μg/m³ PM$_{10}$		Mean (S.D.) 59.2 (16.8) and 31.9 (10.7) for PM$_{2.5}$ and PM$_{10}$, respectively over study subjects
		M + F	1.06	0.4–2.6	per 10 μg/m³ Black Smoke	Residential history of last 1–4 residences up to 1986 matched by GIS to regional and urban background estimates from National Air Quality Monitoring Network 1987–1990 plus local exposure based on distance to major road	Estimated range 9.6–35.8 μg/m³ Black Smoke and 14.7–67.2 μg/m³ NO$_2$ over study subjects
			1.25	0.4–3.7	per 30 μg/m³ NO$_2$		
Case control studies							
USA, Erie County, NY, 1957–65	Vena[23]	M	1.7	1.0–2.9	≥50 versus 0–49 years residence in areas with high (50–200 μg/m³) TSP	Lifetime residential history. Pollutant average 1961–63	Estimate for Erie county all-life residents
USA, Denver, 1979–82	Brownson et al[24]	M	1.66	0.7–4.2	≥100 exposure years versus 0–99	Lifetime residential history. TSP measurements by census tract	Exposure-years = residential years weighted (multiplied) by ranking index (1–10) based on census tract TSP levels
Japan, Osaka, 1965	Hitosugi[22]	F	1.51			Present residence. Pollutant average 1965	
		M	1.8	$P < 0.05$	SPM 390 versus 190 μg/m³, BaP 79 versus 26 μg/m³		
		F	1.2	$P > 0.05$			
Poland, Krakow, 1980–85	Jedrychowski et al[25]	M	1.46	1.1–2.0	TSP > 150 & SO$_2$ > 104 versus TSP < 150 & SO$_2$ < 104	Last residence. Pollutant average 1973–80	
		F	1.17	0.7–2.0	TSP > 150 or SO$_2$ > 104 versus TSP < 150 & SO$_2$ < 104		
Greece, Athens, 1987–89	Katsouyanni et al[26]	F	0.81		High versus low quartile, never-smokers	Lifetime residential and work address histories. Air pollution averages 1983–85 were used to classify areas into five categories. Range of BS from >400 in urban areas to <100 or less in rural areas	

(Continued on next page)

Table 2 (continued) Relative risk of lung cancer and outdoor air pollution measurements in some studies with quantitative or semi-quantitative exposure assessment

Location, study period	Reference	Sex	RR	95% CI	Exposure contrast[a]	Basis for exposure assessment for individuals and/or areas	Comments
Germany, 5 cities, 1984–88	Jöckel et al[28]	F	1.35		Same, smokers 1–29 years		
		F	2.23		Same, smokers 30+ years		
		M	1.16	0.6–2.1	High versus low time-weighted semi-quantitative index	Lifetime residential history. BaP, TSP, SO_2 data + local energy, SO_2 and coal use and degree of industrialization 1900–1980 classified areas in 10-year intervals 1895–1984	
		M	1.82		Same, using 20 years lag		
Italy, Trieste, 1979–81, 1985–86	Barbone et al[30]	M	1.4	1.1–1.8	>294 µg/m²/day versus <175 of particulate deposition	Last residence. Deposition measured 1972–77	
Russia, Moscow, 1991–93	Zaridze et al[32]	F	2.6	1.2–5.6	>200 SO_2, >60 NO_2, >2400 CO, >300 BS versus <50, <48, <1300, <60	Final = 20 years of residential history. Pollutant average 1971–75	Never-smokers only
Poland, Krakow, 1992–94	Pawlega et al[34]	M	0.24	0.1–0.5	TSP 142 & SO_2 110 versus 78 & 71	Last residence. Pollutant average 1973–80	Residual confounding by occupation possible despite some adjustment
Sweden, Stockholm, 1985–90	Nyberg et al[65]	M	1.6	1.1–2.4	NO_2 ≥ 29.26 versus NO_2 < 12.78 (Top decile versus lowest quartile)	30-year residential history. NO_2 estimates based on historical emission data and dispersion modelling (1960s, 70s and 80s)	

[a] µg/m³ if not otherwise indicated.

The authors noted that vehicle density generally levels off in more densely populated areas when other means of transportation are available, which might explain this inconsistency.

To pinpoint possible industrial emissions responsible for the suggested urban excess, populations living near point sources of air pollution have also been studied. Increased risks have been reported for living close to industries such as smelters, foundries, chemical industries, and others with various emissions[14,30,32,67–69], with up to doubled risk, although confidence intervals were mostly wide. Scottish ecological studies of residence near steel and iron foundries suggested a temporal relation between emission reductions and decreased lung cancer rates, with relatively short latency[70–77]. Other studies showed no relationship, however. For example, a recent epidemiological study could not show any association between heavy community exposure to sulphuric acid and lung cancer risk[78].

A number of studies concern sources of inorganic arsenic in air. Ecological studies suggested an increased lung cancer risk[79–84], which three early US case–control studies of residential exposure failed to substantiate[85–87]. In two subsequent case–controls studies with better control for smoking and occupation[88,89], however, estimated RRs for living near the smelter were around 2 and statistically significant. In a third study, the RR in the top quintile of exposure was 1.6 (P value for trend 0.07)[90]. Recent studies of 10 smelter towns in Arizona did not, however, observe any clear association or dose–response, but the arsenic content of the ore used was comparatively low[91,92].

The results of cohort and case–control studies regarding air pollution and lung cancer are too heterogeneous for a formal meta-analysis assessment of attributable cancers. The exposed proportion of the population in industrialized countries appears to lie in the range 15–75%, depending on the definition used[93–95]. As a conservative estimate, 20% of the population have low exposure (RR 1.1), 4% medium exposure (RR 1.3) and 1% high exposure (RR 1.5). In this scenario, the attributable proportion in the EU is approximately 3.6%, corresponding to some 7000 lung cancer cases per year.

Limited results are available for cancers other than the lung. In ecological studies, many individual cancers have shown elevated urban/rural ratios, including cancers of the mouth & throat, nasopharynx, oesophagus, stomach, colon, rectum, larynx, female breast, bladder and prostate[96,97]. Stronger associations were often reported for smoking-related cancers such as oesophagus, larynx and bladder cancer. However, the urban/rural ratio is often higher in men, suggesting that residual confounding by smoking or occupational exposures may be involved. Other ecological studies have related cancer rates to air pollutant measurements, or emission indexes or figures for fuel consumption. For example, stomach cancer was related to SO_2, particulates or fuel consumption in several studies[98–102] but not all[103], and prostate cancer was associated with measured particulate air pollution in two studies[99,104]. Occupational benzene exposure is a recognized cause of leukaemia[105,106]. A recent ecological study in 19 European countries found an inverse temporal association between gasoline use and leukaemia mortality or morbidity, but a weak positive spatial association[107]. Overall, it was not very supportive of an association to environmental benzene exposure. Previous ecological and case–control studies also provide unclear evidence.

Two cohort studies also provided data on cancers in organs other than the lung. For major sites, a Swedish study[46] found significant urban/rural ratios only for cancers of the bladder and uterine cervix, among smokers and non-smokers. In a Finnish cohort[50], there was some excess, for leukaemia and prostate cancer in particular, in urbanized but not in lifetime-urban men (mainly unmarried)[50].

One case–control study[108] examined the relationship between air pollution and childhood cancer, using residential traffic density as exposure proxy. An odds ratio of 1.7 (95% CI 1.0–2.8) was found for total childhood cancers and an odds ratio of 2.1 (95% CI 1.1–4.0) for leukaemias in a comparison of high- and low-traffic density addresses.

In summary, evidence concerning adult cancers other than lung cancer comes mainly from ecological studies, is not consistent, and is insufficient to attempt any estimate of a possible cancer burden. Likewise, no conclusion is possible for childhood cancer.

Cancer risk from exposure to environmental tobacco smoke

Environmental tobacco smoke is composed of sidestream and mainstream smoke, in which known, probable or possible human carcinogens are present. The International Agency for Research on Cancer has evaluated the evidence of a carcinogenic risk from exposure to environmental tobacco smoke, and has classified it as an established human carcinogen[109]. Confounding by dietary, occupational and social class-related factors can be reasonably excluded, and bias from misclassification of smokers is not likely to explain the results. On that occasion, a meta-analysis of epidemiological studies of lung cancer and adult exposure to environmental tobacco smoke was conducted, resulting in RRs of 1.22 (95% CI 1.12–1.32) in women and 1.36 (95% CI 1.02–1.82) in men from spousal exposure, and of 1.15 (95% CI 1.05–1.26) in women and 1.28 (95% CI 0.88–1.84) in men from workplace exposure. Other meta-analyses have reached very similar conclusions[110,111].

Table 3 presents our estimates of the numbers of lung cancers attributable to ETS exposure from the spouse and at the workplace in the European Union based on results of multicentre studies. In a study of lung cancer and ETS exposure, the proportion of never-smokers ever exposed to ETS

Table 3 Number of cases of lung cancer attributable to exposure to environmental tobacco smoke (ETS)

Study	Men					Women				
	RR	PE%	AF%	N^a	NA	RR	PE%	AF%	N^b	NA
Spousal ETS										
EPA[112]	1.17		0.66		29	1.17		6.0		436
Hackshaw et al[110]	1.24	3.9[c]	0.93	4400	41	1.24	37.8[d]	8.3	7220	601
Boffetta et al[113]	1.47		1.8		79	1.11		4.0		288
Workplace ETS										
Boffetta et al[113]	1.13	21.3[e]	2.7	4400	119	1.19	28.2[f]	5.1	7220	367

RR, relative risk from exposure to ETS; PE%, proportion of exposed cases; AF%, fraction of cases attributable to ETS; N, number of cases of lung cancer in the European Union in 1990; NA, number of cases attributable to ETS.
[a]4400 = 3% of lung cancers.
[b]7220 = 20% of lung cancers.
[c]0.13 (from European study) × 0.30 (prevalence of non-smokers).
[d]0.63 (from European study) × 0.60 (prevalence of non-smokers).
[e]0.71 (from European study) × 0.30 (prevalence of non-smokers).
[f]0.47 (from European study) × 0.60 (prevalence of non-smokers).

among controls was 13% in men and 63% in women for spousal ETS, and 71% in men and 47% in women for workplace ETS[113]. The average prevalence of never-smokers in Europe was estimated from a pooled analysis of case–control studies conducted in six European countries[114]: the overall prevalence among controls was 65% in women and 24% in men. Finally, from the same pooled analysis it was estimated that 29% of lung cancers in women and 2% in men occur among never-smokers[114].

The annual number of cases attributable to spousal ETS is in the order of 50 in men and over 500 in women. The corresponding estimates for ETS exposure at the workplace are about 200 cases among men and 270 cases among women.

Estimates made by the US Environmental Protection Agency[112] for the US population, which considered spousal and background sources of ETS, resulted in 1930 cases among women and 1130 cases among men. With respect to previous estimates, our exercise has the advantage of being based on actual measurements of exposure and risk derived from European populations.

The evidence of a causal association between ETS exposure and cancers in organs other than the lung is inconclusive[109].

Cancer risk from residential radon exposure

The carcinogenicity of radon decay products has been widely studied in occupationally exposed populations, in particular underground miners. This agent causes lung cancer in humans, but the evidence for an effect on other neoplasms is not conclusive[115]. The excess RR estimated from occupational cohorts, which included over 2500 cases of lung cancer occurring among over 60,000 miners, has been estimated in the order of 0.0049 per working level month of exposure[116]. Further refinements of this estimate took into account age at exposure and time since first exposure[117] as well as smoking status, with a stronger effect being shown among never-smokers than among smokers.

Although exposure levels in the houses are one order of magnitude smaller than in underground mines, the duration of exposure and the number of exposed individuals stress the importance of residence as a source of exposure to radon decay products. Several case–control studies of lung cancer from residential radon exposure have been reported in the literature, and their results have been reviewed and summarized[115,118,119]. A pooled RR of 1.06 (95% CI 1.01–1.10) has been calculated for individuals exposed at 100 Bq/m^3 *versus* unexposed[119], which is in agreement with the extrapolation from the results of occupationally exposed populations. Results of studies reported after these pooled analyses confirm these conclusions[120–123].

Most studies of residential radon rely on the historical reconstruction of exposure levels *via* household measurements. This approach is subject to substantial misclassification, most likely resulting in an underestimate of the risk. In a few studies, attempts were made to correct such biases, and the estimated RR increased by about 50%[124,125]. Furthermore, in one study[126], in which cumulative radon exposure was estimated from surface monitors rather than measurement in houses, the RR was higher (1.63, 95% CI 1.07–2.93 for exposure at 100 Bq/m^3).

Several estimates have been proposed of the number of lung cancers attributable to residential radon exposure. In one of the most detailed exercises, Darby and colleagues[119] estimated that this agent is responsible for 6.5% of all deaths from lung cancer in the UK, including 5.5% attributable to the joint effect of radon and smoking and 1% to residential radon alone. The figure of 1% corresponds to 349 deaths in the UK in 1998, or 9.4% of lung cancer deaths not due to tobacco smoking. If these figures are applied to other European countries, the number of lung cancer cases attributable to indoor radon exposure is in the order of 2000 per year.

Cancer risk from other sources of indoor air pollution

Based on the observation of very high lung cancer rates in some regions of China and elsewhere among women who spend much of their time at home, exposure to indoor air pollution from combustion sources used for heating and cooking, as well as high levels of cooking oil vapours resulting from some cooking methods, have been identified as risk factors for lung cancer. Table 4 presents a summary of results from relevant studies and illustrates the great variability in exposure measures across the case–control studies carried out in Asian populations.

Three main groups of factors influencing indoor air pollution ('smokiness') have been studied: (i) heating fuel: type of fuel, type of stove or central heating, ventilation, living area, subjective smokiness; (ii) cooking fuel: type of fuel, type of stove or open pit, ventilation of kitchen, location of cooking area in residence, frequency of cooking, smokiness; and (iii) fumes from frying oils: type of oil, frequency of frying, eye irritation when cooking. Many of the results are inconclusive, and the interpretation is difficult since the exposure measures used vary considerably. Nonetheless, strong and significant increases in risk have repeatedly been reported and merit consideration. In a recent review, it was concluded that the epidemiological findings regarding cooking oil vapours (group iii) from Chinese-style cooking are clearly suggestive of an effect and have some support from experimental data[140].

Limited data supporting a similar effect of exposure to cooking-derived indoor air pollution are available from other regions of the world. In a case control study from the Northern Province of South Africa[4], the odds ratio of lung cancer among women using wood or coal as main fuel was 1.4 (95% CI 0.6–3.2). A study conducted among white women in Los Angeles in 1981–82 reported that coal use for cooking and heating in the home during childhood and adolescence was associated with an odds ratio of 2.3 (95% CI 1.0–5.5) for adenocarcinoma and 1.9 (95% CI 0.5–6.5) for squamous cell cancer[141].

It appears plausible that indoor air pollution from combustion or cooking products (oil vapours in particular) could play a role in the causation of lung cancer. The relevance of the risks estimated in China for present-day conditions in Europe and North America is, however, somewhat questionable. Frying is less common in most parts of Europe than in China and kitchens are generally larger, better ventilated and separated from the living quarters. Central heating is increasingly common, and open combustion sources indoors are infrequent. However, given that lung cancer induction may span several decades, earlier living conditions may still play a role today in the risk of lung cancer among the middle-aged and older generations in Europe, although its importance should be waning.

Cancer risk from inorganic arsenic in drinking water

Inorganic arsenic causes cancer at various sites in humans[106]. The main source of environmental exposure to arsenic for the general population

Table 4 Case–control studies in Asian women of lung cancer and indoor air pollution from cooking habits

Study	Location[a], years	Ca	Co	Exposure variable[b]	RR	95% CI
Sobue[127]	Japan, Osaka, 1985	144	713	Wood/straw for cooking	1.8	1.1–2.9
				Combustion heating	1.2	0.8–1.7
Koo and Ho[128]	Hong Kong, 1981–83	200	200	41+ years of cooking	0.4	0.1–1.0
Ko et al[68]	Taiwan, Kaohsiung, 1992–93	105	105	Coal/anthracite for cooking	1.3	0.3–5.8
				Wood/charcoal for cooking	2.7	0.9–8.9
				Kitchen fume extractor	8.3	3.1–22.7
Wu-Williams et al[129]	Shenyang, 1985–87; Harbin, 1985–87	965	959	41+ years coal heating stoves	1.3	1.0–1.7
				50+ years kang use	1.6	0.9–2.8
Xu and co-workers[14,130,131]	Shenyang, 1985–87	520	557	50+ years coal heating stoves	1.2	$P > 0.05$
				50+ years kang use	3.4	$P < 0.05$
				20+ years direct burning kang	2.3	$P < 0.05$
				Indoor air pollution index	1.5	1.0–2.4
				Subjective smokiness	2.0	1.4–2.8
Wang et al[132]	Shenyang, 1992–93	135	135	Kang use	1.0	0.6–1.5
				Coal for heating	0.8	0.4–1.3
				Coal smoke when cooking	2.4	1.4–3.9
Dai et al[133]	Harbin, 1992–93	120	120	30+ years coal stove in bedroom	18.8	3.9–29
				25+ years coal heating	4.7	1.3–17
He et al[134]	Xuanwei, 1985–86	52	202	45+ years cooking	8.4	$P < 0.05$
Lan et al[135]	Xuanwei, 1988–90	139	139	Use of smoky coal	7.5	3.3–17.2
Lan et al[136]	Xuanwei, 1976–92	684	N/A	Use of chimney instead of stove	0.5	0.4–0.6
Liu et al[137]	Guangzhou, 1983–84	92	92	No separate kitchen	5.9	2.1–16
Du et al[138]	Guangzhou, 1985	283	283	Coal fumes exposure	2.2	1.2–4.2
Gao et al[139]	Shanghai, 1984–86	672	735	Coal for cooking	0.9	0.7–1.3
				Gas for cooking	1.1	0.7–1.5
				Wood for cooking	1.0	0.6–1.8
				Eye irritation	1.6	$P < 0.05$
				House smokiness	1.6	$P < 0.05$
Zhong et al[140]	Shanghai, 1992–94	504	601	No separate kitchen	1.24	0.94–1.62
				High smokiness	2.36	1.53–3.62
				Frying	1.79	0.94–3.41

Ca, number of cases; Co, number of controls; RR, relative risk; CI, confidence interval.
[a]Country is China if not otherwise specified.
[b]Reference category includes unexposed or low-exposure categories.

is through ingestion of contaminated water. A high level of arsenic in groundwater (up to 2–5000 μg/l) is found in areas of Argentina, Bangladesh, Bolivia, Chile, China (Xinjiang, Shanxi), India (West Bengal), Mexico, Mongolia, Taiwan, Thailand, the USA (Arizona, California, Nevada) and Vietnam. The most significant exposures, in terms of levels and populations, occur around the Gulf of Bengal, in South America and in Taiwan. In Europe, intermediate levels (not higher than 200 μg/l) are found in areas of Hungary and Romania in the Danube basin, as well as in Spain, Greece and Germany. There is strong evidence of an increased risk of bladder, skin and lung cancers following consumption of water with high arsenic contamination[109]. The evidence for an increased risk of other cancers, such as those of the liver, colon and kidney, are weaker but suggestive of a systemic effect. Most of the available studies have been conducted in areas with elevated arsenic content (typically above 200 μg/l). The results of studies of bladder cancer conducted in areas with low or intermediate contamination are suggestive of a possible increased risk. In an ecological study from Finland, the RR for concentrations of arsenic above 0.5 μg/l *versus* less than 0.1 μg/l was 2.44 (95% CI 0.95–1.96) with 3–9 years of latency, and it was 1.51 (95% CI 0.67–3.38) with 10 or more years of latency[142]. In a study from the USA, the RR for a cumulative dose of 53 mg or more, compared to less than 19 mg, was 1.14 (95% CI 0.7–2.9) overall, but it was 3.3 (95% CI 1.1–10.3) among smokers[143]. Very limited data are available on the risk of other neoplasms at low or intermediate exposure levels.

Few data are available on the proportion of the population in Europe exposed to arsenic in drinking water. In the study from Finland mentioned above[142], 5% of the study population consumed water with concentrations above 5 μg/l, including 1% with concentrations above the WHO guideline of 10 μg/l.

Cancer risk from water chlorination by-products

Access to unpolluted water is one of the requirements of human health. Water quality is influenced by seasons, geology and discharges of agriculture and industry. Microbiological contamination of water is controlled by disinfection methods based on oxidants like chlorine, hypochlorite, chloramine, chlorine dioxine and ozone. Drinking water may contain a variety of potentially carcinogenic agents, including chlorination by-products[144]. Showering and bathing represent another important source of exposure to chlorination by-products.

Chlorination by-products result from the interaction of chlorine with organic chemicals, whose level determines the concentration of the by-products. Among the many halogenated compounds that may be formed, trihalomethanes are those most commonly found. Trihalomethanes include chloroform, bromodichloromethane, chlorodibromomethane and bromoform. Brominated by-products are formed from the reaction of chlorinated by-products with bromide, present at low levels in drinking water. Concentrations of trihalomethanes show a wide range, mainly as a result of the occurrence of water contamination by organic chemicals: average measurements from the USA[144] are in the order of 10 µg/l for chloroform, bromodichloromethane and chlorodibromomethane, whereas those for bromoform are close to 5 µg/l. In 1992, Morris and colleagues[145] carried out a meta-analysis of cancer risk from consumption of chlorinated drinking water. They estimated an RR of 1.21 (95% CI 1.09–1.34) for bladder cancer, based on seven studies. This estimate was not modified after adjusting for smoking. Two further studies of bladder cancer risk have been published[146,147], and their results are in line with the meta-analysis by Morris and colleagues. In three studies, information on duration of exposure was reported[146–148]: a meta-analysis of the results for the category with longest exposure (more than 30 or 40 years, depending on the study) resulted in a pooled RR of 1.68 (95% CI 1.25–2.27). Results based on estimated intake of trihalomethanes and on other disinfection by-products are too sparse to allow a conclusion.

The interpretation of these data is complicated by several factors. The concentration of by-products in water varies depending on the presence of organic contaminants, which differs by geographical area and by season. In addition, people consume water outside their homes, which is seldom considered in the assessment of exposure in epidemiological studies. Furthermore, although the possible confounding effect of smoking has been taken into account in several studies, confounding by other risk factors such as diet remains a possibility. Despite the good consistency of the available studies on bladder cancer, the uncertainties in exposure assessment caution against the conclusion that a causal link has been established between consumption of chlorinated drinking water and increased risk of bladder cancer[149].

The evidence for an association between chlorination by-products and cancers in organs other than the bladder is inconclusive[149], although some of them are considered possible human carcinogens because of evidence of carcinogenicity in experimental systems[109].

Cancer risk from other drinking water pollutants

Several other groups of pollutants of drinking water have been investigated as possible sources of cancer risk in humans[149,150]. They include organic compounds derived from industrial, commercial and agricultural activities, and in particular from waste sites, nitrites & nitrates,

Table 5 Epidemiological studies of nitrate in drinking water (NDW) and risk of stomach cancer

Country	Period	Design	Exposure	Results	Study
Colombia	1968–72	CC, I	NDW in areas of residence	+	Cuello et al[151]
Hungary	1960–79	E, I	NDW and soil type in a county	+	Juhasz et al[152]
UK	1969–73	E, Mo	NDW in 32 rural districts	+ Male; – Female	Fraser and Chilvers[153]
Denmark	1943–72	E, I	NDW in two towns	+	Jensen[154]
Italy	1976–79	E, I	NDW (20+ mg/l) in 1199 communities	+	Gilli et al[155]
UK	1969–73	E, Mo	NDW of 253 urban areas	–	Beresford[156]
USA	1982–85	E, Mo	NDW in Wisconsin areas	–	Rademacher et al[157]

CC, case–control study; E, ecological study; I, incidence; Mo, mortality.

radionuclides and asbestos. Most of the studies were based on ecological comparisons and did not provide a quantitative risk estimate. Several cancer sites were analysed in these studies and selective reporting of positive results, resulting in an over-estimate of the risk, is a possibility.

Despite these limitations, three sets of results are particularly interesting. Firstly, an increased risk of stomach cancer has been repeatedly reported in areas with high nitrate levels in drinking water (Table 5). However, the two most recent studies did not confirm these findings[156,157].

Secondly, two studies are available on the association between nitrate level in drinking water and risk of non-Hodgkin lymphoma in the USA. Weisenburger[158] found a higher rate of lymphoma in eastern counties of Nebraska with more than 20% of wells with nitrate levels exceeding the standard, as compared to counties with less than 10% of such wells. In a case–control study in the same region, Ward et al[159] found an increased risk for high cumulative intake of nitrates in drinking water. A further case–control study of bladder cancer reported an association with high nitrate level in drinking water[160].

Finally, two ecological studies from the USA reported an increased risk of leukaemia in adults among residents in areas with elevated levels of radium in drinking water[161,162]. A third study reported a similar association between radon levels and childhood leukaemia[163].

Conclusions

A number of circumstances of environmental exposure to carcinogens has definitely been linked with an increased risk of cancer in humans. For some of them, the available data allow an attempt to quantify the burden of cancer. Uncertainties in all the components of such quantifications, however, suggest great caution in their interpretation: they should be considered as indicating the likely order of magnitude of the risk based on current knowledge.

It is noteworthy, however, that despite the relatively small relative risks of cancer following exposure to environmental carcinogens, the number of cases that might be caused, assuming a causal relationship, is relatively large, as a result of the high prevalence of exposure. This emphasizes the need for a better understanding of the actual risk of cancer posed by environmental factors, and of the effect of measurements aimed at controlling exposure to environmental carcinogens.

References

1. Tomatis L, Aitio A, Day NE et al. *Cancer: Causes, Occurrence and Control* (IARC Scientific Publications No 100). Lyon: International Agency for Research on Cancer, 1990
2. INSERM. *Effets sur la santé des principaux types d'exposition à l'amiante* (Expertise Collective INSERM). Paris: Institut National de la Santé et de la Recherche Médicale, 1997
3. Botha JL, Irwig LM, Strebel P. Excess mortality from stomach cancer, lung cancer, and asbestosis and/or mesothelioma in crocidolite mining districts in South Africa. *Am J Epidemiol* 1986; **123**: 30–40
4. Mzileni O, Sitas F, Steyn K, Carrara H, Bekker P. Lung cancer, tobacco, and environmental factors in the African population of the Northern Province, South Africa. *Tobacco Control* 1999; **8**: 398–401
5. Theriault GP, Grand-Bois L. Mesothelioma and asbestos in the province of Quebec, 1969–1972. *Arch Environ Health* 1978; **33**: 15–9
6. Camus M, Siemiatycki J, Meek B. Nonoccupational exposure to chrysotile asbestos and the risk of lung cancer. *N Engl J Med* 1998; **338**: 1565–71
7. Hammond EC, Garfinkel L, Selikoff IJ, Nicholson WT. Mortality experience of residents in the neighbourhood of an asbestos factory. *Ann NY Acad Sci* 1979; **330**: 417–22
8. Neuberger M, Kundi M, Friedl HP. Environmental asbestos exposure and cancer mortality. *Arch Environ Health* 1984; **39**: 261–5
9. Magnani C, Terracini B, Ivaldi C, Botta M, Mancini A, Andrion A. Pleural malignant mesothelioma and non-occupational exposure to asbestos in Casale Monferrato, Italy. *Occup Environ Med* 1995; **52**: 362–7
10. Magnani C, Dalmasso P, Biggeri A, Ivaldi C, Mirabelli D, Terracini B. Increased risk of malignant mesothelioma of the pleura after residential or domestic exposure to asbestos: a case–control study in Casale Monferrato, Italy. *Environ Health Perspect* 2001; **109**: 915–9
11. Magnani C, Agudo A, Gonzalez CA et al. Multicentric study on malignant pleural mesothelioma and non-occupational exposure to asbestos. *Br J Cancer* 2000; **83**: 104–11
12. Newhouse M, Thompson H. Mesothelioma of pleura and peritoneum following exposure to asbestos in the London area. *Br J Ind Med* 1965; **2**: 261–9
13. Howel D, Arblaster L, Swinburne L, Schweiger M, Renvoize E, Hatton P. Routes of asbestos exposure and the development of mesothelioma in an English region. *Occup Environ Med* 1997; **54**: 403–9
14. Xu ZY, Blot WJ, Xiao HP et al. Smoking, air pollution, and the high rates of lung cancer in Shenyang, China. *J Natl Cancer Inst* 1989; **81**: 1800–6
15. Luo S, Zhang Y, Mu S, Zhang C, Ma T, Liu X. The risk of lung cancer and mesothelioma in farmers exposed to crocidolite in the environment. *Hua Xi Yi Ke Da Xue Xue Bao* 1998; **29**: 63–5 [in Chinese]
16. Luce D, Bugel I, Goldberg P et al. Environmental exposure to tremolite and respiratory cancer in New Caledonia: a case–control study. *Am J Epidemiol* 2000; **151**: 259–65
17. Hansen J, de Klerk NH, Musk AW, Hobbs MST. Environmental exposure to crocidolite and mesothelioma: exposure–response relationships. *Am J Respir Crit Care Med* 1998; **157**: 69–75
18. WHO. *Air Quality Guidelines for Europe* (WHO Regional Publications, European Series No 23). Copenhagen: World Health Organization Regional Office for Europe, 1987; 182–99

19 Bourdes V, Boffetta P, Pisani P. Environmental exposure to asbestos and risk of pleural mesothelioma: review and meta-analysis. *Eur J Epidemiol* 2000; **16**: 411–7

20 Sakellariou K, Malamou-Mitsi V, Haritou A, Koumpaniou C, Stachouli C, Dimoliatis ID, Constantopoulos SH. Malignant pleural mesothelioma from nonoccupational asbestos exposure in Metsovo (north-west Greece): slow end of an epidemic? *Eur Respir J* 1996; **9**: 1206–10

21 Katsouyanni K, Pershagen G. Ambient air pollution exposure and cancer. *Cancer Causes Control* 1997; **8**: 284–91

22 Hitosugi M. Epidemiological study of lung cancer with special reference to the effect of air pollution and smoking habits. *Bull Inst Public Health* 1968; **17**: 237–56

23 Vena JE. Air pollution as a risk factor in lung cancer. *Am J Epidemiol* 1982; **116**: 42–56

24 Brownson RC, Reif JS, Keefe TJ, Ferguson SW, Pritzl JA. Risk factors for adenocarcinoma of the lung. *Am J Epidemiol* 1987; **125**: 25–34

25 Jedrychowski W, Becher H, Wahrendorf J, Basa-Cierpialek Z. A case–control study of lung cancer with special reference to the effect of air pollution in Poland. *J Epidemiol Community Health* 1990; **44**: 114–20

26 Katsouyanni K, Trichopoulos D, Kalandidi A, Tomos P, Riboli E. A case–control study of air pollution and tobacco smoking in lung cancer among women in Athens. *Prev Med* 1991; **20**: 271–8

27 Mills PK, Abbey DE, Beeson WL, Petersen F. Ambient air pollution and cancer in California seventh-day adventists. *Arch Environ Health* 1991; **46**: 271–80

28 Jöckel KH, Ahrens W, Wichmann HE *et al.* Occupational and environmental hazards associated with lung cancer. *Int J Epidemiol* 1992; **21**: 202–13

29 Dockery DW, Pope 3rd CA, Xu X *et al.* An association between air pollution and mortality in six U.S. cities. *N Engl J Med* 1993; **329**: 1753–9

30 Barbone F, Bovenzi M, Cavallieri F, Stanta G. Air pollution and lung cancer in Trieste, Italy. *Am J Epidemiol* 1995; **141**: 1161–9

31 Pope 3rd CA, Thun MJ, Namboodiri MM, Dockery DW, Evans JS, Speizer FE, Heath Jr CW. Particulate air pollution as a predictor of mortality in a prospective study of U.S. adults. *Am J Respir Crit Care Med* 1995; **151**: 669–74

32 Zaridze DG, Zemlianaia GM, Aitakov ZN. Role of outdoor and indoor air pollution in the etiology of lung cancer. *Vestnik Rossiiskoi Akademii Meditsinskikh Nauk* 1995; **4**: 6–10 [in Russian]

33 Biggeri A, Barbone F, Lagazio C, Bovenzi M, Stanta G. Air pollution and lung cancer in Trieste, Italy: spatial analysis of risk as a function of distance from sources. *Environ Health Perspect* 1996; **104**: 750–4

34 Pawlega J, Rachtan J, Dyba T. Evaluation of certain risk factors for lung cancer in Cracow (Poland)—a case–control study. *Acta Oncol* 1997; **36**: 471–6

35 Abbey DE, Nishino N, McDonnell WF, Burchette RJ, Knutsen SF, Beeson WL, Yang JX. Long-term inhalable particles and other air pollutants related to mortality in nonsmokers. *Am J Respir Crit Care Med* 1999; **159**: 373–82

36 Pike MC, Jing JS, Rosario IP, Henderson BE, Menck HR. Occupation: 'explanation' of an apparent air pollution related excess of lung cancer in Los Angeles County. In: Breslow N, Whitemore A (eds) *Energy and Health*. Philadelphia, PA: SIAM, 1979; 3–16

37 Pershagen G, Simonato L. Epidemiological evidence on outdoor air pollution and cancer. In: Tomatis L (ed.) *Indoor and Outdoor Air Pollution and Human Cancer*. Berlin: Springer-Verlag, 1993; 135–48

38 Hemminki K, Pershagen G. Cancer risk of air pollution: epidemiological evidence. *Environ Health Perspect* 1994; **102 (Suppl. 4)**: 187–92

39 Cohen AJ, Pope CA. Lung cancer and air pollution. *Environ Health Perspect* 1995; **103 (Suppl. 8)**: 219–24

40 Cohen AJ. Outdoor air pollution and lung cancer. *Environ Health Perspect* 2000; **108 (Suppl. 4)**: 743–50

41 Hammond EC, Horn D. Smoking and death rates—Report on forty-four months of follow-up of 187,783 men, I: Total mortality. *JAMA* 1958; **166**: 1159–72

42 Hammond EC, Horn D. Smoking and death rates—Report on forty-four months of follow-up of 187,783 men, II: Death rates by cause. *JAMA* 1958; **166**: 1294–308

43 Buell P, Dunn JE. Relative impact of smoking and air pollution on lung cancer. *Arch Environ Health* 1967; **15**: 291–7

44 Buell P, Dunn JE, Breslow L. Cancer of the lung and Los Angeles-type air pollution: prospective study. *Cancer* 1967; **20**: 2139–47

45 Hammond EC. Smoking habits and air pollution in relation to lung cancer. In: Lee DH (ed.) *Environmental Factors in Respiratory Disease*. New York, NY: Academic Press, 1972; 177–98

46 Cederlöf R, Friberg L, Hrubek Z, Lorich U. *The Relationship of Smoking and Some Social Covariables to Mortality and Cancer Morbidity: A Ten Year Follow-up in a Probability Sample of 55,000 Swedish Subjects Age 18 to 69* (2 volumes). Stockholm: Department of Environmental Hygiene, Karolinska Institute, 1975

47 Hammond EC, Garfinkel L. General air pollution and cancer in the United States. *Prev Med* 1980; **9**: 206–11

48 Doll R, Peto R. The causes of cancer: quantitative estimates of avoidable risks of cancer in the United States today. *J Natl Cancer Inst* 1981 **66**: 1191–308

49 Tenkanen L, Hakulinen T, Hakama M, Saxen E. Sauna, dust and migration as risk factors in lung cancer among smoking and non-smoking males in Finland. *Int J Cancer* 1985; **35**: 637–42

50 Tenkanen L, Teppo L. Migration, marital status and smoking as risk determinants of cancer. *Scand J Soc Med* 1987; **15**: 67–72

51 Tenkanen L. Migration to towns, occupation, smoking, and lung cancer: experience from the Finnish-Norwegian lung cancer study. *Cancer Causes Control* 1993; **4**: 133–41

52 Engholm G, Palmgren F, Lynge E. Lung cancer, smoking, and environment: a cohort study of the Danish population. *BMJ* 1996; **312**: 1259–63

53 Beeson WL, Abbey DE, Knutsen SF. Long-term concentrations of ambient air pollutants and incident lung cancer in California adults: results from the AHSMOG study. *Environ Health Perspect* 1998; **106**: 813–22

54 McDonnell WF, Nishino-Ishikawa N, Petersen FF, Chen LH, Abbey DE. Relationships of mortality with the fine and coarse fractions of long-term ambient PM10 concentrations in nonsmokers. *J Expos Anal Environ Epidemiol* 2000; **10**: 427–36

55 Hoek G, Brunekreef B, Goldbohm S, Fischer P, van den Brandt PA. Association between mortality and indicators of traffic-related air pollution in the Netherlands: a cohort study. *Lancet* 2002; **360**: 1203–9

56 Pope 3rd CA, Burnett RT, Thun MJ, Calle EE, Krewski D, Ito K, Thurston GD. Lung cancer, cardiopulmonary mortality, and long-term exposure to fine particulate air pollution. *JAMA* 2002; **287**: 1132–41

57 Stocks P, Campbell JM. Lung cancer death rates among non-smokers and pipe and cigarette smokers: an evaluation in relation to air pollution by benzopyrene and other substances. *BMJ* 1955; **2**: 923–9

58 Haenszel W, Loveland DB, Sirken MG. Lung cancer mortality as related to residence and smoking histories, I: White males. *J Natl Cancer Inst* 1962; **28**: 947–1001

59 Haenszel W, Taeuber KE. Lung cancer mortality as related to residence and smoking histories, II: White females. *J Natl Cancer Inst* 1964; **32**: 803–38

60 Dean G. Lung cancer and bronchitis in Northern Ireland, 1960–62. *BMJ* 1966; **5502**: 1506–14

61 Dean G, Lee PN, Todd GF, Wicken AJ. *Report on a Second Retrospective Mortality Study in North-East England, Part I: Factors Related to Mortality from Lung Cancer, Bronchitis, Heart Disease and Stroke in Cleveland County, with Particular Emphasis on the Relative Risks Associated with Smoking Filter and Plain Cigarettes*. London: Tobacco Research Council, 1977

62 Dean G, Lee PN, Todd GF, Wicken AJ. *Report on a Second Retrospective Mortality Study in North-East England, Part II: Changes in Lung Cancer and Bronchitis Mortality and in Other Relevant Factors Occurring in Areas of North-east England, 1963–72*. London: Tobacco Research Council, 1978

63 Samet JM, Humble CG, Skipper BE, Pathak DR. History of residence and lung cancer risk in New Mexico. *Am J Epidemiol* 1987; **125**: 800–11

64 Holowaty EJ, Risch HA, Miller AB, Burch JD. Lung cancer in women in the Niagara Region, Ontario: a case–control study. *Can J Public Health* 1991; **82**: 304–9

65 Nyberg F, Gustavsson P, Jarup L, Bellander T, Berglind N, Jakobsson R, Pershagen G. Urban air pollution and lung cancer in Stockholm. *Epidemiology* 2000; **11**: 487–95

66 Robertson LS. Environmental correlates of intercity variation in age-adjusted cancer mortality rates. *Environ Health Perspect* 1980; **36**: 197–203

67 Shear CL, Seale DB, Gottlieb MS. Evidence for space-time clustering of lung cancer deaths. *Arch Environ Health* 1980; **35**: 335–43
68 Ko YC, Lee CH, Chen MJ *et al.* Risk factors for primary lung cancer among non-smoking women in Taiwan. *Int J Epidemiol* 1997; **26**: 24–31
69 Bhopal RS, Moffatt S, Pless-Mulloli T, Phillimore PR, Foy C, Dunn CE, Tate JA. Does living near a constellation of petrochemical, steel, and other industries impair health? *Occup Environ Med* 1998; **55**: 812–22
70 Lloyd OL. Respiratory-cancer clustering associated with localised industrial air pollution. *Lancet* 1978; **1**: 318–20
71 Gailey FA, Lloyd OL. A wind-tunnel study of the flow of air pollution in Armadale, central Scotland. *Ecol Dis* 1983; **2**: 419–31
72 Lloyd OL, Smith G, Lloyd MM, Holland Y, Gailey F. Raised mortality from lung cancer and high sex ratios of births associated with industrial pollution. *Br J Ind Med* 1985; **42**: 475–80
73 Lloyd OL, Williams FL, Gailey FA. Is the Armadale epidemic over? Air pollution and mortality from lung cancer and other diseases, 1961–82. *Br J Ind Med* 1985; **42**: 815–23
74 Gailey FA, Lloyd OL. Atmospheric metal pollution monitored by spherical moss bags: a case study of Armadale. *Environ Health Perspect* 1986; **68**: 187–96
75 Lloyd OL, Ireland E, Tyrrell H, Williams F. Respiratory cancer in a Scottish industrial community: a retrospective case–control study. *J Soc Occup Med* 1986; **36**: 2–8
76 Smith GH, Williams FL, Lloyd OL. Respiratory cancer and air pollution from iron foundries in a Scottish town: an epidemiological and environmental study. *Br J Ind Med* 1987; **44**: 795–802
77 Williams FL, Lloyd OL. The epidemic of respiratory cancer in the town of Armadale: the use of long-term epidemiological surveillance to test a causal hypothesis. *Public Health* 1988; **102**: 531–8
78 Petrauskaite R, Pershagen G, Gurevicius R. Lung cancer near an industrial site in Lithuania with major emissions of airway irritants. *Int J Cancer* 2002; **99**: 106–11
79 Blot WJ, Fraumeni Jr JF. Arsenical air pollution and lung cancer. *Lancet* 1975; **2**: 142–4
80 Newman JA, Archer VE, Saccomanno G, Kuschner M, Auerbach O, Grondahl RD, Wilson JC. Histologic types of bronchogenic carcinoma among members of copper-mining and smelting communities. *Ann NY Acad Sci* 1976; **271**: 260–8
81 Pershagen G, Elinder CG, Bolander AM. Mortality in a region surrounding an arsenic emitting plant. *Environ Health Perspect* 1977; **19**: 133–7
82 Matanoski GM, Landau E, Tonascia J, Lazar C, Elliott EA, McEnroe W, King K. Cancer mortality in an industrial area of Baltimore. *Environ Res* 1981; **25**: 8–28
83 Cordier S, Theriault G, Iturra H. Mortality patterns in a population living near a copper smelter. *Environ Res* 1983; **31**: 311–22
84 Xiao HP, Xu ZY. Air pollution and lung cancer in Liaoning Province, People's Republic of China. *NCI Monogr* 1985; **69**: 53–8
85 Lyon JL, Fillmore JL, Klauber MR. Arsenical air pollution and lung cancer. *Lancet* 1977; **2**: 869
86 Greaves WW, Rom WN, Lyon JL, Varley G, Wright DD, Chiu G. Relationship between lung cancer and distance of residence from nonferrous smelter stack effluent. *Am J Ind Med* 1981; **2**: 15–23
87 Rom WN, Varley G, Lyon JL, Shopkow S. Lung cancer mortality among residents living near the El Paso smelter. *Br J Ind Med* 1982; **39**: 269–72
88 Brown LM, Pottern LM, Blot WJ. Lung cancer in relation to environmental pollutants emitted from industrial sources. *Environ Res* 1984; **34**: 250–61
89 Pershagen G. Lung cancer mortality among men living near an arsenic-emitting smelter. *Am J Epidemiol* 1985; **122**: 684–94
90 Frost F, Harter L, Milham S, Royce R, Smith AH, Hartley J, Enterline P. Lung cancer among women residing close to an arsenic emitting copper smelter. *Arch Environ Health* 1987; **42**: 148–52
91 Marsh GM, Stone RA, Esmen NA *et al.* A case–control study of lung cancer mortality in six Gila Basin, Arizona smelter towns. *Environ Res* 1997; **75**: 56–72
92 Marsh GM, Stone RA, Esmen NA *et al.* A case–control study of lung cancer mortality in four rural Arizona smelter towns. *Arch Environ Health* 1998; **53**: 15–28
93 WHO European Centre for Environment. Air pollution. In: *Concern for Europe's Tomorrow: Health and the Environment in the WHO European Region.* Stuttgart: Wissenschaftliche Verlagsgesellschaft, 1995; 139–75

94 Dreyer L, Andersen A, Pukkala E. Avoidable cancers in the Nordic countries: external environment. *APMIS Suppl.* 1997; **76**: 80–2

95 EEA. *Europe's Environment: The Second Assessment*. Luxembourg: Office for Official Publications of the European Communities, 1998

96 Levin ML, Haenszel W, Carroll BE, Gerhardt PR, Handy VH, Ingraham SC. Cancer incidence in urban and rural areas of New York State. *J Natl Cancer Inst* 1960; **24**: 1243–57

97 Goldsmith JR. The 'urban factor' in cancer: smoking, industrial exposures, and air pollution as possible explanations. *J Environ Pathol Toxicol* 1980; **3**: 205–17

98 Stocks P. On the relations between atmospheric pollution in urban and rural localities and mortality from cancer, bronchitis and pneumonia, with particular reference to 3:4 benzopyrene, beryllium, molybdenum, vanadium and arsenic. *Br J Cancer* 1960; **14**: 397–418

99 Hagstrom RM, Sprague HA, Landau E. The Nashville Air Pollution Study, VII: Mortality from cancer in relation to air pollution. *Arch Environ Health* 1967; **15**: 237–48

100 Gardner MJ, Crawford MD, Morris JN. Patterns of mortality in middle and early old age in the county boroughs of England and Wales. *Br J Prev Soc Med* 1969; **23**: 133–40

101 Winkelstein Jr W, Kantor S. Stomach cancer: positive association with suspended particulate air pollution. *Arch Environ Health* 1969; **18**: 544–7

102 Lave LB, Seskin EP. Air pollution and human health. *Science* 1970; **169**: 723–33

103 Chinn S, Florey CD, Baldwin IG, Gorgol M. The relation of mortality in England and Wales 1969–1973 to measurements of air pollution. *J Epidemiol Community Health* 1981; **35**: 174–9

104 Winkelstein Jr W, Kantor S. Prostatic cancer: relationship to suspended particulate air pollution. *Am J Public Health* 1969; **59**: 1134–8

105 IARC. *Some Industrial Chemicals and Dyestuffs*. IARC Monographs on the Evaluation of the Carcinogenic Risk of Chemicals to Humans, vol. 29. Lyon: International Agency for Research on Cancer, 1982

106 IARC. *Overall Evaluations of Carcinogenicity: An Updating of IARC Monographs Volumes 1 to 42*. IARC Monographs on the Evaluation of Carcinogenic Risks to Humans, **Suppl.** 7. Lyon: International Agency for Research on Cancer, 1987

107 Swaen GM, Slangen JJ. Gasoline consumption and leukemia mortality and morbidity in 19 European countries: an ecological study. *Int Arch Occup Environ Health* 1995; **67**: 85–93

108 Savitz DA, Feingold L. Association of childhood cancer with residential traffic density. *Scand J Work Environ Health* 1989; **15**: 360–3

109 IARC. *Tobacco Smoking and Involuntary Tobacco Smoke*. IARC Monographs on the Evaluation of the Carcinogenic Risk of Chemicals to Humans, vol. 83. Lyon: International Agency for Research on Cancer; In press

110 Hackshaw AK, Law MR, Wald NJ. The accumulated evidence on lung cancer and environmental tobacco smoke. *BMJ* 1997; **315**: 980–8

111 Boffetta P. Involuntary smoking and lung cancer. *Scand J Work Environ Health* 2002; **28** (**Suppl. 2**): 30–40

112 US EPA. *Respiratory Health Effects of Passive Smoking: Lung Cancer and Other Disorders*. Washington, DC: US Environmental Protection Agency, 1992

113 Boffetta P, Agudo A, Ahrens W *et al*. Multicenter case–control study of exposure to environmental tobacco smoke and lung cancer in Europe. *J Natl Cancer Inst* 1998; **90**: 1440–50

114 Simonato L, Roesch F, Gaborieau V, Sartorel S. *A Multicentric Case-Control Study of the Major Risk Factors for Lung Cancer in Europe with Particular Emphasis on Intercountry Comparison*. Padua: Venetial Cancer Registry, 1997

115 IARC. *Ionizing Radiation, Part 2, Some Internally Deposited Radionuclides*. IARC Monographs on the Evaluation of Carcinogenic Risks to Humans, vol. 77. Lyon: International Agency for Research on Cancer, 2001

116 Lubin JH, Boice Jr JD, Edling C *et al*. Lung cancer in radon-exposed miners and estimation of risk from indoor exposure. *J Natl Cancer Inst* 1995; **87**: 817–27

117 National Research Council. *Committee on Health Risks of Exposure to Radon: BEIR VI, Health Effects of Exposure to Radon*. Washington, DC: National Academy Press, 1999

118 Lubin JH, Boice Jr JD. Lung cancer risk from residential radon: meta-analysis of eight epidemiologic studies. *J Natl Cancer Inst* 1997; **89**: 49–57

119 Darby S, Hill D, Doll R. Radon: a likely carcinogen at all exposures. *Ann Oncol* 2001; **12**: 1341–51

120 Pisa FE, Barbone F, Betta A, Bonomi M, Alessandrini B, Bovenzi M. Residential radon and risk of lung cancer in an Italian alpine area. *Arch Environ Health* 2001; **56**: 208–15
121 Tomasek L, Kunz E, Müller T et al. Radon exposure and lung cancer risk—Czech cohort study on residential radon. *Sci Total Environ* 2001; **272**: 43–51
122 Barros-Dios JM, Barreiro MA, Ruano-Ravina A, Figueiras A. Exposure to residential radon and lung cancer in Spain: a population-based case–control study. *Am J Epidemiol* 2002; **156**: 548–55
123 Wang Z, Lubin JH, Wang L et al. Residential radon and lung cancer risk in a high-exposure area of Gansu Province, China. *Am J Epidemiol* 2002; **155**: 554–64
124 Lagarde F, Pershagen G, Akerblom G et al. Residential radon and lung cancer in Sweden: risk analysis accounting for random error in the exposure assessment. *Health Phys* 1997; **72**: 269–76
125 Darby S, Whitley E, Silcocks P et al. Risk of lung cancer associated with residential radon exposure in south-west England: a case–control study. *Br J Cancer* 1998; **78**: 394–408
126 Alavanja MC, Lubin JH, Mahaffey JA, Brownson RC. Residential radon exposure and risk of lung cancer in Missouri. *Am J Public Health* 1999; **89**: 1042–8
127 Sobue T. Association of indoor air pollution and lifestyle with lung cancer in Osaka, Japan. *Int J Epidemiol* 1990; 19 (**Suppl. 1**): 62–6
128 Koo LC, Ho JH. Diet as a confounder of the association between air pollution and female lung cancer: Hong Kong studies on exposures to environmental tobacco smoke, incense, and cooking fumes as examples. *Lung Cancer* 1996; **14** (**Suppl. 1**): 47–61
129 Wu-Williams AH, Dai XD, Blot W et al. Lung cancer among women in north-east China. *Br J Cancer* 1990; **62**: 982–7
130 Xu ZY, Blot WJ, Li G et al. Environmental determinants of lung cancer in Shenyang, China. In: O'Neill IK, Chen J, Bartsch H (eds) *Relevance to Human Cancer of N-nitroso Compounds, Tobacco Smoke and Mycotoxins* (IARC Scientific Publications No 105). Lyon: International Agency for Research on Cancer, 1991; 460–5
131 Xu ZY, Brown L, Pan GW et al. Lifestyle, environmental pollution and lung cancer in cities of Liaoning in northeastern China. *Lung Cancer* 1996; **14** (**Suppl. 1**): 149–60
132 Wang SY, Hu YL, Wu YL, Li X, Chi GB, Chen Y, Dai WS. A comparative study of the risk factors for lung cancer in Guangdong, China. *Lung Cancer* 1996; **14** (**Suppl. 1**): 99–105
133 Dai XD, Lin CY, Sun XW, Shi YB, Lin YJ. The etiology of lung cancer in nonsmoking females in Harbin, China. *Lung Cancer* 1996; **14** (**Suppl. 1**): 85–91
134 He XZ, Chen W, Liu ZY, Chapman RS. An epidemiological study of lung cancer in Xuan Wei County, China: Current progress—Case–control study on lung cancer and cooking fuel. *Environ Health Perspect* 1991; **94**: 9–13
135 Lan Q, Chen W, Chen H, He XZ. Risk factors for lung cancer in non-smokers in Xuanwei County of China. *Biomed Environ Sci* 1993; **6**: 112–8
136 Lan Q, Chapman RS, Schreinemachers DM, Tian L, He X. Household stove improvement and risk of lung cancer in Xuanwei, China. *J Natl Cancer Inst* 2002; **94**: 826–35
137 Liu Q, Sasco AJ, Riboli E, Hu MX. Indoor air pollution and lung cancer in Guangzhou, People's Republic of China. *Am J Epidemiol* 1993; **137**: 145–54
138 Du YX, Cha Q, Chen XW et al. An epidemiological study of risk factors for lung cancer in Guangzhou, China. *Lung Cancer* 1996; **14** (**Suppl. 1**): 9–37
139 Gao YT, Blot WJ, Zheng W et al. Lung cancer among Chinese women. *Int J Cancer* 1987; **40**: 604–9
140 Zhong LJ, Goldberg MS, Parent ME, Hanley JA. Risk of developing lung cancer in relation to exposure to fumes from Chinese-style cooking. *Scand J Work Environ Health* 1999; **25**: 309–16
141 Wu AH, Henderson BE, Pike MC, Yu MC. Smoking and other risk factors for lung cancer in women. *J Natl Cancer Inst* 1985; **74**: 747–51
142 Kurttio P, Pukkala E, Kahelin H, Auvinen A, Pekkanen J. Arsenic concentrations in well water and risk of bladder and kidney cancer in Finland. *Environ Health Perspect* 1999; **107**: 705–10
143 Bates MN, Smith AH, Cantor KP. Case–control study of bladder cancer and arsenic in drinking water. *Am J Epidemiol* 1995; **141**: 523–30
144 IARC. *Chlorinated Drinking Water; Chlorination By-products; Some Other Halogenated Compounds; Cobalt and Cobalt Compounds*. IARC Monographs on the Evaluation of Carcinogenic Risks to Humans, vol. **52**. Lyon: International Agency for Research on Cancer, 1991
145 Morris RD, Audet AM, Angelillo IF, Chalmers TC, Mosteller F. Chlorination, chlorination by-products, and cancer: a meta-analysis. *Am J Public Health* 1992; **82**: 955–63

146 McGeehin MA, Reif JS, Becher JC, Mangione EJ. Case–control study of bladder cancer and water disinfection methods in Colorado. *Am J Epidemiol* 1993; **138**: 492–501
147 King WD, Marrett LD. Case–control study of bladder cancer and chlorination by-products in treated water (Ontario, Canada). *Cancer Causes Control* 1996; **7**: 596–604
148 Cantor K, Hoover R, Hartge P *et al*. Bladder cancer, drinking water source and tap water consumption: a case–control study. *J Natl Cancer Inst* 1987; **79**: 1269–79
149 Cantor KP. Drinking water and cancer. *Cancer Causes Control* 1997; **8**: 292–308
150 Cantor KP. Arsenic in drinking water: how much is too much? *Epidemiology* 1996; **7**: 113–5
151 Cuello C, Correa P, Haenszel W, Gordillo G, Brown C, Archer M, Tannenbaum S. Gastric cancer in Colombia, I: Cancer risk and suspect environmental agents. *J Natl Cancer Inst* 1976; **57**: 1015–20
152 Juhasz L, Hill MJ, Nagy G. Possible relationship between nitrate in drinking water and incidence of stomach cancer. In: Walker EA, Griciute L, Castegnaro M, Börzsönyi M (eds) *N-nitroso Compounds: Analysis, Formation and Occurrence* (IARC Scientific Publications No 31). Lyon: International Agency for Research on Cancer, 1980; 619–23
153 Fraser P, Chilvers C. Health aspects of nitrate in drinking water. *Sci Total Environ* 1981; **18**: 103–16
154 Jensen OM. Nitrate in drinking water and cancer in northern Jutland, Denmark, with special reference to stomach cancer. *Ecotoxicol Environ Safety* 1982; **6**: 258–67
155 Gilli G, Corrao G, Favilli S. Concentrations of nitrates in drinking water and incidence of gastric carcinomas: first descriptive study of the Piemonte Region, Italy. *Sci Total Environ* 1984; **34**: 35–48
156 Beresford SA. Is nitrate in the drinking water associated with the risk of cancer in the urban UK? *Int J Epidemiol* 1985; **14**: 57–63
157 Rademacher JJ, Young TB, Kanarek MS. Gastric cancer mortality and nitrate levels in Wisconsin drinking water. *Arch Environ Health* 1992; **47**: 292–4
158 Weisenburger DD. Environmental epidemiology of non-Hodgkin's lymphoma in eastern Nebraska. *Am J Ind Med* 1990; **18**: 303–5
159 Ward MH, Mark SD, Cantor KP, Weisenburger DD, Correa-Vallasenor A, Zahm SH. Drinking water nitrate and the risk of non-Hodgkin's lymphoma. *Epidemiology* 1996; **7**: 465–71
160 Ward MH, Cantor KP, Riley D, Merkle S, Lynch CF. Nitrate in public water supplies and risk of bladder cancer. *Epidemiology* 2003; **14**: 183–90
161 Lyman GH, Lyman CG, Johnson W. Association of leukemia with radium groundwater contamination. *JAMA* 1985; **254**: 621–6
162 Fuortes L, McNutt LA, Lynch C. Leukemia incidence and radioactivity in drinking water in 59 Iowa towns. *Am J Public Health* 1990; **80**: 1261–2
163 Collman GW, Loomis DP, Sandler DP. Childhood cancer mortality and radon concentration in drinking water in North Carolina. *Br J Cancer* 1991; **63**: 626–9

Air pollution and infection in respiratory illness

Anoop J Chauhan* and **Sebastian L Johnston**[†]

Department of Respiratory Medicine, St Mary's Hospital, Portsmouth and [†]National Heart and Lung Institute & Wright Fleming Institute of Infection & Immunity, Faculty of Medicine, Imperial College London, London, UK

The detrimental effects of air pollution on health have been recognized for most of the last century. Effective legislation has led to a change in the nature of the air pollutants in outdoor air in developed countries, while combustion of raw fuels in the indoor environment remains a major health hazard in developing countries. The mechanisms of how these pollutants exert their effects are likely to be different, but there is emerging evidence that the toxic effects of new photochemical pollutants such as nitrogen dioxide are likely to be related to infection. This review discusses the relationship between air pollution and infection and will explore some of the mechanisms of how both could act synergistically to cause respiratory illnesses especially in exacerbating symptoms in individuals with pre-existing respiratory conditions such as asthma and chronic obstructive pulmonary disease.

A lesson from history

Correspondence to:
Anoop J Chauhan,
Consultant Respiratory
Physician, Department of
Respiratory Medicine,
St Mary's Hospital,
Milton Road,
Portsmouth PO3 6AD, UK.
E-mail: anoop.chauhan@
porthosp.nhs.uk

It is recognized that severe air pollution episodes have occurred in the UK since the early 17th century, but with rapid industrialization, such episodes became more severe and more frequent towards the 19th century. Legislation at this time led only to a reduction in smoke from industry, and difficulties in implementation gave little improvement in air quality until the early 20th century. Severe smog episodes through industrial and domestic combustion of solid fuels (coal) occurred during calm, winter weather. Between 1948 and 1962, eight air pollution episodes occurred in London, but the now well described 'Great Smog' episode in December 1952 was the most significant. Smoke concentration rose >50 times above the average limit, and at the National Gallery visibility was poor enough for individuals not to see their own feet. An estimated excess death toll of 4000 occurred during this period, a three-fold increase over the expected mortality rates for the time of year. Most deaths were among infants and elderly people. Similar episodes of severe air pollution in the United States after the Second World War and in Belgium in 1930 had aroused public concerns about the health effects of air pollutants.

Increased public and parliamentary concern led to effective legislation. In the UK, subsequent Clean Air Acts of 1956 and 1968 and similarly the Clean Air Act of 1970 in the USA had considerably reduced air pollution from stationary sources from homes, commerce and industry. Both the emission of smoke then sulphur dioxide (SO_2) fell dramatically. However, it is now accepted that any air quality benefits have been at least partially offset by increasing emissions of other visible and photochemical pollutants from mobile 'sources' such as car exhaust fumes. There is increasing awareness that new stationary sources in homes are also sources of the newer photochemical pollutants such as the oxides of nitrogen from unflued gas cookers and heaters[1].

The historical perspective of traditional air pollutants continues to interest contemporary air quality researchers. During the 1952 smog episode, there was a sharp rise in hospital admissions and visits to primary care physicians, but recent re-analysis[2] of the mortality data indicates the number of deaths was underestimated and was nearer 12,000. While the majority of deaths occurred in individuals with chronic respiratory disease, the steep rise in mortality was sustained for several months and did not drop immediately after the resolution of the smog episode (as might be expected if it were simply a triggering phenomenon). This raises issues of the mechanisms of how these pollutants may have caused the deaths perhaps through acidification with sulphuric acid that overwhelmed the buffering capacity of the lungs, interaction with other pollutants, toxic effects at subsequently lower and hitherto 'normal' levels of pollutants or perhaps more importantly, interaction with other co-factors such as infections. It has been assumed that many of the exacerbations of pre-existing respiratory illnesses that occurred during the smog episodes were the result of acute respiratory infections (ARIs). Recent evidence from developing countries confirms the direct exposure–response relationship between indoor air pollution by combustion of biomass fuels generating high levels of particulates and sulphur dioxide with increased acute respiratory infections in adults and children[3]. These observations have renewed interest in how air pollutants may exert their toxic effects through interaction with infection. This brief historical review of air pollution events confirms that air pollution by whatever mechanism can kill and equally importantly, that with sufficient public interest and the political will, effective legislation can reduce the mortality effects of such pollutants.

Combustion of biomass fuels and infections

Acute respiratory infections are the leading cause of the global burden of disease accounting for more than 6% of worldwide disease and mortality in developing countries[4,5]. Acute lower respiratory infections were attributed to have caused up to 4 million deaths worldwide from 1997 to 1999[4]. In developing countries, smoking amongst women is rare and outdoor pollution is restricted to larger cities yet both women and children suffer a huge burden of respiratory illness, largely through ARIs, which are responsible for nearly a third of all deaths in children under 5 years old[6]. Poor sanitation, low birth weight and poverty contribute to the causes of infections but the combustion of biomass fuels remains a rapidly growing problem. The main source of indoor air pollution globally is the combustion of biomass fuels (wood, dung, charcoal) as this is the main source of domestic energy. They produce small amounts of energy but large amounts of indoor pollutants, often emitting 50 times more pollutant concentrations than energy equivalent natural gas[7]. The often poor ventilation and dispersion characteristics of dwellings in developing countries allow pollutant concentrations to rise further and indoor concentrations of particulates between 500 and 100,000 $\mu g/m^3$ are not uncommon[3,8], and levels of indoor particulates 20 times greater than those due to cigarette smoking have been described[8]. Combustion of biomass fuels emits a variety of pollutants including particulates (PM_{10}), nitrogen dioxide (NO_2), carbon monoxide (CO), sulphur dioxide (SO_2) and hydrocarbons. Several studies from different parts of the world provide an indication of indoor pollutant concentrations, and whereas there are no internationally recognized standards for indoor air quality, the WHO estimates the number of people exposed to unacceptable levels indoors to exceed the number exposed to unacceptable levels of outdoor pollutants in all of the world's cities collectively.

The evidence for a relationship between indoor pollution and respiratory infections in developing countries is clear and has been recognized for at least two decades. The elderly, cooking mothers and the very young who spend most time indoors are likely to be at highest risk. Infant girls in Gambia whose mothers carried them on their back while cooking had more respiratory infections than girls not carried on their backs[9], children in Zimbabwe with recurrent pneumonia were more likely to come from homes where wood was used for fuel[10], and infants and children in Nepal who reported more time spent next to stoves suffered more life threatening episodes of acute respiratory infection[11]. The effects of these indoor pollutants on chronic respiratory illness in cooking women has also been reported with rates of chronic bronchitis of 18% in women who have never smoked[12] and rates of COPD of >30% in women over 50 years of age in Kashmir[13].

The effect of biomass fuel combustion has also been studied in developed countries. Children from Arizona, USA from homes using woodstoves for heating and cooking were four times more likely to suffer physician confirmed ARIs[14]. The Harvard Six Cities Study reported the use of wood stoves to be associated with a 30% increase in respiratory symptoms ranging from chronic cough to asthma in children aged 7–10 years[15]. The importance of indoor pollution in children has recently been reviewed[7].

Contemporary indoor and outdoor pollutants and infections

The combustion of fuels indoors in homes from developed countries produces a variety of pollutants. The main (and most consistently reported) pollutant among them is nitrogen dioxide which is produced in both indoor and outdoor air from sources such as unvented gas stoves and motor cars, respectively. There is an increasing body of evidence to suggest that exposure to this pollutant is linked to respiratory disease in children and adults in the developed world, although until recently, less has been known about the potential mechanisms involved. Many of the epidemiological studies of outdoor NO_2 exposure have found associations between exposure to the pollutant and health effects, often at levels well below current WHO guidelines. These health effects have included accident and emergency room visits[16,17], hospital admissions[18,19], mortality[20,21], increased symptoms[22,23] and reduced lung function[24,25]. The APHEA project (Air Pollution on Health: European Approach) incorporated data from 15 European cities, with a total population in excess of 25 million people. An increase of 50 μg/m^3 NO_2 (1 h maximum) was associated with a 2.6% increase in asthma admissions and a 1.3% increase in daily all-cause mortality[26]. There are many methodological issues that complicate interpretation of outdoor exposure studies such as exposure misclassification, confounding, co-linearity and insensitive measures of health effects. Nonetheless, this study strongly suggested a role for NO_2, or other pollutants for which NO_2 is a marker, in precipitating acute exacerbations of respiratory disease, although it has been difficult to separate the individual effects of the pollutant from the complex mixture of pollutants found in outdoor air.

With the recognition that indoor combustion sources contributed significantly to NO_2 levels inside homes and to personal NO_2 exposure, a number of studies have focused on indoor exposures. In most of the investigations, exposure to indoor pollution or NO_2 has been classified on the basis of the presence of a gas stove or other major source of fuel combustion in the home, and there have been fewer studies where actual levels of NO_2 or other pollutants indoors have been recorded, and even fewer where personal exposure has been estimated or directly measured, consequently some of the evidence has been inconsistent. Methodological limitations of the studies can readily explain the lack of definitive data, in particular in terms of misclassification of exposure and health outcomes and the incorrect (and therefore inaccurate) grouping of subjects with regard to exposure and disease. Although some studies have measured indoor NO_2 levels at fixed points, our own observations confirm that neither fixed site monitoring nor use of categorical variables such as gas or electric cooking accurately predict personal NO_2 exposure[27]. We have also confirmed that personal NO_2 exposure (at least in children) is not consistently related to either outdoor or indoor NO_2 exposure[28]. In considering the effects of indoor NO_2 exposure, information from a wide range of cross-sectional and longitudinal studies was brought together in a meta-analysis that estimated that the odds ratio of respiratory illness in children exposed to a long-term increase of 30 µg/m³ in NO_2 exposure (comparable to the presence of a gas cooker) was 1.20[29], and up to 1.29 in children aged 5–6 years and 1.60 in infants[30].

The link with infection

There has been a series of studies alluding to the link between infection and air pollution particularly by nitrogen dioxide. Two studies investigated the incidence of croup (laryngo-tracheo-bronchitis usually due to influenza or parainfluenza viruses) and air pollution. The effects of NO_2, TSP (total suspended particulates) and SO_2 exposure were investigated in more than 6000 paediatrician-reported cases of croup and more than 4500 cases of 'obstructive bronchitis' in five German cities over 3 years[31]. Increases in TSP and NO_2 levels of 10 and 70 µg/m³ were associated with a 27% and 28% increase in cases of croup, respectively. In illustration of the point made earlier, the close relationship between TSP and NO_2 levels did not allow their causal effects to be separated. In three other German towns[32], levels of SO_2, NO_2, CO, O_3 and dust were analysed with 875 cases of croup over a 24-month period. There was a significant association between croup frequency and the daily means of NO_2 for the peak period between September and March (the peak virus season), and with daily NO_2 and CO for the whole year. Another three cities in Finland from 'high' and 'low' pollution levels were studied over a 12-month period including 679 cases and 759 controls[33]. The annual mean concentration of NO_2 in the more polluted city was 15 µg/m³ higher. The odds ratios for one or more upper respiratory infections in children in the polluted city *versus* those in the less polluted cities were 2.0 in the younger age group and 1.6 in the older age group. The authors did not separate the effects of the individual pollutants in the analyses. In another case-control study of outdoor NO_2 in Stockholm, 197 children admitted to hospital because of wheezing bronchitis (reasonably assumed to be of infective aetiology) were compared with 350 controls. Time-weighted personal NO_2 exposures were estimated based on outdoor levels. The risk of wheezing bronchitis was significantly related to outdoor NO_2 exposure in girls ($P = 0.02$) but not in boys, and presence of a gas stove in the home appeared to be a risk factor only for girls.

More recently, several studies have suggested that air pollution may modify symptoms in individuals who may already be infected. Short-term effects of SO_2 and particulate matter (but with low acidity) were studied in 89 children with asthma for 7 months. Exposure to elevated levels of air pollution was associated with decreased peak expiratory flow rates, increased respiratory symptoms, increased school absence and fever, and increased medication use. Furthermore, there was evidence that exposure to air pollution might have enhanced the respiratory symptoms while children were experiencing respiratory infections[34]. In another study, the impact of air pollution on 321 non-smoking adults was investigated recording the daily incidence of respiratory symptoms over a 6-month period. The incidence of lower respiratory tract symptoms was related to the 1-h daily maximum ozone levels (OR = 1.22). The use of a gas stove in the home was also associated with lower respiratory tract symptoms (OR = 1.23), and the effects of ozone were significantly greater in individuals with a pre-existing respiratory infection[35]. Another study investigated the relationship between use of anti-inflammatory asthma medication and susceptibility to the effects of air pollution in 22 asthmatic children from non-smoking households. The associations between air pollution (mainly particulates) and symptoms were notably stronger in 12 asthmatic children who were not taking anti-inflammatory medications *versus* 10 subjects who were[36]. Another study examined specifically short-term indoor NO_2 exposures at school and in the home in 388 children aged 6–11 years. Exposure to NO_2 at hourly peak levels of the order of <160 µg/m^3 compared with background levels of 40 µg/m^3 was associated with a significant increase in sore throat, colds and absences from school, although infection was not confirmed[37]. Also, significant dose–response relationships were demonstrated for these four measures with increasing levels of NO_2 exposure.

The body of evidence thus strongly suggests a link between air pollution and severity of illness associated with respiratory infection, and that individuals with pre-existing lung disease may be at greater risk. The majority of the studies of air pollution and infection have been conducted in children, although until recently studies had not clarified the exact nature of the respiratory illness and infection. There now is good evidence of such a link in children with pre-existing asthma. However, the link with chronic obstructive pulmonary disease in adults has been less clear. Time-series analyses from Australia, Europe and North America suggest an association between air pollution and admissions to hospital for adults with COPD. An increase in daily maximum 1-h concentration of NO_2 and particulates was associated with a 4.60% and 3.01% increase, respectively in admissions for COPD in Sydney[38]. Another study of hospital admissions in Birmingham, UK, reported associations between PM_{10} and all respiratory admissions, pneumonia and deaths from COPD without a threshold effect[39]. A recent time series study from Rome investigated the relationship between air pollution levels and admissions for respiratory diseases over 3 years. A same day increase in outdoor NO_2 was associated with a 2.5% increase in all respiratory admissions per interquartile range (IQR) change. The effect of NO_2 was stronger on acute respiratory infections (4.0% increase) and on asthma among children (10.7% increase). A similar increase in carbon monoxide was also associated with a 2.8% increase in admissions for all respiratory admissions, a 5.5% increase for asthma and a 4.3% increase for COPD[40]. Further data from the APHEA project in six European cities of differing climates, has also shown[41] that a variety of pollutants including SO_2, black smoke, NO_2, total suspended particulates and ozone were associated with daily admissions for COPD with low relative risks ranging from 1.02 to 1.04. Another study from Minneapolis-St. Paul reported that an increase of 100 µg/m³ in daily PM_{10} was a risk factor for admissions for pneumonia (RR = 1.17) and admissions for COPD (RR = 1.57). Similarly, an increase of 100 µg/m³ in daily ozone concentration was associated with admissions for pneumonia (RR = 1.15)[42].

This array of studies of indoor or outdoor pollutants and acute respiratory infections (or respiratory symptoms in general) in children and adults suggests a relationship but none have confirmed infection microbiologically but instead relied on clinical criteria based on symptoms or use of health services. Consequently little is known of the spectrum of infectious agents or whether these pollutants increase susceptibility to infection or whether they exacerbate pre-existing morbidity following infection.

We have confirmed that respiratory tract viral infections are the major precipitants of acute exacerbations of asthma in children[43] and that episodes of viral infection are strongly associated in time with increases in hospital admissions for asthma[44]. In the former study human rhinoviruses alone accounted for 50% of exacerbations. If adequate virus detection methods are used, rhinoviruses and respiratory syncytial virus are detected in the vast majority of children (>80%) and infants (up to 100%) admitted to hospital with acute upper and lower respiratory illnesses. These data demonstrate that acute episodes of upper and lower respiratory illnesses in predisposed children both in the community and in those admitted to hospital are related to virus infections. Most of the epidemiological indoor pollution studies have alluded to the presence of respiratory infection as the health effect following NO_2 exposure, although very few studies (including those from the developing world) had until recently attempted to confirm the presence of infection microbiologically.

Our own recent observations provide the first direct evidence of a link between upper respiratory virus infections, personal NO_2 exposure and the severity of asthma exacerbations in children. This study is unique so far, in that it confirmed the presence of infection microbiologically. A cohort of 114 asthmatic children aged 8–11 years prospectively recorded daily upper and lower respiratory-tract symptoms, peak expiratory flow (PEF), and measured personal NO_2 exposures every week for up to 13 months. Nasal aspirates were taken during reported episodes of upper respiratory-tract illness and tested for infection by common respiratory viruses and atypical bacteria with RT-PCR assays. Severity of associated asthma exacerbations was analysed in relation to high *versus* low NO_2 exposures in the week before the viral infection. There were significant increases in the severity of asthma symptoms with 60% increased severity for all virus and >200% for respiratory syncytial virus infections for high compared with low NO_2 exposure in the week before the start of the virus-induced exacerbation. The highest category of NO_2 exposure was also associated with more severe falls in peak expiratory flow with virus infection by up to 75%. These effects were observed at levels within current air quality standards[45].

Controlled exposure and infection studies

There is a wealth of literature on the effects of controlled photochemical pollutant exposure (O_3, NO_2, SO_2) alone and lung function effects in healthy individuals and those with pre-existing lung disease. Inhaled O_3 provokes a dose-dependent fall in lung function and an increase in bronchial hyper-responsiveness to histamine. These responses vary considerably between individuals and surprisingly, little difference has been observed between subjects with asthma and healthy individuals after O_3 exposure. Inhalation of high concentrations of SO_2 provokes acute airway bronchoconstriction in normal subjects whereas lower concentrations provoke the same response in asthmatic subjects. The response is usually rapid, and maximal effects are seen within 5 min and usually resolution is within 15–30 min. Given that SO_2 was one of the main pollutants implicated in the London smog episodes, these short lived broncho-provocant effects are unlikely to explain why the deaths associated with the worst smogs were maintained for several months after resolution of the pollution episodes. Inhalation of NO_2 at higher concentrations induced no change in resting lung function in either normal or asthmatic subjects exposed to concentrations of 1880 μg/m^3 or less[46]. Bronchial hyper-responsiveness has been shown to increase only modestly in normal subjects exposed to >1880 μg/m^3 NO$_2$, and in about a third of studies in asthmatic subjects. In short, controlled exposure studies show largely modest effects on lung function and the pattern of response does not explain the epidemiological evidence, or the mechanisms of how these pollutants may interact with infectious agents.

There have been only a limited number of studies on the effects of controlled pollutant exposure and infection in humans largely through methodological difficulties and ethical constraints. There are a few studies that have suggested an interaction with infection. Three studies have investigated alveolar macrophage function after exposure to oxidant pollutant exposure *in vivo* and infection *in vitro*. In one study, nine volunteers were exposed to >1000 μg/m^3 NO$_2$ continuously or interspersed with three 15 min peak levels of >3500 μg/m^3 NO$_2$ in a sequential double blind randomized fashion (each subject served as their own control)[47]. Alveolar macrophages obtained by broncho-alveolar lavage after exposure were incubated with influenza virus. Two patterns of response were found after continuous exposure; alveolar macrophages from four of the nine subjects showed depressed inactivation of the virus [cells from these four subjects demonstrated an increase in interleukin-1 (IL-1) production after NO$_2$ *versus* air], whereas five showed no difference compared with control values. There were no differences from the intermittent peak exposure group. This indicates that differences in susceptibility to infection after oxidant pollutant exposure could occur in a larger population. Another study investigated the effects of high NO$_2$ exposure (9400, 18,800 or 28,200 μg/m^3) *in vitro* on alveolar macrophages obtained from 15 subjects[48]. After stimulation with influenza virus, both air and NO$_2$ exposed cells released increased amounts of IL-1 (indicating macrophage activation), but there was no significant difference between groups. The results indicated that human alveolar macrophages are resistant to injury by NO$_2$ *in vitro* and that toxicity effects of NO$_2$ may require local factors in the lung. In another 10 subjects exposed to >3500 μg/m^3 NO$_2$, alveolar macrophages obtained by lavage showed a 42% reduction in ability to phagocytose *Candida albicans* and a 72% decrease in superoxide production, superoxide production being important in phagocytosis[49].

A study using 152 young volunteers reported exposure to 1880 or 3760 μg/m^3 NO$_2$, for 2 h a day for 3 days and intra-nasal challenge with influenza A virus immediately after exposure[50]. Despite only one of the volunteers developing any symptoms, 91% were infected in the NO$_2$ exposed subjects as determined by virus recovery and/or antibody titres compared with 71% of air-exposed controls. This difference was not, however, statistically significant. As infected subjects did not become ill, the study could address the effect of NO$_2$ exposure on infection but not on the severity of illness. These findings indicate, but do not prove, that oxidant pollutants may play a role in increasing susceptibility to respiratory virus infection.

Mechanisms of interaction between infection and air pollution

The lung defence mechanisms against inhaled particles and gaseous pollutants include innate mechanisms such as aerodynamic filtration, mucociliary clearance, particle transport and detoxification by alveolar macrophages, as well as local and systemic innate and acquired antiviral immunity. In particular, alveolar macrophages provide an innate defence mechanism against bacteria and viruses. There is increasing experimental evidence that pollution exposure adversely affects lung defence mechanisms. Virus particles are ingested by phagocytosis and, in common with epithelial and other virus-infected cells, macrophages produce interferons which potently inhibit viral replication. Macrophages will also contribute to the neutralization of viral infections by removing the debris of the destroyed, virus-containing, cells and by presenting viral antigens to T-lymphocytes. In addition to the resulting humoral immune response, cell-mediated responses such as the development of cytotoxic T-lymphocytes (capable of destroying cells infected with virus) play an important role in the control of many viral infections of the respiratory tract. Many of these functions can be modulated by exposure to NO_2 and other pollutants in experimental models.

Animal infectivity models

It has now been almost 40 years since the 'infectivity model' was developed. This linked the effects of pollutants on pulmonary antibacterial activity following pollutant exposure with disease and mortality as end-points in animals. The majority of studies have been carried out using rodents, particularly mice, on the basis that the lung defence mechanisms of rodents and humans are sufficiently similar to permit their use as a surrogate. Some investigated various alveolar macrophage functions *ex vivo* but the great majority have used acute exposures to determine the concentrations of NO_2 at which antibacterial defences are overwhelmed. Different exposure regimes whereby animals have been challenged with the infectious organisms either before or after exposure to varying concentrations of NO_2, have also been used. For example, exposure to NO_2 occurred before infectious challenge in mice, whereas others were exposed to NO_2 at 1, 6 and 24 h *after* an infectious challenge. In all instances, mice exposed to NO_2 either before or after the bacterial challenge showed increased mortality[51]. In another study, mice were challenged with *Staphylococcus aureus* and then exposed to NO_2 for 4 h: pulmonary bactericidal capacity was progressively impaired in groups exposed to 7100 µg/m³ NO_2 or higher[52]. Short-term exposures to NO_2 and murine cytomegalovirus (MCMV) infection were investigated by Rose *et al*[53] studying a number of parameters including macrophage viability, *in vivo* and *in vitro* phagocytosis by macrophages and systemic cell-mediated and humoral responses. The minimum inoculum of viral particles capable of reliably producing infection was much lower in mice exposed to NO_2 compared with air-exposed animals. Another study addressed the risk of re-infection after NO_2 exposure. Animals were exposed to high doses of NO_2 or clean air during primary infection with MCMV. This produced infection in all animals. Thirty days later, animals were re-inoculated with the virus. The number of re-infected animals in the group which had been exposed to NO_2 was 11/20 compared with 1/22 amongst air-exposed animals, suggesting that NO_2 exposure damages the development of virus-specific immunity following a primary infection[54]. The finding is supported by the observation that splenic lymphocytes obtained from NO_2-exposed animals 30 days after primary infection failed (in contrast to lymphocytes from air-exposed animals) to proliferate in response to MCMV antigen.

Impaired bronchial immunity

Collective findings from both animal and human models provide some evidence of alterations in local bronchial immunity; acute exposure to oxidant pollutants results in ciliostasis in both the upper[55] and lower airways[56] which may prevent the nasal and bronchial mucosa filtering inhaled particles such as aero-allergens, bacteria or viruses delivered to the airway. In the studies by Devalia and colleagues (carried out in the absence of anti-oxidant protection), exposure of cultured human bronchial epithelial cells to NO_2 concentrations of >1000 µg/m³ reduced ciliary beat frequency and caused ciliary dyskinesis. The reduction in muco-ciliary activity following NO_2 exposure up to 4700 µg/m³ has been confirmed in a study of 24 healthy subjects *in vivo*[57]. Studies investigating single exposures to NO_2 have found increases in mast cells, lymphocytes and natural killer (NK) cells in BAL fluid[58]. Studies of repeated exposure to 7520 µg/m³ NO_2 (on six occasions, alternate days) have shown evidence of impaired local bronchial immunity by a reduction in total macrophages, B cells, NK lymphocytes, peripheral blood lymphocytes and a reduction in the T-helper-inducer/T-cytotoxic-suppressor ratio in alveolar lavage[59].

Alveolar macrophage function

Evidence from animal and human studies suggests that alterations in alveolar macrophage function are important in the increased risks of infection. Activated macrophages provide protection against bacterial and viral infections by a variety of mechanisms including oxygen dependent pathways involving superoxide radical-anion mechanisms (*e.g.* myeloperoxidase) and cytokine production [particularly IL-1, -6 and -8, interferons α and β and tumour necrosis factor α (TNF-α)] which are important in mounting an immune response to infection. Exposure of human alveolar macrophages (AMs) to concentrations of NO_2 ranging from 188 to 940 $\mu g/m^3$ for short durations resulted in a significant reduction of lipo-polysaccharide (LPS)-stimulated IL-1β, IL-6, IL-8 and TNF-α (but not TGF-β)[60]. Cytotoxicity of AMs remained unaffected. The results indicate that at concentrations relevant to human exposure, there is a functional impairment of AM without significant alterations in cytotoxicity. More recent work investigated exposure to particulate pollution of three size ranges from PM_{10} to $PM_{0.1}$ and the effect on antigen presenting cells by evaluating the expression of surface receptors involved in T-cell interaction on both human AMs and blood-derived monocytes (Mo). Mo up-regulated the expression of all four receptors in response to each of the particle fractions, whereas expression was unaffected in AM. However, when Mo and AM were separately exposed to the three PM size fractions and assessed for T-helper lymphocyte chemoattraction (by production of IL-16), AM alone (and not Mo) produced IL-16, and this chemoattractant was released only in response to $PM_{2.5-10}$. This suggests that a wide size range of pollution particles contains materials that may promote antigen presentation by Mo, whereas the capability to specifically recruit T-helper lymphocytes is contained in AM stimulated with the 'coarse' PM fraction[61]. When the experiments were extended to different particle sizes and compositions, the response of AM was highly variable, leading the authors to conclude that composition rather than size was responsible for the oxidant response and that oxidant activation by various sources of particulate matter is cell specific[62]. The ability of different pollutants and particle fractions to cause variable defects in bronchial immunity may also determine the risk of symptoms following pollutant exposure and infection.

Epithelial interaction

Viral infections can cause severe pathological abnormalities in both the upper and lower respiratory tract, although the extent of epithelial damage varies between viruses. Whereas Influenza A infection causes intense inflammation of the bronchi, trachea and larynx with desquamation of ciliated epithelial cells, rhinoviruses cause little or only patchy epithelial damage. Experimental studies of oxidant pollutant exposure show that the lower airway epithelium (particularly the transitional zone between bronchial epithelium and proximal alveolar regions) is particularly susceptible. In contrast, nasal, laryngeal and tracheal regions are not readily damaged by oxidant pollutants possibly because they have a thicker, extra-cellular anti-oxidant hypophase[63]. It is possible that the penetration of allergen into the epithelium would be facilitated both by epithelial shedding, and by reduced ciliary clearance, resulting in easier access of allergen to antigen presenting cells and therefore increased inflammation. The epithelium is also an important source of regulatory proteins or mediators with protective roles, such as nitric oxide or bronchodilator prostaglandins E_2 and I_2, which may play a role in maintaining bronchial patency.

The epithelium may also interact directly with virus infections and air pollutants. Spannhake reported experiments infecting primary human (nasal) epithelial cells and cells of the bronchial epithelial BEAS-2B line with human rhinovirus type 16 (RV16) and exposed them to 3880 µg/m³ NO_2 or 400 µg/m³ O_3 for 3 h. Infection with rhinovirus, NO_2 and O_3 independently increased release of IL-8 through oxidant-dependent mechanisms. The combined effect of RV16 and oxidant ranged from 42% to 250% greater than additive for NO_2 and from 41% to 67% for O_3. Both individual and combined effects were inhibited by anti-oxidant treatment. Perhaps the most interesting observation is that the surface expression of intercellular adhesion molecule 1 (ICAM-1) underwent additive enhancement in response to *combined* stimulation. These data indicate that oxidant pollutants can amplify the generation of pro-inflammatory cytokines by RV16-infected cells and suggest that virus-induced inflammation in upper and lower airways may be exacerbated by NO_2 and O_3. Given that ICAM-1 is also the receptor for the major group of rhinoviruses, a potential mechanism of how oxidant pollutants could increase susceptibility to rhinovirus infection is also suggested[64]. Another study reported by Becker investigated respiratory syncytial (RS) viral replication and virus-induced IL-6 and IL-8 production in BEAS-2B cells following exposure to approximately 1000, 2000 and 2500 mg/m³ NO_2. The internalization, release of infectious virus, and virus-induced cytokine production were all significantly reduced at the highest category of NO_2 exposure. This led the authors to conclude that increases in viral clinical symptoms associated with NO_2 may not be caused by increased susceptibility of the epithelial cells to infection but may result from effects of NO_2 on other aspects of antiviral host defences[65].

The mechanisms underlying the relationship between infection and the development of lower airway symptoms after air pollution exposure are

Table 1 Putative interaction between air pollution and infection in individuals with pre-existing lung disease

Effect	Air pollution	Infection
Bronchoconstriction	++	+++
Bronchial hyperresponsiveness	+	+++
Inflammatory mediator release	++	+++
Ciliary dyskinesis	++	+++
Inflammatory cell activation	++	+++
Epithelial damage	++	++/±
T-lymphocyte function	++	+++
Alveolar macrophage function	+++	++
Interaction with allergens	++	+++
↑ Epithelial derived cytokines	++	+++
↓ Macrophage derived cytokines	++	+

Strength of effect based on an arbitrary scale: +, mild; ++, moderate; +++, severe.

not fully understood. Oxidant pollutant exposures have the potential to exacerbate the inflammatory effects of virus infections in the lower airway, especially in individuals with pre-existing lung disease. Table 1 summarizes some of the mechanisms that may be involved in the synergistic interaction.

Will improving air quality improve health?

The data discussed above strongly suggests that the vast majority of serious morbidity and mortality related to air pollution occurs *via* interactions with respiratory infection. However, a discussion of the role of air pollutants and infection is not complete without considering whether improving air quality can reduce the burden of respiratory disease. The success of the Clean Air Legislation Acts described earlier in eradicating the severe smog episodes in London, Europe and the USA provides clear proof that improved air quality can reduce adverse health effects. The clearest demonstration of this has also been observed in the Utah Valley following closure of a steel mill for 14 months in 1987 as a result of a labour strike. Outdoor particulate concentrations and respiratory hospital admissions both fell dramatically during the period of mill closure, but returned to the pre-closure levels as the mill re-opened[66]. More recently the introduction of use of low sulphur fuels in Hong Kong was associated with between 2.01% and 3.9% of reductions in cardiovascular, respiratory and all-cause deaths[67]. This provided direct evidence that control of sulphur-rich pollution has immediate and long-term health benefits and has immediate implications for how air quality may be controlled in developing countries. Similar recent descriptions of the reduction in particulate pollution in Dublin for up to 5 years after the ban of coal sales showed a reduction of up to 15.5% for respiratory deaths and 10.3% for cardiovascular deaths, equating to 116 fewer respiratory and 243 fewer cardiovascular deaths in Dublin after the ban[68]. Whereas it is accepted that reducing air pollution is associated with beneficial effects for the traditional pollutants from stationary sources, there is now also evidence that control of traffic related pollution may have equally beneficial effects. Asthma admissions were reduced in Atlanta in parallel with traffic measures taken to reduce traffic density during the 2000 Olympic Games[69].

The future

These analyses of air pollution provide real evidence that air pollution is significantly associated with mortality and morbidity, provide evidence for possible mechanisms and interaction with infection and confirm that reducing pollutants could improve public health. Further work is still needed on the effect of indoor pollutants, especially in developing countries where the public health impact is likely to be dramatic. While there is emerging evidence confirming that air pollutants are intimately related to infections, the challenges for epidemiologists and clinical scientists remain to go beyond the short-term triggering phenomena and consider how the increase in susceptibility to air pollutants reflects fundamental interactions with co-factors such as allergens, domestic biomass fuel combustion, diet and virus infections. The challenges remain to unravel the mechanisms that drive such effects of air pollution.

References

1   POST. *Air Quality in the UK*. London: The Parliamentary Office of Science and Technology, 2002
2   Bell ML, Davis DL. Reassessment of the lethal London fog of 1952: novel indicators of acute and chronic consequences of acute exposure to air pollution. *Environ Health Perspect* 2001; **109 (Suppl. 3)**: 389–94
3   Ezzati M, Kammen D. Indoor air pollution from biomass combustion and acute respiratory infections in Kenya: an exposure–response study [erratum appears in *Lancet* 2001; 358: 1104]. *Lancet* 2001; **358**: 619–24
4   World Health Report. Geneva: World Health Organization, 1998
5   WHO. *World Health Report 2000. Health Systems: Improving Performance*. Geneva: World Health Organization, 2000
6   WHO. *World Health Report 1997. Acute Respiratory Infections in Children: Respiratory Infections Programme and the Programme Support Service*. Geneva: World Health Organization, 1997
7   Smith KR, Samet JM, Romieu I, Bruce N. Indoor air pollution in developing countries and acute lower respiratory infections in children. *Thorax* 2000; **55**: 518–32
8   Pandey MR, Smith KR, Boleij JSM, Wafula EM. Indoor air pollution in developing countries and acute respiratory infection in children. *Lancet* 1989; **1**: 427–9
9   Armstrong JR, Campbell H. Indoor air pollution exposure and lower respiratory infections in young Gambian children. *Int J Epidemiol* 1991; **20**: 424–9

10 Collings DA, Sithole SD, Martin KS. Indoor woodsmoke pollution causing lower respiratory disease in children. *Trop Doct* 1990; **20**: 151–5
11 Pandey MR, Neupane RP, Gautam A, Shrestha IB. Domestic smoke pollution and acute respiratory infections in a rural community of the hill region of Nepal. *Environ Int* 1989; **15**: 337–40
12 Pandey MR. Prevalence of chronic bronchitis in a rural community of the hill region of Nepal. *Thorax* 1984; **39**: 331–6
13 Chen BH, Hong CJ, Pandey MR, Smith KR. Indoor air pollution in developing countries. *World Health Statistics Quarterly—Rapport Trimestriel de Statistiques Sanitaires Mondiales* 1990; **43**: 127–38
14 Morris K, Morgenlander M, Coulehan JL. Wood burning stoves and lower respiratory tract infection in American Indian children. *Am J Dis Child* 1990; **144**: 105–8
15 Dockery D, Speizer F, Stram D, Ware J, Spengler J, Ferris B. Effects of inhalable particles on respiratory health of children. *Am Rev Respir Dis* 1989; **139**: 587–94
16 Kesten S, Szalai J, Dzyngel B. Air quality and the frequency of emergency room visits for asthma. *Ann Allergy Asthma Immunol* 1995; **74**: 269–73
17 Buchdahl R, Parker A, Stebbings T, Babiker A. Association between air pollution and acute childhood wheezy episodes: prospective observational study [see comments]. *BMJ* 1996; **312**: 661–5
18 Bates DV, Sizto R. Air pollution and hospital admissions in Southern Ontario: the acid summer haze effect. *Environ Res* 1987; **43**: 317–31
19 Ponka A, Virtanen M. Chronic bronchitis, emphysema, and low-level air pollution in Helsinki, 1987–1989. *Environ Res* 1994; **65**: 207–17
20 Saldiva PH, Lichtenfels AJ, Paiva PS, Barone IA, Martins MA, Massad E *et al*. Association between air pollution and mortality due to respiratory diseases in children in Sao Paulo, Brazil: a preliminary report. *Environ Res* 1994; **65**: 218–25
21 Anderson HR, Limb ES, Bland JM, Ponce de Leon A, Strachan DP, Bower JS. Health effects of an air pollution episode in London, December 1991. *Thorax* 1995; **50**: 1188–93
22 Braun-Fahrlander C, Ackermann-Liebrich U, Schwartz J, Gnehm HP, Rutishauser M, Wanner HU. Air pollution and respiratory symptoms in preschool children. *Am Rev Respir Dis* 1992; **145**: 42–7
23 Mukala K, Pekkanen J, Tiittanen P, Alm S, Salonen RO, Jantunen M *et al*. Seasonal exposure to NO2 and respiratory symptoms in preschool children. *J Expos Anal Environ Epidemiol* 1996; **6**: 197–210
24 Frischer TM, Kuehr J, Pullwitt A, Meinert R, Forster J, Studnicka M *et al*. Ambient ozone causes upper airways inflammation in children. *Am Rev Respir Dis* 1993; **148**: 961–4
25 Scarlett JF, Abbott KJ, Peacock JL, Strachan DP, Anderson HR. Acute effects of summer air pollution on respiratory function in primary school children in southern England. *Thorax* 1996; **51**: 1109–14
26 Touloumi G, Katsouyanni K, Zmirou D, Schwartz J, Spix C, de Leon AP *et al*. Short-term effects of ambient oxidant exposure on mortality: a combined analysis within the APHEA project. Air Pollution and Health: a European Approach. *Am J Epidemiol* 1997; **146**: 177–85
27 Linaker CH, Chauhan AJ, Inskip H, Frew AJ, Sillence A, Coggon D *et al*. Distribution and determinants of personal exposure to nitrogen dioxide in school children. *Occup Environ Med* 1996; **53**: 200–3
28 Linaker CH, Chauhan AJ, Inskip HM, Holgate ST, Coggon D. Personal exposures of children to nitrogen dioxide relative to concentrations in outdoor air. *Occup Environ Med* 2000; **57**: 472–6
29 Hasselblad V, Eddy DM, Kotchmar DJ. Synthesis of environmental evidence: nitrogen dioxide epidemiology studies. *J Air Waste Manage Assoc* 1992; **42**: 662–71
30 Li Y, Powers TE, Roth HD. Random-effects linear regression meta-analysis models with application to the nitrogen dioxide health effects studies. *J Air Waste Manage Assoc* 1994; **44**: 261–70
31 Schwartz J, Spix C, Wichmann HE, Malin E. Air pollution and acute respiratory illness in five German communities. *Environ Res* 1991; **56**: 1–14
32 Rebmann H, Huenges R, Wichmann HE, Malin EM, Hubner HR, Roll A. Croup and air pollutants: results of a two-year prospective longitudinal study. *Zentralbl Hyg Umweltmed* 1991; **192**: 104–15

33   Jaakkola JJ, Paunio M, Virtanen M, Heinonen OP. Low-level air pollution and upper respiratory infections in children [see comments]. *Am J Public Health* 1991; **81**: 1060–3
34   Peters A, Dockery DW, Heinrich J, Wichmann HE. Short-term effects of particulate air pollution on respiratory morbidity in asthmatic children. *Eur Respir J* 1997; **10**: 872–9
35   Ostro BD, Lipsett MJ, Mann JK, Krupnick A, Harrington W. Air pollution and respiratory morbidity among adults in southern California. *Am J Epidemiol* 1993; **137**: 691–700
36   Delfino RJ, Zeiger RS, Seltzer JM, Street DH, McLaren CE. Association of asthma symptoms with peak particulate air pollution and effect modification by anti-inflammatory medication use. *Environ Health Perspect* 2002; **110**: A607–17
37   Pilotto LS, Douglas RM, Attewell RG, Wilson SR. Respiratory effects associated with indoor nitrogen dioxide exposure in children. *Int J Epidemiol* 1997; **26**: 788–96
38   Morgan G, Corbett S, Wlodarczyk J. Air pollution and hospital admissions in Sydney, Australia, 1990 to 1994 [see comments]. *Am J Public Health* 1998; **88**: 1761–6
39   Wordley J, Walters S, Ayres JG. Short term variations in hospital admissions and mortality and particulate air pollution. *Occup Environ Med* 1997; **54**: 108–16
40   Fusco D, Forastiere F, Michelozzi P, Spadea T, Ostro B, Arca M *et al*. Air pollution and hospital admissions for respiratory conditions in Rome, Italy. *Eur Respir J* 2001; **17**: 1143–50
41   Anderson HR, Spix C, Medina S, Schouten JP, Castellsague J, Rossi G *et al*. Air pollution and daily admissions for chronic obstructive pulmonary disease in 6 European cities: results from the APHEA project. *Eur Respir J* 1997; **10**: 1064–71
42   Schwartz J. PM10, ozone, and hospital admissions for the elderly in Minneapolis-St. Paul, Minnesota. *Arch Environ Health* 1994; **49**: 366–74
43   Johnston SL, Pattemore PK, Sanderson G, Smith S, Lampe F, Josephs L *et al*. Community study of role of viral infections in exacerbations of asthma in 9–11 year old children [see comments]. *BMJ* 1995; **310**: 1225–8
44   Johnston SL, Pattemore PK, Sanderson G, Smith S, Campbell MJ, Josephs LK *et al*. The relationship between upper respiratory infections and hospital admissions for asthma: a time-trend analysis. *Am J Respir Crit Care Med* 1996; **154**: 654–60
45   Chauhan AJ, Inskip HM, Linaker CH, Smith S, Schreiber J, Johnston SL *et al*. Personal exposure to nitrogen dioxide (NO2) and the severity of virus-induced asthma in children. *Lancet* 2003; **361**: 1939–44
46   Department of Health. *Oxides of Nitrogen*. Advisory Group on the Medical Aspects of Air Pollution Episodes (AGMAAPE), 3rd Report. Department of Health, 1993
47   Frampton MW, Smeglin AM, Roberts NJ, Finkelstein JN, Morrow PE, Utell MJ. Nitrogen dioxide exposure in vivo and human alveolar macrophage inactivation of influenza virus in vitro. *Environ Res* 1989; **48**: 179–92
48   Pinkston P, Smeglin A, Roberts Jr NJ, Gibb FR, Morrow PE, Utell MJ. Effects of in vitro exposure to nitrogen dioxide on human alveolar macrophage release of neutrophil chemotactic factor and interleukin-1. *Environ Res* 1988; **47**: 48–58
49   Becker S, Devlin R, Horstman D, Gerrity T, Madden M, Biscardi F *et al*. Evidence for mild inflammation and change in alveolar macrophage function in humans exposed to 2 ppm NO2. *Proceedings of Indoor Air '93* 1993; **1**: 471–6
50   Goings SA, Kulle TJ, Bascom R, Sauder LR, Green DJ, Hebel JR *et al*. Effect of nitrogen dioxide exposure on susceptibility to influenza A virus infection in healthy adults. *Am Rev Respir Dis* 1989; **139**: 1075–81
51   Ehrlich R. Effect of nitrogen dioxide on resistance to respiratory infection. *Bacteriol Rev* 1966; **30**: 604–14
52   Jakab GJ. Modulation of pulmonary defense mechanisms by acute exposures to nitrogen dioxide. *Environ Res* 1987; **42**: 215–28
53   Rose RM, Fuglestad JM, Skornik WA, Hammer SM, Wolfthal SF, Beck BD *et al*. The pathophysiology of enhanced susceptibility to murine cytomegalovirus respiratory infection during short-term exposure to 5 ppm nitrogen dioxide. *Am Rev Respir Dis* 1988; **137**: 912–7
54   Rose RM, Pinkston P, Skornik WA. Altered susceptibility to viral respiratory infection during short-term exposure to nitrogen dioxide. *Research Report—Health Effects Institute* 1989; **24**: 1–24
55   Carson JL, Collier AM, Hu SC, Devlin RB. Effect of nitrogen dioxide on human nasal epithelium. *Am J Respir Cell Mol Biol* 1993; **9**: 264–70

56 Devalia JL, Campbell AM, Sapsford RJ, Rusznak C, Quint D, Godard P et al. Effect of nitrogen dioxide on synthesis of inflammatory cytokines expressed by human bronchial epithelial cells in vitro. *Am J Respir Cell Mol Biol* 1993; **9**: 271–8

57 Helleday R, Huberman D, Blomberg A, Stjernberg N, Sandstrom T. Nitrogen dioxide exposure impairs the frequency of the mucociliary activity in healthy subjects. *Eur Respir J* 1995; **8**: 1664–8

58 Sandstrom T, Andersson MC, Kolmodin-Hedman B, Stjernberg N, Angstrom T. Bronchoalveolar mastocytosis and lymphocytosis after nitrogen dioxide exposure in man: a time-kinetic study. *Eur Respir J* 1990; **3**: 138–43

59 Sandstrom T, Helleday R, Bjermer L, Stjernberg N. Effects of repeated exposure to 4 ppm nitrogen dioxide on bronchoalveolar lymphocyte subsets and macrophages in healthy men. *Eur Respir J* 1992; **5**: 1092–6

60 Kienast K, Knorst M, Muller-Quernheim J, Ferlinz R. Modulation of IL-1 beta, IL-6, IL-8, TNF-alpha, and TGF-beta secretions by alveolar macrophages under NO2 exposure. *Lung* 1996; **174**: 57–67

61 Becker S, Soukup J. Coarse(PM(2.5–10)), fine(PM(2.5)), and ultrafine air pollution particles induce/increase immune costimulatory receptors on human blood-derived monocytes but not on alveolar macrophages. *J Toxicol Environ Health A* 2003; **66**: 847–59

62 Becker S, Soukup JM, Gallagher JE. Differential particulate air pollution induced oxidant stress in human granulocytes, monocytes and alveolar macrophages. *Toxicol In Vitro* 2002; **16**: 209–18

63 Gordon RE, Solano D, Kleinerman J. Tight junction alterations of respiratory epithelium following long-term NO2 exposure and recovery. *Exp Lung Res* 1986; **11**: 179–93

64 Spannhake EW, Reddy SPM, Jacob DB, Yu XY, Saatian B, Tian J. Synergism between rhinovirus infection and oxidant pollutant exposure enhances airway epithelial cell cytokine production. *Environ Health Perspect* 2002; **110**: 665–70

65 Becker S, Soukup JM. Effect of nitrogen dioxide on respiratory viral infection in airway epithelial cells. *Environ Res* 1999; **81**: 159–66

66 Pope CA. Respiratory disease associated with community air pollution and a steel mill, Utah Valley. *Am J Public Health* 1989; **79**: 623–8

67 Hedley AJ, Wong CM, Thach TQ, Ma S, Lam TH, Anderson HR. Cardiorespiratory and all cause mortality after restrictions on sulphur content of fuel in Hong Kong: an intervention study. *Lancet* 2002; **360**: 1646–52

68 Clancy L, Goodman P, Sinclair H, Dockery DW. Effect of air-pollution control on death rates in Dublin, Ireland: an intervention study [comment]. *Lancet* 2002; **360**: 1210–4

69 Friedman MS, Powell KE, Hutwagner L, Graham LM, Teague WG. Impact of changes in transportation and commuting behaviors during the 1996 Summer Olympic Games in Atlanta on air quality and childhood asthma. *JAMA* 2001; **285**: 897–905

Evaluating evidence on environmental health risks

Lesley Rushton* and **Paul Elliott**[†]

MRC Institute for Environment and Health, Leicester, UK, and [†]Small Area Health Statistics Unit, Department of Epidemiology and Public Health, Faculty of Medicine, Imperial College, London, UK

The assessment of adverse health effects from environmental hazards involves integration of evidence from a variety of sources, including experimental studies, both in animals and humans, *in vitro* studies, and epidemiological research. It requires an understanding of the sources, nature and levels of exposure to which humans may be subjected, the nature of the health outcome or toxic effect and the mechanisms by which this might occur, the relationship between dose and response, and a knowledge of the variability and susceptibility of potentially exposed populations. After outlining the process of risk assessment, this paper gives an overview of the most relevant human study methods used to investigate environment and health effects and discusses issues such as confounding and effect modification, that are important to consider when interpreting the results from such studies. Future challenges are outlined, such as increasing responsibility required by scientists to the sensitive issue of data protection and confidentiality, and also new opportunities, such as the increasing availability of computerized data, the incorporation of molecular epidemiological methods to aid the investigation of mechanistic pathways and gene–environment interactions, and the development and utilization of sophisticated statistical approaches.

Introduction

Correspondence to:
Lesley Rushton, MRC
Institute for Environment
and Health, 94 Regent
Road, Leicester LE1 7DD,
UK. E-mail: lr@le.ac.uk

The assessment of risks of adverse health effects from environmental hazards is an area of increasing public interest and a topic of a large body of research within many different disciplines. It involves both knowledge of the source and nature of the environmental hazard and an understanding of the relationship of the exposure to the disease. The primary aims of environmental research vary but can include:

- the identification of causal relationships between environmental hazards and ill health in general populations and specific subgroups;
- the evaluation of changes in health with environmental changes;
- the provision of evidence for the setting of 'acceptable' standards for known environmental contaminants.

In the broadest sense, the environment can be defined as external conditions influencing the development of people, animals or plants. A distinction is sometimes made between conditions from which individuals may have no or only partial control, for example, exposures encountered at work and to substances in ambient air, and those for which some element of personal choice exists, for example, 'lifestyle' factors such as smoking, eating a high fat diet and drinking alcohol. In the public mind, environmentally caused ill-health is often limited to the former situation, particularly hazards such as chemical exposures, and even more so if these are man-made. The research community, however, would generally consider a much broader set of factors, encompassing all that might impact on human health.

Different kinds of research contribute relevant information for assessment of environmental risks. Toxicological studies conducted on experimental animals can rigorously assess the biological response to a range of levels of exposure to a given substance or combination of substances. This has the advantage of specificity but presents a problem when extrapolating to humans, both because of species differences, and because high doses are often used in an experimental setting in animals, whereas typically humans will be exposed to low doses of pollutants. This can be partially overcome through clinical studies employing human volunteers, for example in food challenge studies or chamber studies of air pollutants. However, for ethical reasons these are generally limited to a narrow range of low exposure levels, which cause a mild and reversible health effect. Epidemiological approaches are thus the most widely used methods to test for potentially causal associations between environmental exposures and human health, both in the general population and for specific groups, defined for example, by occupation or location. Epidemiological studies are observational rather than experimental in design and thus have a number of limitations, especially where excess risks of any adverse health effect are small, as will be the case with most environmental exposures encountered today in the developed world. Despite these limitations, a large number of risks that have been recognized so far have been established using these methods. This paper gives an overview of the most common epidemiological methods and their advantages and limitations, and a discussion of the future challenges, for investigating the influence of environment on health. Toxicological methods are not addressed in this chapter, nor are hazards relating to personal choice.

General considerations

Assessment of the impact of a potential adverse health effect from an environmental pollutant is dependent on an understanding of several important issues, including:

- the hypothesized health outcome or toxic effect;
- the nature of the exposure;
- the relationship between dose and response;
- the variability and susceptibility of the potentially exposed population, for example regarding sub-groups of the population that might be at especial risk due either to the pattern and distribution of exposures in the population, or to non-environmental factors that might influence the risk of disease.

It is important to have a clear view of these aspects both when designing a study and when interpreting information from the published literature.

Health outcomes

There are various ways of classifying a toxic response to a hazard including:

- acute or chronic effects;
- reversible or irreversible effects;
- local or systemic effects;
- immediate or delayed effects.

An acute effect is generally caused by a high single exposure (occurring in seconds to hours) through accidental or gross over-exposure. The adverse condition develops quickly (within minutes or days) with obvious signs and symptoms, as was the case with chloracne, for example, following exposure to 2,3,7,8-tetrachlorodibenzo-p-dioxin (TCDD) as a consequence of the Seveso accident in northern Italy in July 1976[1]. Unless the effect is fatal, damage is often reversible. Chronic effects usually occur after lower exposures repeated over months or years. The adverse condition may develop slowly over months or years with incremental damage and is often irreversible; an example in an occupational setting is lung fibrosis secondary to inhalation of dusts contaminated with silicone dioxide (silicosis)[2]. Local effects are those which occur at the site of first contact between the organism and pollutant whereas systemic effects occur only after a substance has been absorbed and distributed within the body. Immediate effects are those which develop shortly after exposure, whereas delayed effects develop some considerable time afterwards.

Early changes as a result of low-level exposure to an environmental pollutant will often remain undetected. Human cancer, for example, may take many years to become clinically apparent following an initial environmental exposure as part of a multistage process in carcinogenesis[3]. Damage may be arrested if the exposure is removed, as happens after quitting tobacco smoking where eventually risks approach that of non-smokers, especially at younger ages[4]; however with other conditions, for example following asbestos exposure, there may be progressive deterioration.

In environmental epidemiology, concern usually centres on chronic effects from low-level exposures. Although at the individual level the excess *relative* risks associated with such exposures may be low, they may nonetheless be associated with a substantial burden of disease in the population, that is *attributable* risk is high. This is because of the ubiquity of exposure to many low-level pollutants, such that large numbers of people, or even whole populations, are exposed (toxicants in outdoor air would be an example). According to Rose[5] 'a large number of people exposed to a small risk may generate many more cases than a small number of people exposed to a high risk'.

The nature of exposure

Much of the information regarding the kinetics and metabolism of a substance, *i.e.* the uptake, distribution, accumulation, excretion and biotransformation, is determined through animal and *in vitro* studies. An understanding of these and their relevance to humans is essential in assessing the totality of evidence on the potential risk of an environmental pollutant on human health.

Knowledge of the toxicokinetics of a substance can also be very useful in determining the most appropriate measure of exposure to an environmental contaminant that should be used when assessing potential human risk. Human exposure represents a dynamic process from the source of the substance in question, through intake, uptake and transport to a critical organ. A lack of adequate exposure data has been reported to be the major limiting factor in preventing the identification of causal associations[6]. Critical issues include:

- the type of the assessment method used. For example, has direct measurement been carried out using personal monitoring or biomarkers or have indirect methods been used, such as exposure modelling or the use of monitoring and time-activity data?

- the characteristic patterns of exposure over time. For example, for how long does exposure take place, how is it distributed (continuously or intermittently) and are there any critical time windows of exposure which are relevant to the health effect of interest?

- the metric used to represent exposure. Does this represent the relevant exposure patterns? For example is cumulative exposure assessed when some measure of short-term higher intermittent exposures would be more appropriate?

- inclusion of all key sources of exposure *via* all possible routes and all media to give an aggregate exposure level. There is a tendency for many studies to investigate adverse health effects related to a restricted, though often substantial, potential source of exposure and to ignore other sources. For example, very few studies of occupational groups attempt to assess exposure from non-occupational sources. Figure 1 illustrates several of the potential sources, media and exposure routes through which exposure to environmental pesticides could occur.

The relationship between dose and response

The notion that there may be a causal relationship between an environmental exposure and a health effect is strengthened if a clear dose–response relationship can be shown. If the data are suitable, mathematical models can be developed to describe the dose–response relationship. In interpreting and using these models, a distinction is often drawn between situations in which the critical toxic effect is considered to have a threshold, *i.e.* a level below which no adverse health effect would be expected to occur, *e.g.* copper in water, or is considered to have no threshold, *i.e.* it is assumed that as little as one molecule could theoretically cause an effect, *e.g.* ionizing radiation or inorganic arsenic. The former is usually derived from consideration of known metabolic and mechanistic information and empirical observations in animals and humans of an absence of effect below certain doses (though statistical power to detect such effects at low doses is often limited). The latter has been traditionally applied to mutagenic and genotoxic substances. Although in practice studies may indicate an apparent threshold, the uncertainty owing to the inability to detect effects at very low levels of exposure both in animal and human studies limits the use of these models in assessing risk[7].

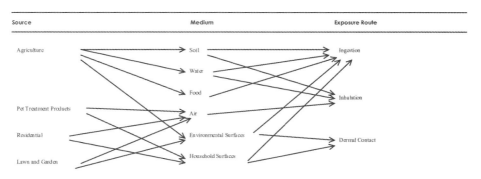

Fig. 1 Factors to consider in an aggregate exposure assessment of a pesticide.

Risk assessment methodology

In carrying out a risk assessment for an environmental hazard the totality of evidence is evaluated. The formal process of risk assessment shown in Figure 2 is now accepted internationally and incorporates:

- knowledge of the source of exposure to a substance, methods of release and transport of the substance and an assessment of exposure to human populations of concern;

- hazard identification, *i.e.* identification from animal and human studies of the potential adverse health effects and hazard characterization, in particular determination of the dose–response relationship;

- integration of hazard identification and characterization and human exposure assessment to give a risk characterization which evaluates the nature and severity of adverse effects at given levels of exposure.

The results of a risk assessment inform the setting of environmental standards in water, air, food and soil. The methods used for deciding the values to be used in standard setting depend on whether a substance is thought to have a threshold or is a non-threshold substance. With a threshold substance, many countries set standards by selecting a pivotal or key study (often an animal study) to identify the most sensitive endpoint—generally the health effect that occurs at the lowest dose level. From this, the No Observed Adverse Effect Level (NOAEL) is obtained, the highest dose level at which

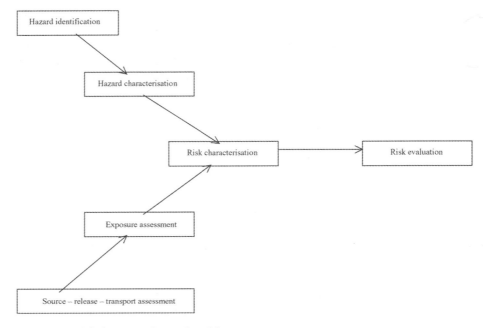

Fig. 2 Essential elements of assessing risk.

no adverse health effect is observed. This level is then adjusted by dividing by 'uncertainty' factors to account for interspecies and intraspecies differences. Additional uncertainty factors are also sometimes added, for example if it is necessary to extrapolate from a study that gives only a lowest observed adverse effect level to obtain a NOAEL, if the exposure route differs between the study and the typical human exposure situation, or if the severity of the effect is deemed to justify additional protection. The magnitude of each adjustment factor ranges from 1 to 10. In practice the first two are usually sufficient, and the figure of 10 is used, so that typically the permitted concentration of a chemical is one-hundredth that of the highest concentration associated with no observed adverse effect in the key animal study. Alternative approaches used by a few countries, notably the USA, include the use of benchmark doses and probabilistic modelling (including Bayesian).

There are three approaches taken for non-threshold substances: banning or non-approval of substances to which humans would be deliberately exposed, *e.g.* a water treatment chemical; the use of probabilistic methods to extrapolate from animal studies to the low levels of exposure experienced environmentally by humans to set a level at which risk is regarded as very low, *e.g.* 1 in 100,000 excess incidence of cancer over a lifetime of exposure; setting of an exposure level which is as low as technically achievable or reasonably practicable, *e.g.* the setting of a maximum exposure limit for occupational settings.

There are several comprehensive textbooks which explain in detail the scientific methodology employed to carry out both animal and human studies and interpret the results from them[8–15] and it is not within the remit of this paper to repeat this. However, as the following papers in this volume summarize and evaluate the evidence on various environmental issues, particularly from studies on humans, a brief overview follows of the most relevant human study methods.

Human study methods

Studies of humans can be categorized as (i) experimental or (ii) observational, with several different types of study within these two categories as illustrated in Figure 3.

Essential elements of an experimental study include the control of intervention assignment by the investigator, comparison of interventions with standard or control treatment and, usually, some form of random allocation to different interventions. Of the experimental designs, formal clinical trials which in most cases involve patients, *i.e.* sick individuals, are the least used for environmental issues.

Field trials involve individuals without any particular disease, *i.e.* who are not patients, and are recruited from the general population. An example

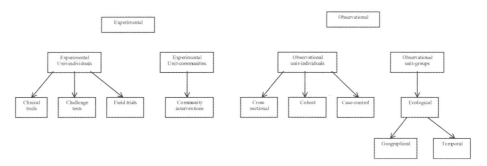

Fig. 3 Epidemiological study designs.

of a field trial is testing the effectiveness of large doses of vitamin C in prevention of the common cold. Community intervention trials differ from clinical and field trials in that the interventions are implemented for whole communities, for example, administering fluoride to a community water supply. Challenge tests or controlled human exposure studies, such as air pollution chamber studies and food challenge tests can potentially provide good evidence between exposure and effect as exposures take place in a controlled setting, and potential confounders can be systematically identified and controlled for. However, for ethical reasons, these studies are limited to studies of mild and reversible health effects, and usually low levels of exposure. In addition, a laboratory setting may not always simulate the real world.

The majority of human studies of environmental influences on health are carried out using observational study designs. The three most common, in which observations are recorded on individuals, are cross-sectional, cohort and case-control studies.

Cross-sectional studies describe the frequency of the disease of interest in a population at a particular period of time and the variation between subgroups defined in terms of personal characteristics, time, place, and where available, relevant exposures. They represent a 'snapshot' of a population and are useful for generating hypotheses about the aetiology of a disease. They are particularly useful when dealing with health data that are continuously distributed in the population, such as blood pressure or serum cholesterol. For categorical health outcomes, the cross-sectional study is in fact similar to the case-control study (see below) except that sampling is based on *prevalent* rather than *incident* cases[15]. Their drawback is that information on exposure and effect are collected simultaneously, thus making it more difficult to attribute causation.

Cohort (or prospective) studies follow over time a group of people, the cohort, with particular characteristics in common, including levels of exposure, to observe the development of disease. The rate at which the disease develops in the exposed people in the cohort is compared with the rate in the non-exposed or in a standard group such as the national population; this comparison of rates is known as the *relative risk* of the disease for a particular exposure. In case-control studies, individuals with a given disease (the cases) are compared with a group of individuals without the disease (the controls). Information on past exposure to possible risk factors is then obtained for both cases and controls and compared, giving the *odds ratio* (an estimate of the relative risk) of disease associated with a particular exposure[15].

Given the large numbers of people potentially exposed to many environmental pollutants and the likelihood that for many pollutants the excess risk of disease at the level of the individual is small, then in order to detect any excess risk, many thousands, hundreds of thousands or even millions of people may need to be studied, over a prolonged period, if the cohort approach is adopted. This is especially true for rare diseases such as congenital anomalies, childhood and many adult cancers that may be thought to be of particular interest with regard to the possible health effects of environmental pollutants. An alternative is to use the case-control approach which is much more efficient as only cases and a relatively small number of controls need to be assembled and studied. However, because of the retrospective nature of the exposure assessment, and possible selection effects, such studies are prone to bias that can seriously distort risk estimates.

Where individual-level follow-up studies are infeasible, and case-control studies are considered too expensive or impracticable, group-level (or ecological) studies are often undertaken. Ecological studies are generally thought to be of weaker design than individual-level studies because inferences made at the group level may not pertain at individual level, the so-called *ecological fallacy*[16]. However, such concerns notwithstanding, ecological studies have played a major role in the investigation of aetiological associations of public health importance, such as the relationship of type and amount of dietary fat to heart disease[17]. Also, it is not always understood that in many common study designs, people are *grouped* for the purposes of exposure assessment—such studies are in fact ecological in nature. Examples include the use of job exposure matrices for occupational studies[18] and studies of health effects associated with differing measures of social class[19]. Because of the difficulty in studying low-level risks from environmental exposures in large numbers of people, much of our knowledge concerning risk of chemicals in humans is in fact derived from occupational studies. Features of occupational and environmental epidemiological studies are compared in Table 1.

In the context of environmental epidemiology, two ecological designs are commonly employed, as indicated in Figure 3—those that group people in time, and those that group in space. Temporal studies are exemplified by studies on the health effects of outdoor air pollutants. In these studies, time-varying data on concentrations of air pollutants are available from

Table 1 Comparison of occupational and environmental epidemiological studies

	Occupational	Environmental
Exposure	High	Low
	Well characterized	Poorly characterized
	Local	Widespread
Risk	Large excess relative risk	Small excess relative risk
	Attributable risk low	Attributable risk high
Control for confounding	++	+/–

one or a small number of monitoring stations at city level[20]. It is assumed that exposure to a range of outdoor pollutants of whole city populations can be characterized by such data; patterns of daily mortality or morbidity (such as hospital admissions for asthma) are then compared for the city in relation to daily fluctuations in air pollution. This is a powerful study design since the daily mortality and exposure to outdoor air pollution for many thousands or millions of people can readily be captured using routine data sources, and problems of confounding by individual characteristics such as smoking are minimized, as effectively people act as their own controls. Such studies have been mostly consistent in finding higher rates of mortality associated with days with higher levels of air pollutants (allowing for appropriate lag periods), especially particulate matter and sulphur dioxide. These effects have been reported at much lower concentration levels of pollutants than those traditionally associated with air pollution episodes, such as the London smog of 1952[21].

The second type of ecological study involves using location as a proxy for exposure. Often, proximity (to a factory or polluting industry) is used as the marker of exposure; examples include the incidence of certain cancers near municipal solid waste incinerators[22], and the occurrence of congenital anomalies and other birth effects near landfill sites in Great Britain[23]. In other instances, some kind of exposure modelling is done to define areas considered to have high levels of exposure[24]. Semi-ecologic designs offer an attractive means of reducing the possible biases that may affect ecological studies[25]. In these studies, data on the exposure of interest (such as air pollution) is measured at the ecological level, but other data, including major confounders (see below), are collected at individual level, *e.g.* for a representative sample of individuals in the study regions[26].

Confounding and interaction

An important issue in the interpretation of results of an observational study is whether they might be explained by a factor or factors other than that under investigation. This is not usually a problem in an experimental setting (such as a randomized controlled clinical trial) since the process of randomization, provided that the sample size is sufficient, should ensure that the groups under study are similar with respect to other potentially causative factors. In epidemiological studies, the term *confounding* is used to describe the situation where an association between the factor of interest and the disease outcome is explained by the association of both these factors with another variable, the *confounder*, which itself is either a cause or closely related to a cause of the disease. Age and social class, for example, are commonly regarded as confounders as they are strongly related to disease occurrence and are also related to a wide range of environmental exposures. The effects of confounding variables can be at least partially removed, either by *matching* (in a case-control setting) or by statistical *adjustment* in the analysis. Either way, in the presence of possible confounding, judgments as to causality for the factor of interest have to be tempered by the possibility that the results from a study may be affected by potential confounding effects, unless the association is so large (relative risk > 5 or so) that it cannot readily be ascribed to other factors. In studying the effects of low-level environmental pollutants, risks are likely to be much lower than this, so that unmeasured or inadequately controlled (*residual*) confounding in the observational data should always be considered. This is one of the most important reasons why it is essential to examine all the evidence, from animal and human experimental studies as well as epidemiological studies, when making judgments as to causality.

Another important issue to consider is whether the effect on disease outcome of one factor is modified by levels of another factor, so-called *effect modification* or *interaction*. An example is the effect of cigarette smoking on risk of cardiovascular disease, which is strongly modified by age. At 40 years follow up of the British doctors study[4], below the age of 65, the relative risk of cardiovascular disease among smokers compared with non-smokers was 2.1, whereas it was 1.2 at ages 85 and over. Interaction effects are increasingly thought to be important, and may lead to new ideas about aetiology and mechanisms of disease; however, they are difficult to investigate as much larger sample sizes are required than studies examining only 'main' effects. In particular, recent interest has focused on potential *gene–environment interactions* in determining the combined genetic and environmental influences on disease risk (see section on molecular epidemiology, below).

Challenges and future developments

Data availability and quality

In order to make robust inferences about potential environmental causes of disease, it is imperative to obtain valid data on health outcomes and environmental exposures. For purpose-designed environmental epidemiological studies such as case-control or cohort studies, evaluation of the health data will often involve a detailed validation exercise and assessment of the quality of the diagnostic information (for example, case note and histology review). In contrast, the quality of the exposure data has been regarded as the 'Achilles heel' of such investigations, as often only group data or proxy information on environmental or occupational exposures has been available. Poorly measured exposure data will lead to misclassification which in turn leads to bias toward no effect—the so-called 'regression dilution' problem[27]. Recent developments in exposure biomarkers[28] and molecular epidemiology (see below) should in the future lead to improved exposure assessment methods, with increased specificity, and hence improved ability to detect true differences in disease risk.

Such purpose-designed studies are, however, time-consuming and expensive to carry out, and often recourse is made to routine sources of data, especially in geographical studies where location is used as a proxy for exposure, so-called *spatial epidemiology*[11]. The health data may be available with associated point locations, or as aggregated counts for areas, and will potentially be subject to a number of inaccuracies. For any health event there is always the potential for diagnostic error or misclassification, especially at older ages where diagnostic tests and post-mortem examinations are carried out less frequently than at younger ages. Some events may be captured poorly, if at all, in routine registers (*e.g.* miscarriages). For others, such as cancers, case registers may be subject to double counting and under-registration as well as diagnostic inaccuracies. Some assessment of the basic quality of the routine data is therefore essential to inform their use in spatial analyses[29].

For the future, the increasing use and availability of computerization in medical care means that potentially large new databases of morbidity (for example from general practitioner consultations) will become available. The quality of such data will need careful evaluation and no doubt will vary across specialties, practices and over time and space. Nonetheless they promise exciting new opportunities for carrying out environmental epidemiology, which hitherto has been limited with respect to the range of outcome data available from routine sources.

Data protection and confidentiality

The current climate of legislation in the European Union and elsewhere is moving toward greater recognition of the rights of individuals to confidentiality of personal data, including health data, and the need for consent for medical research. It therefore behoves the research community to be sensitive to their responsibilities for the protection of the privacy and confidentiality of individuals. In purpose-designed individual-level studies, this will involve obtaining proper consent to participate, the right to withdraw consent at any time, and the right of privacy and confidentiality of data obtained on individuals. Secondary use of routine data for epidemiology where the data were originally collected for other purposes (*e.g.* health care management or delivery) is a more complex issue, because here in many instances it can be argued that the public good (in terms of describing disease patterns, potentially identifying new causes of disease in the population, *etc.*) outweighs the individual right to privacy of data. In these circumstances, often it will be impossible or impracticable to obtain consent, and in any case data on the entire population are required to avoid possible biases as a result of selection effects. In the UK, recent legislation has made it possible to use such routinely collected data without consent provided that certain conditions and safeguards are met. It is imperative for the future of epidemiological research that such uses of the data are allowed to continue, given that appropriate safeguards are in place.

Molecular epidemiology

Advances in molecular epidemiology offer the opportunity to combine the scientific disciplines of epidemiology and molecular toxicology to investigate the interactions between genetic and environmental factors in the cause of disease. Many studies have demonstrated that exposure to relatively high levels of substances, such as carcinogens, does not affect all individuals equally, supporting the theory that genetic factors may influence an individual's susceptibility and resistance to disease[30]. Future research will need to incorporate measurement of susceptibility to aid the investigation of mechanistic pathways and to detect gene–environment interactions.

Dealing with uncertainties

Uncertainty and variability permeate all aspects of assessment of risks from environmental pollutants. Barnett and O'Hagan[31] in discussing the setting of environmental standards draw attention to several sources of uncertainty inherent in the investigation of a pollutant–effect relationship. These include an imperfect understanding of the mechanisms by which the pollutant influences the effect, variation in the way individuals react to a given level of a pollutant, variation in the levels of exposure received by individuals, imprecise exposure assessment methods, and other causes of the disease of interest. Barnett and O'Hagan[31] suggest that the quantification of uncertainty and variation should be addressed using statistical inference procedures and the use of probabilistic modelling. In particular, these methods facilitate the incorporation of sensitivity analyses to examine the robustness of models to input assumptions and to identify those parameters to which the output is most sensitive. In addition, Bayesian modelling has been successfully used to analyse risks when available data have been inadequate for use in the classical statistical approaches to risk assessment and when there is the desire to incorporate the opinions of experts. The use of modelling approaches to the assessment of risk has been viewed with some scepticism by some governments, particularly in Europe. Effort is needed by scientists to ensure that adequate testing of the robustness of results from modelling is carried out and that they are able to interpret and communicate these results to non-specialists.

The role of systematic review and meta-analysis

Systematic review and meta-analysis methods are well developed in the area of clinical trials, and involve the collation of the literature on a particular area of interest, assessment of the extent and quality of the studies found and provision of a compilation of the results, often including quantitative estimation of risk estimates from the combined studies. These methods facilitate transparency and reproducibility of the methodology and results, and ease of updating. They can identify gaps in the knowledge base and areas for future research. Quantitative meta-analyses can give greater statistical power than single studies and provide a framework for investigation of possible sources of heterogeneity between studies[32]. In spite of controversy over the opportunity for bias and other sources of heterogeneity compared with clinical trials, these techniques are being increasingly used in epidemiological research, and a number of guidelines have been produced recently on the topic[33,34]. To date there has been very little use either of systematic review methodology or of meta-analysis techniques in toxicology or in environmental epidemiology.

Within the past 10 years, there has also been a growing interest in cross-design synthesis, *i.e.* the quantitative combination of results from different study designs[35]. In particular, the potential for the use of meta-analysis techniques to combine data from animal studies with that of human studies is beginning to be explored, which will have an important impact on the development of risk assessments and the setting of environmental standards.

Conclusion

Integration of the evidence available from a variety of sources for the assessment of adverse heath effects from environmental hazards is a complex process. The methods used in the individual scientific components contributing to the totality of information are, themselves, also often complex. This paper has given an overview of the commonest methods used for human studies and discussed issues important for interpretation of results. A difficult challenge will continue to be that of communication, both of complex methodology and the interpretation of results, and a need to acknowledge and address the concerns of the public whom the scientific community and regulatory authorities are aiming to protect. Risk communication is itself a developing scientific discipline and involves interaction, both co-operative and contending, between a large number of scientific, social and political institutions, media organizations, the public and the decision-makers. The problems widely encountered are founded in a lack of trust and confidence in risk information provided by 'experts' and in the processes by which decisions are made. A cultural shift is required by all bodies concerned to develop appropriate ways of establishing more open and transparent methodologies for assessing risks and to increase the contribution of the public to the development of risk management and risk reduction options.

References

1 Caramaschi R, Del Corno G, Favaretti C, Giambelluca SE, Montesarchio E, Fara GM. Chloracne following environmental contamination by TCDD in Seveso, Italy. *Int J Epidemiol* 1981; **10**: 135–43
2 American Thoracic Society. Adverse effects of crystalline silica exposure. *Am J Respir Crit Care Med* 1997; **155**: 761–8
3 Moolgavkar SH. Stochastic models for estimation and prediction of cancer risk. In: Barnett V, Stein A, Turkman KF (eds) *Statistics for the Environment 4: Health and the Environment*. Chichester: Wiley, 1999; 237–57
4 Doll R, Peto R, Wheatley K, Gray R, Sutherland I. Mortality in relation to smoking: 40 years' observations on male British doctors. *BMJ* 1994; **309**: 901–11
5 Rose G. *The Strategy of Preventive Medicine*. Oxford: Oxford University Press, 1992; 24
6 Checkoway H. Rapporteur's Summary: Epidemiologic evaluation of exposure–effect relationships. In: Rappaport SM, Smith TJ (eds) *Exposure Assessment for Epidemiology and Hazard Control*. MI: Lewis, 1991; 67–73
7 Institute for Environment and Health. *Risk Assessment Approaches Used by UK Government for Evaluating Human Health Effects of Chemicals*. Leicester: Institute for Environment and Health, 1999
8 Altman DG. *Practical Statistics for Medical Research*. London: Chapman & Hall, 1991
9 Bertollini R, Lebowitz MD, Saracci R, Savitz DA (eds) *Environmental Epidemiology: Exposure and Disease*. Boca Raton, FL: CRC Press, 1996
10 Ballantyne B, Marrs T, Turner P. *General & Applied Toxicology Abridged Edition*. Basingstoke: Macmillan Press Ltd, 1995
11 Elliott P, Wakefield JC, Best NG, Briggs DJ (eds) *Spatial Epidemiology: Methods and Applications*. Oxford: Oxford University Press, 2000
12 Fletcher RH, Fletcher SW, Wagner EH. *Clinical Epidemiology: The Essentials*. Baltimore: Williams & Wilkins, 1996
13 Pocock S. *Clinical Trials: A Practical Approach*. Chichester: Wiley, 1983
14 Hayes WA. *Principles and Methods of Toxicology*, 4th edn. Philadelphia: Taylor & Francis, 2001

15 Rothman KJ, Greenland S. *Modern Epidemiology*. Philadelphia: Lippincott-Raven, 1997
16 Piantadosi S, Byar DP, Green SB. *Am J Epidemiol* 1988; **127**: 893–904
17 Keys A (ed) *Coronary Heart Disease in Seven Countries*. American Heart Association Monograph no. 29. New York: American Heart Association, 1970
18 Goldberg M, Kromhout H, Guenel P, Fletcher AC, Gerin M, Glass DC, Heederik D, Kauppinen T, Point A. Job exposure matrices in industry. *Int J Epidemiol* 1993; **22** (**Suppl. 2**): S10–15
19 Marmot MG, Shipley MJ, Rose G. Inequalities in death—specific explanations of a general pattern? *Lancet* 1984; **1**: 1003–6
20 Katsouyanni K, Touloumi G, Spix C, Schwartz J, Balducci F, Medina S, Rossi G, Wojtyniak B, Sunyer J, Bacharova L, Schouten JP, Ponka A, Anderson HR. Short-term effects of ambient sulphur dioxide and particulate matter on mortality in 12 European cities: results from time series data from the APHEA project. Air Pollution and Health: a European Approach. *BMJ* 1997; **314**: 1658–63
21 Wilson R, Spengler J. *Particles in Our Air. Concentrations and Health Effects*. Boston: Harvard University Press, 1996
22 Elliott P, Shaddick G, Kleinschmidt I, Jolley D, Walls P, Beresford J, Grundy C. Cancer incidence near municipal solid waste incinerators in Great Britain. *Br J Cancer* 1996; **73**: 702–10
23 Elliott P, Briggs D, Morris S, de Hoogh C, Hurt C, Jensen TK, Maitland I, Richardson S, Wakefield J, Jarup L. Risk of adverse birth outcomes in populations living near landfill sites. *BMJ* 2001; **323**: 363–8
24 Nyberg F, Gustavsson P, Jarup L, Bellander T, Berglind N, Jakobsson R, Pershagen G. Urban air pollution and lung cancer in Stockholm. *Epidemiology* 2000; **11**: 487–95
25 Prentice RL, Sheppard L. Aggregate data studies of disease risk factors. *Biometrika* 1995; **82**: 113–25
26 Dockery DW, Pope III CA, Xu X *et al*. An association between air pollution and mortality in six US cities. *N Engl J Med* 1993; **329**: 1753–9
27 MacMahon S, Peto R, Cutler J, Collins R, Sorlie P, Neaton J, Abbott R, Godwin J, Dyer A, Stamler J. Blood pressure, stroke, and coronary heart disease. Part 1, Prolonged differences in blood pressure: prospective observational studies corrected for the regression dilution bias. *Lancet* 1990; **335**: 765–74
28 Hulka BS, Wilcosky TC, Griffith JD. *Biological Markers in Epidemiology*. Oxford: Oxford University Press, 1990
29 Best NG, Wakefield JC. Accounting for inaccuracies in population counts and case registration in cancer mapping studies. *J R Stat Soc A* 1999; **162**: 363–82
30 Schulte PA, Perera FP. *Molecular Epidemiology Principles and Practices*. London: Academic Press, Inc., 1993
31 Barnett V, O'Hagan A. *The Statistical Approach to Handling Uncertainty and Variation*. London: Chapman and Hall, 1997
32 Blettner M, Sauerbrei W, Schlehofer B, Scheuchenpflug T, Friedenreich C. Traditional reviews, meta-analyses and pooled analyses in epidemiology. *Int J Epidemiol* 1999; **28**: 1–9
33 Egger M, Schneider M, Smith GD. Meta-analysis: spurious precision? Meta-analysis of observational studies. *BMJ* 1998; **316**: 140–4
34 Sutton AJ, Abrams KR, Jones DR, Sheldon TA, Song F. *Methods for Meta-analysis in Medical Research*. Chichester: John Wiley & Sons Ltd, 2000
35 Piegorsch WW, Cox LH. Combining environmental information II: Environmental epidemiology and toxicology. *Environmetrics* 1996; **7**: 309–24

Environmental effects and skin disease

JSC English*, RS Dawe[†] and J Ferguson[†]

Department of Dermatology, Queen's Medical Centre, University Hospital, Nottingham and [†]Department of Dermatology, Ninewells Hospital, Dundee, UK

The skin is the largest organ in the body and one of its main functions is to protect the body from noxious substances, whether they are ultraviolet radiation, toxic chemicals or prolonged/repeated exposure to water. It is the level of exposure that determines if damage to the organism will result. The harm that can occur to the skin with sufficient exposure will be considered. Contact dermatitis, halogen acne, chemical depigmentation, connective tissue diseases and skin cancer are the conditions that will be covered in this chapter, as environmental exposure is important in their aetiologies. Systemic absorption will not be dealt with. Most environmental exposure to harmful substances will occur at work, but exposure may occur at home or during normal day-to-day activities.

Dermatitis

Contact dermatitis is an eczematous eruption caused by external agents. The causes can be broadly divided into irritant substances that have a direct toxic effect on the skin, and allergic chemicals where immune hypersensitivity reactions occur.

Irritant contact dermatitis

Correspondence to: Dr JSC English, Department of Dermatology, Queen's Medical Centre, University Hospital, Nottingham NG7 2UH, UK. E-mail:john.english@ mail.qmcuh-tr.trent.nhs.uk

Irritant contact dermatitis (ICD) is caused by direct damage of the skin by the irritant substance. It develops if the exposure of the substance or substances is sufficient or the skin is particularly susceptible as is found in atopic dermatitis (AD) patients. The commonest irritants found in the environment are: soaps, detergents, water, solvents and a dry atmosphere. The occurrence of irritant contact dermatitis depends upon the degree of exposure. Most of us do not develop ICD when washing our own hair but an apprentice hairdresser is at high risk of developing dermatitis because the exposure to shampoo and water is so great. Irritant contact dermatitis is very common with at least 10% of the population suffering from hand dermatitis[1]. An infant's skin is more susceptible to irritant damage than adult skin.

Allergic skin reactions

There are two allergic mechanisms that can commonly affect the skin. They are immediate (IgE driven) and delayed (cell mediated) allergic reactions. The T cell lymphocytes involved in these allergic mechanisms can be subdivided into two functional subsets by their differing cytokine profiles. Immediate allergic reactions involve the central immune organs releasing B lymphocytes which are then stimulated to differentiate into plasma cells producing allergen specific IgE. This is the T helper 2 (Th2) response with cytokine profile of interleukin 4 and 5. Delayed allergic reactions involve lymphocyte-mediated mechanisms called the T helper 1 (Th1) response with cytokine profile of interleukin 2 and gamma interferon.

Hay fever, asthma and contact urticaria are examples of immediate allergic reactions. Common causes are: house dust mite, grasses, pollens, natural rubber latex, dairy products and peanuts. These allergenic proteins do not penetrate normal skin particularly well but can penetrate damaged skin such as AD and hence aggravate it. Sometimes exposure to these allergens can lead to anaphylaxis. Peanuts and natural rubber latex are good examples.

Allergic contact dermatitis (ACD) is caused by delayed allergenic mechanisms Th1 response. It takes the immune system a few hours to react. Common causes are: nickel, fragrances, rubber additives, preservatives, plants and medicaments (Table 1). The immune system of atopics tends towards Th2 responses and so is less likely to develop ACD even though their exposure to certain delayed allergens will be greater than for non-atopics.

Atopic dermatitis

There has been a three-fold increase in patients suffering from AD in the past 30 years. The reason for this is not clear but the hygiene hypothesis seems to be the best explanation[2]. The incidence of childhood AD in the UK is at least between 15 and 20% and some countries even higher[3]. There has been an increase in frequency in first generation immigrants to western industrialized countries and there was a substantial increase in prevalence of atopic disease in Eastern Germany following reunification of Germany in 1990[4]. It would therefore appear that exposure to pollutants in some way protects against the development of atopy or that the exposure to infectious diseases associated with poorer, less developed nations protects against atopy. This is the basis of the hygiene hypothesis.

Table 1 Commonest allergens causing ACD and their sources

Allergen	Source
Potassium dichromate	Leather, cement, *etc.*
Neomycin	Antibiotic
Thiruam mix	Rubber additive
Paraphenylenediamine	Permanent hair dye
Cobalt	Metal
Caine mix III	Local anaesthetic
Formaldehyde	Biocide
Rosin	Colophony, resin from spruce trees
Balsam of Peru	Fragrance
Isopropyl-phenyl-paraphenylenediamine	Industrial rubber additive
Wool alcohols	Lanolin
Mercapto mix	Rubber additive
Epoxy resin	Two part adhesive
Paraben mix	Preservative
Paratertiarybutyl phenol formaldehyde resin	Adhesives
Fragrance mix	Fragrance
Quaternium 15	Biocide
Nickel	Metal
Methylchloroiso thiazolinone + methylisothiazolinone	Biocide
Mercaptobenzothiazole	Rubber additive
Primin	Primula
Sesquiterpene lactone mix	Compositae plant allergy

Chloracne (halogen acne)

Acne is a very common skin disease in adolescence but on rare occasions environmental pollution can result in a variant of acne called chloracne. Environmental acne results from various chemical exposures and the eruption may be mild, involving localized exposure, or covered areas of the body or severe, explosive and disseminated with involvement of almost every follicular orifice. Chloracne almost always represents a cutaneous sign of systemic exposure to highly toxic chemicals. Chloracne results from environmental exposure to certain halogenated aromatic hydrocarbons and is considered to be one of the most sensitive indicators of systemic poisoning by these compounds.

Chloracne was first observed by Von Bettman in 1897 and by Herxheimer in 1899. Since that time, a number of chloracnegenic chemicals have been identified. Before World War II, most cases were thought to be caused by chloronaphthalenes and polychlorinated biphenyls (PCBs)[5]. More recently, trace contaminants formed during the manufacture of PCBs and other polychlorinated compounds, especially herbicides, have been causally linked to chloracne development[6]. These include polyhalogenated dibenzofurans in association with PCBs, polychlorinated dibenzo-*p*-dioxins, and chlorinated azo- and azoxybenzenes which are contaminants of 3,4-dichloraniline and related herbicides.

The chloracnegenic compounds are structurally similar, all sharing relative molecular planarity and containing two benzene rings with halogen atoms occupying at least three of the lateral ring positions. The position of halogen substitution appears to be critical, as substitution to positions that lead to molecular nonplanarity reduces biological activity. Stereo-specific binding of these compounds to a receptor is implicated in their toxicity.

Most cases of chloracne have resulted from environmental exposure in chemical manufacturing or rarely from end-use of products. Exposure is usually through direct contact, but inhalation and ingestion may also be operative in some cases. Non-occupational chloracne has resulted from industrial accidents, contaminated industrial waste, and contaminated food products. A widely publicized example was the extensive environmental contamination with 2,3,7,8-tetrachlorodibenzo-*p*-dioxin (TCDD) (Fig. 1), which occurred on July 10, 1976, at the ICMESA chemical plant near Seveso, Italy. During production of trichlorophenol, an explosion resulted in the formation and ultimate discharge into the atmosphere of an estimated 2 kg of TCDD. The contaminated area encompassed more than 200 acres of land, and 135 cases of chloracne, mostly in children, were confirmed among some 2000 inhabitants. Other examples were the widespread ingestion of tainted rice cooking oil that occurred in Japan in 1968 and in Taiwan in 1979. Popular brands of oil were contaminated with PCBs and dibenzofurans resulting in the largest epidemics of chloracne to date. Over 1000 patients were affected with oil disease called Yusho in Japan, and Yucheng in Taiwan. For a listing of known chloracnegens, see Table 2.

Although chloracne tends to slowly resolve upon cessation of exposure to chloracnegenic compounds, the duration of chloracne correlates with the severity of the disease, which is usually a reflection of the degree and extent of exposure. Thus, in severely exposed victims of the Yusho incident in 1968, lesions characteristic of chloracne continued to develop for as long as 14 years after the initial exposure. Treatment has, in general, been disappointing, but the retinoids seem to hold some promise.

The prevention and control of chloracne require a totally enclosed manufacturing process with no opportunity for direct skin contact or inhalation of the toxic chemicals—a difficult environmental engineering

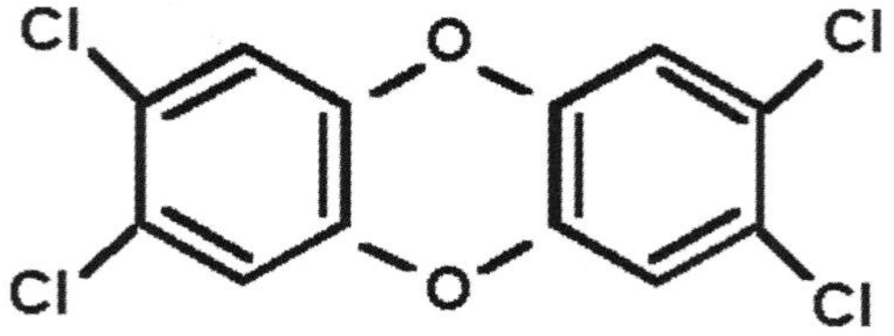

Fig. 1 2,3,7,8-Tetrachlorodibenzo-*p*-dioxin.

Table 2 Chloracne-producing chemicals

Polyhalogenated naphthalenes
  Polychloronaphthalenes
  Polybromonaphthalenes

Polyhalogenated biphenyls
  Polychlorobiphenyls (PCBs)
  Polybromodiphenyls (PBBs)

Polyhalogenated dibenzofurans
  Polychlorodibenzofurans, especially tri-, tetra-, penta- and hexachloro-dibenzofuran
  Polybromodibenzofurans, especially tetra-bromodibenzofuran

Contaminants of polychlorophenol compounds—especially herbicides (2,4,5-T and pentachlorophenol) and herbicide intermediates (2,4,5-trichlorophenol)
  2,3,7,8-Tetrachlorodibenzo-p-dioxin (TCDD)
  Hexachlorodibenzo-p-dioxin
  Tetrachlorodibenzofuran

Contaminants of 3,4-dichloroaniline and related herbicides (propanil, methazole, etc.)
  3,4,3'4'-Tetrachloroazoxybenzene (TCAOB)
  3,4,3'4'-Tetrachloroazobenzene (TCAB)

Other
  1,2,3,4-Tetrachlorobenzene (experimental)
  Dichlobenil (Casoran)—a herbicide (clinical only)
  DDT (crude trichlorobenzene?)

task. The only other alternative is to attempt to alter the chemical synthetic process to eliminate or minimize exposure to the contaminant chloracnegens.

Chemical depigmentation

Chemical leukoderma is defined as pigmentation or hypopigmentation of the skin due to industrial exposure to a chemical or chemicals known to have a destructive effect on epidermal melanocytes[7].

Certain chemicals, particularly the substituted phenols, are destructive to functional melanocytes[8]. Many of these compounds cause permanent depigmentation of the skin, resembling vitiligo. The most commonly implicated chemicals are *para*-tertiary butyl phenol, *para*-tertiary butyl catechol, monobenzyl ether of hydroquinone, hydroquinone and related compounds[7,8]. A list of chemicals known to cause occupational leukoderma is shown in Table 3.

The diagnosis of occupational vitiligo should be suspected if a worker who potentially has been exposed to depigmenting chemicals develops leukoderma on the dorsal aspects of the hands or in a more widespread distribution[9]. There should be particular suspicion if more than one worker is involved. The chemicals to which the worker is exposed should be identified and investigation made to see if it or they are known to cause depigmentation.

Table 3 Chemicals capable of causing occupational leukoderma

Hydroquinone
Monobenzylether of hydroquinone
Monoethylether of hydroquinone (*p*-ethoxyphenol)
Monomethylether of hydroquinone (*p*-methoxyphenol)
p-Cresol
p-Isopropylcatechol
p-Methylcatechol
p-Nonylphenol
p-Octylphenol
p-Phenylphenol
p-tert-Amylphenol
p-tert-Butylcatechol
p-tert-Butylphenol
N,N',N"-Triethylenethiophosphoramide (thio-TEPA)
Mercaptoamines, e.g. N-2-mercaptoethyl-dimethylamine hydrochloride (MEDA)
Physostigmine

There is no specific treatment for chemical leukoderma. Removal of the offending chemical may result in partial repigmentation but this process may take years and may not occur at all. Treatment should be aimed at preventing further exposure. Camouflage cosmetics may be used and the depigmented skin protected from ultraviolet irradiation by sunscreens.

Scleroderma-like disease

Scleroderma-like diseases have been observed increasingly over the last few decades[10]. These comprise diseases that in addition to skin changes similar to those of scleroderma also involve other organ systems, but are not consistent with classical scleroderma. The most important aetiological factors and clinical manifestations are summarized in Table 4. A common factor in these diseases is that they generally show definite clinical improvement when exposure to the relevant agent ceases apart from quartz induced systemic sclerosis.

Quartz induced scleroderma

Quartz is widely distributed in nature (silicon dioxide, SiO_2). Silicon-derived materials predominantly as quartz) account for 27% of the earth's crust. Silicon dioxide is present in over 90% of all minerals. Quartz is chemically virtually inactive and can remain unaltered for decades in human tissues. The principal medical significance of the substance is as the causative agent of pulmonary silicosis.

Table 4 Connective tissue diseases related to environmental exposures

Inducing factors	Symptoms/disease
Occupational agents	
Vinyl chloride	Raynaud's phenomenon, sclerodactyly, acro-osteolysis, hepatic fibrosis, angiosarcoma, plaque-like fibrotic cutaneous lesions, leuco- and thrombocytopaenia
Organic solvents	Skin fibrosis, irritant dermatitis, hepatitis, neurological symptoms
Bis (4-amino-3-methylcyclohexyl) methane—used in epoxy production	Skin sclerosis, erythema, fatique, myalgia, arthralgia
Quartz	Systemic sclerosis
Iatrogenic agents	
Bleomycin	Pulmonary fibrosis, scleroderma-like lesions
Pentazocine	Pigmentary changes, pannicultitis, ulcerations and sclerotic fibrosis on injection sites
L-Tryptophan	Sclerodermatous induration, peripheral eosinophilia, myalgia, arthralgia
Silicon	Systemic sclerosis, Sjogren's syndrome, arthritis
Other substances	
Toxic oil syndrome	Scleroderma-like changes, neuromuscular atrophy, hypertension, Sicca syndrome

The first suggestion of a possible association between occupational exposure and the development of systemic sclerosis came from the Scottish physician Bramwell in 1914. He observed nine patients of whom five had worked as stonemasons. In 1957 Erasmus reported 16 cases of scleroderma amongst 8000 miners in South African gold mines, in comparison to a control group of 25,000 hospital patients without a single case of scleroderma. Since then similar case reports and studies have been published in the USA, France, Italy, Japan, Switzerland, Canada and Germany[11].

Quartz exposure occurs particularly in the following industries and occupations: mining, stone industry, slate industry, foundry industry, fire-clay manufacture and processing, construction, rubber industry (talc is heavily contaminated with silicon dioxide), ceramic and glass industries.

Treatment of scleroderma is unsatisfactory as there is no specific therapy and the course is almost always progressive. Death usually results from general debility or from renal, myocardial or pulmonary sclerosis. In quartz industry-related cases, removal of the patient from exposure does not usually result in clinical improvement.

Vibration white finger

Vibration white finger (VWF) consists of the episodic appearance of white-finger skin patches (Raynaud's phenomenon) in response to environmental cold and is accompanied by secondary loss of sensation caused by vascular ischaemia. It can be part of but is not synonymous with the hand-arm vibration syndrome[12].

The pathogenesis of VWF is poorly understood. Chronic vibration exposure may damage endothelial vasoregulatory mechanisms by disturbing the endothelial-derived relaxing factor-mediated vasodilatory function[13]. Operatives using vibrating tools, such as lumberjacks, coal miners and road and construction workers, are at risk of developing VWF. Affected individuals develop symptoms of Raynaud's phenomenon on exposure to cold or vibration, usually after many years of working with vibrating tools[12]. The diagnosis is usually made by history alone; ice provocation tests are not always reliable in precipitating attacks of VWF[12].

With widespread knowledge of the cause of VWF, controls over duration of use of relevant machinery and improved personal protective equipment have led to a reduction in the incidence of VWF[12]. The treatment of VWF is the same as for Raynaud's phenomenon. It is generally believed that symptoms of VWF regress some time after cessation of exposure[12].

Skin cancer

In 1775, the first cancer of any type to be linked with environmental exposure was scrotal squamous carcinomas in British chimney sweeps reported by Percivall Pott[14]. In the rest of Europe the disease was unknown because of wearing protective clothing and the reduced carcinogenicity of wood soot as opposed to coal predominantly burnt in Britain. Soot formed by burning wood has much lower levels of the polycyclic hydrocarbon, benzo(a)pyrene (Fig. 2), implicated in the aetiology of skin cancer. Skin cancer was still reported in chimney sweeps in Britain in the 1950s. By 1945 in Britain, almost 50% of industrial skin cancer was attributable to exposure to pitch and tar in occupations such as mule spinners and jute workers and in the engineering industry[15].

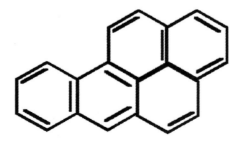

Fig. 2 Benzo(a)pyrene.

In the past, however, there have been several virtual epidemics of skin cancer, which were traceable to environmental exposures. The major environmental carcinogens recognized were polycyclic hydrocarbons, ionizing radiation and arsenic. Ultraviolet radiation is now the most important carcinogen in the aetiology of environmental skin cancer.

Polycyclic hydrocarbons are produced by incomplete combustion and distillation of coal, natural gas and oil shale. These chemicals are contained in tar, fuel oils, lubricating oils and greases, oil shale and bitumen (Table 5)[16].

The diagnosis of skin cancer is similar to that of non-occupational skin cancers. Generally the exposed sites are involved. Previously the scrotum was involved frequently because of continuous exposure to carcinogens and the increased likelihood of skin absorption in that site. There may be co-existing signs of exposure prior to or in addition to evidence of skin cancer. These may include oil folliculitis and hyperkeratoses described in people working with mineral oil, and pitch or tar warts. Oil hyperkeratoses were described as being flat, white, circular, hyperkeratotic smooth plaques, small in diameter and often clustered. In addition there were verrucose pigmented round or oval irregular raised warts. Tar warts were pigmented small papules which were often seen around the face on the eyes, eyelids, cheek, forearms and back of the hands.

Prevention of skin cancers is most important. In the workplace it is important to consider substitution of carcinogens where possible; an example is the declining exposure to polycyclic hydrocarbons in recent decades. Protection of the skin, with either protective clothing or with engineering control such as machine guarding, is important. Daily washing is essential. Since most of the skin cancers are associated with a very long latency

Table 5 Occupations with potential exposure to causative agents in environmental skin cancer

Polycyclic hydrocarbons
  Tar distilling
  Coal gas manufacturing
  Briquettes manufacturing
  Shale oil workers
  Refinery workers

Ultraviolet light
  Outdoor workers
  Welders
  Laser exposure
  Printers

Ionizing radiation
  Nuclear power plant workers
  X-ray technicians
  Uranium mining

period, it is important to have continued surveillance of older or retired workers. Finally, the skin cancers need to be treated as appropriate.

Sunlight effects on the skin

Non-ionizing radiation, predominantly in the ultraviolet region (Fig. 3), from the sun is important in a variety of skin cancers and inflammatory skin diseases.

There is good evidence that sunlight (particularly ultraviolet B, and to a lesser degree ultraviolet A radiation) plays an important role in the aetiology of the commonest skin cancers, basal cell carcinoma[17,18] and squamous cell carcinoma[19] and its precursor lesions actinic keratoses[20] and Bowen's intraepidermal neoplasia. There is more controversy about the role of sunlight in the development of malignant melanoma, but it is probable that sunlight exposure, perhaps particularly intermittent exposure in childhood, is important[21–24].

Non-ionizing radiation from the sun is important in the clinical expression of the idiopathic photodermatoses (Table 6), the cutaneous porphyrias, and drug-induced photosensitivity. It is also important in the rare geno-photodermatoses, such as the many types of xeroderma pigmentosum which share the common feature of impaired ability to repair ultraviolet induced DNA damage. Many other dermatoses can be photoaggravated, including cutaneous lupus erythematosus, Jessner's lymphocytic infiltrate, melasma, pemphigus vulgaris, and actinic lichen planus. The effects of sunlight on the skin are not all adverse, and are beneficial for the majority of those with some common inflammatory skin diseases including psoriasis and atopic dermatitis.

How might environmental pollution alter the transmission of ultraviolet radiation and visible light to the skins of people living in different parts of the world? And what are the possible effects, for good or ill, on the prevalence and severity of common and serious skin diseases?

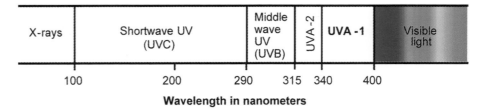

Fig. 3 Ultraviolet and visible rays (and the longer infrared rays) reach Earth. The ultraviolet rays are most relevant in causing, triggering, aggravating and relieving skin diseases.

Table 6 Idiopathic photodermatoses

Disease	Prevalence	Comment
Polymorphic light eruption (PLE)	Very common: prevalence increases at latitudes further from equator	Wide spectrum of severity: ranges from being a nuisance, precluding holidays in sunny climates, to life-ruining severity
Actinic prurigo	Rare in Europe; common amongst Amerindians	Usually more severe than PLE
Idiopathic solar urticaria	Rare	Often severely limiting, particularly for those with severe abnormal visible wavelength photosensitivity
Chronic actinic dermatitis	Rare (1 in 6000 in Tayside, UK)	A chronic severely disabling condition, with particularly severe ultraviolet B photosensitivity in the majority
Hydroa vacciniforme	Very rare	Usually spontaneously resolves after several years, but often severe and difficult to manage while present
Juvenile springtime eruption	Probably common	Childhood only; usually no more than an avoidable nuisance

Possible effects of pollutants: effects on high atmospheric ozone

Depletion of ozone in the upper atmosphere (stratosphere), thought largely to be due to human activities particularly the release of chlorofluorocarbons (CFCs), has resulted in an increase in shorter wavelength ultraviolet B (UVB, see Fig. 3) reaching the Earth's surface. The actual increase in shorter wavelength UVB reaching ground level has been most marked in the southern hemisphere[25]. In the northern regions studied, counteracting effects of air pollutants in the lower atmosphere have generally limited such an effect[26], although an increase in UVB at ground level in Scotland did coincide with reduction in stratospheric ozone[27].

Mathematical modelling has estimated that the actual increase in skin cancers seen in the northern hemisphere as a result of ozone depletion will be small[28,29], although not insignificant when we consider that non-melanoma skin cancers are the commonest cancers in peoples with constitutively pale skin. Yet the effects of ozone depletion might be expected to cause more problems in the South. An increase in patients with sunburn, and other photodermatoses[30], presenting to the dermatologist serving Punta Arenas, the most southerly city in South America, has been noted to coincide with increased ground-level UVB measurements associated with movements of the Antarctic 'ozone hole'. While confounding factors, particularly media attention to the ozone hole likely to increase the probability of someone with sunburn seeking a dermatologist's opinion, could not be excluded in this observational study, this was an important report as it showed for the first time a possible real association of skin disease and changes in stratospheric ozone.

Overall, small changes in stratospheric ozone, increasing shorter ultraviolet wavelengths reaching ground level, particularly in the southern hemisphere, may be expected to increase the frequency of sunburn episodes, and skin cancers, although the evidence that this is actually happening is limited. If there are effects on other common conditions such as polymorphic light eruption (PLE), it is unclear what these will be: possibly PLE will even decrease in frequency as this condition is generally provoked more readily by longer wavelength ultraviolet (UVA) radiation, and it is likely that its relative rarity near the equator is due to greater UVB exposure.

High atmospheric ozone and lower atmospheric pollution

Pollutants, such as the smoke from forest burning in South East Asia[31], would be expected to reduce the frequency of conditions caused or aggravated by ultraviolet exposure.

It is theoretically possible that worry about the possible risks of CFC pollution affecting the ozone layer will lead to increased sunscreen use. Recently, several widely used absorbent sunscreens have been shown to have oestrogenic activity comparable to that of other xenoestrogens released into the environment and entering the food chain[32], raising concern about possible effects on ecology and human health.

Key points for clinical practice

- Irritant contact dermatitis is the commonest dermatosis caused by environmental exposure.
- Chloracne, chemical depigmentation and certain connective tissue diseases can be caused by environmental pollution.
- Ultraviolet radiation is the most important carcinogen in the aetiology of skin cancer.
- Ultraviolet radiation can aggravate certain dermatoses and cause photosensitivity.
- There is as yet little hard evidence that atmospheric pollution has had (or will have) an effect on skin diseases influenced by ultraviolet radiation from the sun.

References

1. Meding B, Swanbeck G. Occupational hand eczema in an industrial city. *Contact Dermatitis* 1990; **22**: 13–23
2. Holt PG. Parasites, atopy, and the hygiene hypothesis: resolution of a paradox? *Lancet* 2000; **356**: 1699–701
3. Kay J, Gawkrodger DJ, Mortimer MJ, Jaron AG. The prevalence of atopic eczema in the general population. *J Am Acad Dermatol* 1994; **30**: 35–9
4. von Mutius E, Weiland SK, Fritzsch C et al. Increasing prevalence of hay fever and atopy among children in Leipzig, East Germany. *Lancet* 1998; **351**: 862–6
5. Taylor JS. Environmental chloracne: Update and review. *Ann NY Acad Sci* 1979; **320**: 295–307
6. Tindall JP. Chloracne and chloracnegens. *J Am Acad Dermatol* 1985; **13**: 539–58
7. Wattanakrai P, Miyamoto L, Taylor JS. Occupational pigmentary disorders. In: Kanerva L, Elsner P, Wahlberg JE, Maibach HI (eds) *Handbook of Occupational Dermatology*. Berlin: Springer, 2000; 280–94
8. Ortonne JP, Mosher DB, Fitzpatrick TB. Hypomelanosis secondary to irradiation and physical trauma, chemical hypomelanosis, hypomelanosis associated with inflammation. In: *Vitiligo and Hypomelanosis of Hair and Skin*. New York and London: Plenum Medical Book Company, 1983; 475–522
9. Gawkrodger DJ. Pigmentary changes due to occupation. In: English JSC (ed.) *A Colour Handbook of Occupational Dermatology*. London: Mansons, 1998; 147–58
10. Black CM, Welsh KI. Occupationally and environmentally induced scleroderma-like illness: etiology, pathogenesis, diagnosis, and treatment. *Int Med Spec* 1988; **9**: 135–54
11. Ziegler V, Haustein UF. Die progressive Sklerodermie—eine quarzinduzierte Berufskrankheit? *Dermatol Monatsschr* 1992: **178**: 34–43
12. Gemne G. Raynaud's phenomena ('white fingers') in workers using hand-held vibrating tools. In: Kanerva L, Elsner P, Wahlberg JE, Maibach HI (eds) *Handbook of Occupational Dermatology*. Berlin: Springer, 2000; 162–6
13. Gemne G. Pathophysiology and pathogenesis of disorders in workers using hand-held vibrating tools. In: Pelmear P, Taylor W, Wassermann D (eds) *Hand-arm Vibration. A Comprehensive Guide for Occupational Health Professionals*. New York: Van Nostrand Reinhold, 1992; 41–76
14. Waldron HA. A brief history of scrotal cancer. *Br J Ind Med* 1983; **40**: 390–401
15. Cruikshank CD, Squire JR. Skin cancer in the engineering industry from use of mineral oil. *Br J Ind Med* 1950; **7**: 1–11
16. Epstein JH, Ormsby A, Adams RM. Occupational skin cancer. In: Adams RM (ed.) *Occupational Skin Disease*, 3rd edn. WB Saunders Company, 1999; 142–64
17. Kricker A, Armstrong BK, English DR, Heenan PJ. Does intermittent sun exposure cause basal cell carcinoma? A case-control study in Western Australia. *Int J Cancer* 1995; **60**: 489–94
18. Gallagher RP, Hill GB, Bajdik CD, Fincham S, Coldman AJ, McLean DI et al. Sunlight exposure, pigmentary factors, and risk of nonmelanocytic skin cancer. I. Basal cell carcinoma. *Arch Dermatol* 1995; **131**: 157–63
19. Gallagher RP, Hill GB, Bajdik CD, Coldman AJ, Fincham S, McLean DI et al. Sunlight exposure, pigmentation factors, and risk of nonmelanocytic skin cancer. II. Squamous cell carcinoma. *Arch Dermatol* 1995; **131**: 164–9
20. Kocsard E. Solar keratoses and their relationship to non-melanoma skin cancers. *Aust J Dermatol* 1997; **38 (Suppl 1)**: S30
21. Nelemans PJ, Rampen FH, Ruiter DJ, Verbeek AL. An addition to the controversy on sunlight exposure and melanoma risk: a meta-analytical approach. *J Clin Epidemiol* 1995; **48**: 1331–42
22. Armstrong BK. Epidemiology of malignant melanoma: intermittent or total accumulated exposure to the sun? *J Dermatol Surg Oncol* 1988; **14**: 835–49
23. Bell CM, Jenkinson CM, Murrells TJ, Skeet RG, Everall JD. Aetiological factors in cutaneous malignant melanomas seen at a UK skin clinic. *J Epidemiol Community Health* 1987; **41**: 306–11
24. Fears TR, Scotto J, Schneiderman MA. Skin cancer, melanoma, and sunlight. *Am J Public Health* 1976; **66**: 461–4
25. Madronich S, McKenzie RL, Björn LO, Caldwell MM. Changes in biologically active ultraviolet radiation reaching the Earth's surface. *J Photochem Photobiol B* 1998; **46**: 5–19

26 Frederick JE. Ultraviolet sunlight reaching the Earth's surface: a review of recent research. *Photochem Photobiol* 1993; **57**: 175–8
27 Moseley H, Mackie RM. Ultraviolet B radiation was increased at ground level in Scotland during a period of ozone depletion. *Br J Dermatol* 1997; **137**: 101–2
28 Diffey BL. Stratospheric ozone depletion and the risk of non-melanoma skin cancer in a British population. *Phys Med Biol* 1992; **37**: 2267–79
29 Diffey BL. Ozone depletion and skin cancers. In: Grob JJ, Stern RS, MacKie RM, Weinstock WA (eds) *Epidemiology, Causes and Prevention of Skin Diseases*. Oxford: Blackwell Science, 1997; 77–84
30 Abarca JF, Casiccia CC, Zamorano FD. Increase in sunburns and photosensitivity disorders at the edge of the Antarctic ozone hole, southern Chile, 1986–2000. *J Am Acad Dermatol* 2002; **46**: 193–9
31 Sastry N. Forest fires, air pollution, and mortality in southeast Asia. *Demography* 2002; **39**: 1–23
32 Schlumpf M, Cotton B, Conscience B, Haller V, Steinmann B, Lichtensteiger W. In vitro and in vivo estrogenicity of UV screens. *Environ Health Perspect* 2001; **109**: 239–44

Ambient air pollution and health

Klea Katsouyanni

Department of Hygiene and Epidemiology, University of Athens Medical School, Athens, Greece

The adverse health effects of air pollution became widely acknowledged after severe pollution episodes occurred in Europe and North America before the 1960s. In these areas, pollutant levels have decreased. During the last 15 years, however, consistent results, mainly from epidemiological studies, have provided evidence that current air pollutant levels have been associated with adverse long- and short-term health effects, including an increase in mortality. These effects have been better studied for ambient particle concentrations but there is also substantial evidence concerning gaseous pollutants such as ozone, NO_2 and CO. Attempts to estimate the impact of air pollution effects on health in terms of the attributable number of events indicate that the ubiquitous nature of the exposure results in a considerable public health burden from relatively weak relative risks.

Introduction

Anthropogenic air pollution (*i.e.* that superimposed on the background of natural pollution originating from plants, radiological decomposition, forest fires, volcanic eruptions, *etc.*) has existed since people learned how to use fire, but has increased rapidly with industrialization.

The well known and severe air pollution episodes in Europe and North America before 1960 provided indisputable evidence that those high levels of air pollution can have very important adverse health effects, including a significant increase in mortality[1]. Since then, legal and other corrective measures have contributed to a reduction in air pollutant concentrations, especially of black smoke (an index of ambient particles) and sulphur dioxide (SO_2), to moderate or low levels in many, though not all, the areas traditionally affected by air pollution[2]. Until the mid 1980s, it was generally thought that ambient pollution levels in Europe did not threaten human health[3]. However, results from epidemiological studies during the last 15 years have consistently shown that moderate and low concentrations of traditional pollutants such as ambient particles can have both short- and long-term effects on health. Furthermore, in Europe and elsewhere, a change in the emission sources (with vehicles increasingly becoming the most important source in many areas) has

Correspondence to:
Klea Katsouyanni, Mikras Asias 75, (Goudi), Athens, Greece. E-mail: kkatsouy@med.uoa.gr

contributed to changes in the air pollution mixture, which is now characterized by high concentrations of nitrogen oxides and photochemical oxidants. Today we are less concerned with very severe air pollution episodes but much more with the consequences of acute and chronic air pollution exposure for excess respiratory and cardiovascular morbidity and mortality[4].

Clean air is considered to be a basic requirement for human health and well being[2,5]. Individual and population exposure to air pollution is caused by both indoor and outdoor sources. Although the components of indoor and outdoor pollution may be the same, and the exposure–response relationship is not affected by the source of a specific pollutant, outdoor and indoor sources can usefully be treated separately as they are determined by different factors and require different management policies. The focus in this chapter is on outdoor (ambient) air pollution; indoor air pollution is discussed in Chapter 11. At the same time, it needs to be noted that, whilst ambient air pollution exposure occurs outdoors, it also penetrates indoors, at a rate which depends on the nature of a particular pollutant. The involuntary and ubiquitous nature of the exposure results in a considerable public health burden of relatively weak adverse health effects.

The air pollutants routinely measured by most organized monitoring systems include various indicators of ambient particle concentrations and gases. Addressing the problem by pollutant is in many ways inadequate since, in the real world, individuals are exposed to mixtures of pollutants which may act in combination or synergistically. The study of mixtures of air pollutants is, however, extremely complex and as yet in its infancy. This chapter will therefore focus on particles and gases, which are most relevant for health according to current scientific knowledge.

Ambient particles

Measurement, sources, distribution and relevant components of the particle mix

Ambient particles are a mixture with various physical and chemical characteristics. Relevant, interrelated, physical characteristics of particles are size, surface and number. Possibly relevant chemical characteristics include the content of transition metals, crustal material, polycyclic aromatic hydrocarbons, carbonaceous material, sulphates and nitrates. Their concentrations may thus be measured using a wide range of different indices. The traditional ambient particle indicator in Europe[6] has been Black Smoke (BS), measured by reflectometry, representing black particles of aerodynamic diameter <4 μm. The reflectometry units are typically transformed to mass using a calibration curve (the OECD curve).

Widespread monitoring has also been made of total particle mass (TSP) concentration. This is dominated by large particles outside the respirable range, thought today not to be so relevant for health. The US EPA in 1979 defined PM_{10} (particles with diameter <10 µm) as the ambient particle indicator to be used for regulatory purposes and in 2000 added $PM_{2.5}$ (those with diameter <2.5 µm). The airborne particle mix in each location has different chemical and physical characteristics and depends on the range of sources and their proportional contribution to the mix.

Particles derived from combustion sources (vehicles, power plants, *etc.*) are generally smaller whilst those coming from abrasion (road dust, wind blown soil) are often larger[4]. Until recently, all regulations have been based on the particle mass per unit volume. Nevertheless, the number of particles (and surface area) to mass ratio increases with decreasing size, and it seems that number of particles may also be relevant to health effects[8]. The particle mix is composed of primary particles (which are emitted) and secondary particles, such as sulphates and nitrates, which are formed in the atmosphere.

Smaller particles tend to be remarkably homogeneously spread over large areas, penetrate effectively indoors and consist to a larger extent of primary and secondary combustion products (containing elemental carbon and PAHs, sulphates and nitrates).

Although there are some results indicating that particular components of the particle mix are responsible for specific health outcomes, the existing evidence is still limited (see below).

Current guidelines and regulations for ambient particles

The current WHO air quality guidelines for Europe[2] accept that available information does not allow a judgment of concentrations below which no effects are to be expected. Thus, only concentration–response tables for acute health effects are provided, based on studies mainly using PM_{10} and a few using $PM_{2.5}$ as the particle indicator, and relative risks of long-term effects. No guideline values are recommended and risk managers are referred to the risk estimates provided.

The European Union adopted a general framework Directive[9] for air pollution in 1996 and a daughter directive[10], including ambient particulate matter regulations in 1999. The standards adopted can be summarized as follows:

- for the 24 h levels, 50 µg/m³ of PM_{10} should not be exceeded more than 35 times per year by 2005 and more than seven times per year by 2010;
- for the annual levels, 40 and 20 µg/m³ should not be exceeded by the years 2005 and 2010, respectively.

There is a planned review of this regulation currently on-going which should result in a decision about any need for revision during 2003.

The US EPA has adopted standards for PM_{10} (not to exceed 150 µg/m³ on a 24-h basis and 50 µg/m³ on an annual basis) complemented by standards for $PM_{2.5}$ (not to exceed 65 µg/m³ on a 24-h basis and 15 µg/m³ on annual basis). These are also currently under the process of a review based on new scientific evidence[11].

In European cities, the mean annual levels of PM_{10} in the 1990s ranged between 14 and approximately 65 µg/m³, whilst black smoke levels ranged between 10 and 65 µg/m³. In several cities, the level of 50 µg/m³ is exceeded for more than 35 days per year[12].

Health effects of ambient particle concentrations

Short-term (acute) effects

The short-term effects of particles have been the main focus for study, especially in time-series studies, in several locations throughout the world. Acute effects are well established for total non-accidental, respiratory, cardiopulmonary and cardiac daily mortality, as well as respiratory hospital admissions[2]. There is also evidence of acute effects on respiratory function, lower respiratory symptoms and increased medication use by asthmatic subjects[2].

Typically, effect estimates are given as an increase in the health outcome associated with a 10 µg/m³ increase in particle concentrations. There is, however, heterogeneity in the effect estimates reported from different studies. In the WHO guidelines, based on studies until 1994, an increase of 0.74% (95% CI 0.62–0.86%) is reported for the daily total number of deaths and 0.80% (95% CI 0.48–1.12%) for the daily hospital respiratory admissions. More recently, two large multi-city studies, the European APHEA (Air Pollution and Health: a European Approach)[13], and the US NMMAPS (National Mortality, Morbidity and Air Pollution Study)[14], have provided estimates based on 29 and 20 cities, respectively. The APHEA estimate for the daily total number of deaths is 0.6% (95% CI 0.4–0.8%) and the NMMMAPS estimate 0.5% (95% CI 0.1–0.9%). Later estimates from NMMAPS were 0.4% based on 90 cities[15]. After problems were discovered with the software used in the original analyses, these estimates were recalculated, using different modelling methods[16]. This gave a revised estimate of 0.2% (still statistically significant) for the 90 cities in NMMAPS. The optimal model to be used to control for confounding effects is still not clear. The US EPA has organized a workshop where several sensitivity analyses have been presented and a report published[17]. It appears that the above reported estimates cover the extremes of likely effects. For hospital admissions, the reported increase in COPD and asthma admissions for the elderly from the APHEA study[18] is 1.0% (95% CI 0.4–1.5%) and from the NMMAPS study[15] 1.5% (95% CI 1.0–1.9%).

It is clear that the above health effects concern to a larger extent the more sensitive population subgroups, but the specific characteristics of these subgroups have not been exactly identified. There is evidence that the socially deprived, the elderly and persons with pre-existing respiratory or cardiac disease or diabetes are more susceptible to the health effects of air pollution[19–21]. It is also apparent that the acute effects of air pollution do not represent only short-term harvesting: analyses using distributed lag models have indicated that the effects persist over a longer period of time (>1.5 month) and the extent of mortality displacement may be considerable, depending on the cause of death[22,23].

Long-term effects
Long-term effects of chronic exposure to ambient particle concentrations have been studied less. The results and calculations of attributable risks and years of life-lost have largely been based on two US cohort studies[24,25]. Relative risk estimates for total mortality reported from Dockery et al[24], per 10 µg/m^3 in long-term average pollutant concentration, were 1.10 for PM_{10}, 1.14 for $PM_{2.5}$ and 1.33 for sulphates. Corresponding estimates from Pope et al[25] were 1.07 for $PM_{2.5}$ and 1.08 for sulphates. Based on these studies, it has been calculated that the expected reduction in life expectancy from air pollution exposure is in the order of a few years[26]. Recently, Pope published a further analysis of the ACS data and evidence from a European cohort study has provided consistent results[27,28]. An estimate for three countries (Austria, France and Switzerland) using the effects reported from the US cohort studies concluded that about 6% of the annual total mortality may be attributed to air pollution exposure[29], whilst in the WHO 'Global Burden of Disease' project about 1,000,000 premature deaths are attributed to high PM concentrations worldwide[30].

Ozone

Measurement, sources and distribution

Ozone is one of a range of photochemical oxidants which are formed as secondary pollutants by the action of solar radiation in the presence of primary pollutants, mainly nitrogen oxides and volatile organic compounds[2]. Tropospheric ozone pollution should be distinguished from the problem of stratospheric ozone depletion, which is linked to global warming and risks of UV radiation. Because of its generation procedure, tropospheric ozone is a more important problem in the summertime and in areas with more prolonged sunshine. In the presence of precursor primary pollutants (especially NO), ozone is 'scavenged'. As a result, low concentrations tend to occur in busy city centres, where NO concentrations are high, whilst higher concentrations are observed downwind in city suburbs to which ozone is transported but where NO and other precursor concentrations are relatively low. Thus the spatial distribution of ozone and resulting personal exposure patterns differ from those of other pollutants. Ozone measurements are often expressed as ppb or $\mu g/m^3$ (1 ppb = 2 $\mu g/m^3$ at 20°C).

Regulations

The WHO guidelines for ozone give a level of 120 $\mu g/m^3$ for an 8-h average[2]. The US EPA regulations[11] comprise an 8-h standard of 157 $\mu g/m^3$ and a 1-h level of 235 $\mu g/m^3$. The EU regulation for ozone for the protection of human health, still under consideration, is 120 $\mu g/m^3$ for an 8-h average not to be exceeded for more than 20 days per year by the year 2010[31]. In several European cities, the 90th percentile of ozone 1-h concentrations currently exceeds 120 $\mu g/m^3$ and the maximum[13] reaches more than 200–300 $\mu g/m^3$.

Health effects

Ozone, as a potent oxidant, may react with a variety of biomolecules[4], potentially causing both short- and long-term effects. Its effects have been assessed in controlled exposure experiments as well as epidemiological studies. Short-term effects are better established. They include an increase in the daily total number of deaths, especially for the warm season, an increase in hospital respiratory admissions, increased respiratory symptoms, pulmonary function changes, increased airway responsiveness and airway inflammation. In the NMMAPS Study, a 0.5% increase in mortality associated with 20 $\mu g/m^3$ (10 ppb) in the daily O_3 concentrations is reported during summertime[15]. The corresponding estimate from the APHEA project is a 2.9% increase in mortality associated with a 50 $\mu g/m^3$ increase in daily ozone[32]. In a study from Montreal, Goldberg et al estimated a 3.3% increase in daily deaths in the warm season, associated with an increase of 21.3 $\mu g/m^3$ in 3-day ozone concentration[33].

Experimental studies have mainly focused on acute exposures of up to a few hours. These show functional decrements in healthy exercising adults at concentrations around 160 $\mu g/m^3$, whilst there are more severe effects at concentrations of 500 $\mu g/m^3$ or more. A number of field studies done in children and young individuals indicate that pulmonary function decrements can occur at levels of 120–240 $\mu g/m^3$. After repeated prolonged exposure, pulmonary function shows adaptation but there is evidence that inflammation continues[34]. There is also evidence for association of short-term peaks in O_3 exposures and lung epithelial damage[35]. It must be noted that individual responsiveness to O_3 exposure varies substantially for reasons which remain largely unexplained.

For long-term effects, the evidence is less consistent. There are a few studies indicating that long-term ozone exposure may be a risk factor for asthma incidence[36,37], lung function growth[38], and lung cancer incidence and mortality[39,40]. In these studies, it is not entirely clear that the ozone effects are not confounded by particle levels. Furthermore, the ACS cohort study found no indication of an association between long-term ozone exposure and either lung-cancer or total mortality[27].

Nitrogen dioxide (NO$_2$)

Measurement, sources and distribution

Nitrogen dioxide is mainly produced as a result of emissions from vehicles and is thus considered a good indicator of ambient, traffic-generated air pollution[41]. Power plants and fossil-fuel burning industries also contribute to NO_2 pollution. There are also significant indoor sources of NO_2, such as gas stoves[2,4], and indoor NO_2 levels may dominate the total personal exposure to NO_2. However, NO_2 typically forms part of a complex pollutant mixture which is different in indoor from outdoor air[4].

During high temperature combustion, nitric oxide (NO), NO_2 and other nitrogen oxides (NO_x) are generated. Part of the NO is converted to NO_2 through oxidation reactions which involve oxygen and ozone. NO_2, in the presence of sunlight, participates with hydrocarbons and oxygen in the formation of ozone and other secondary photochemical oxidants and is therefore an important precursor of O_3 formation. NO_2 also reacts with aerosols to form secondary (often acidic) particles[4,42].

NO_2 is measured routinely by monitoring networks and is expressed either as $\mu g/m^3$ or ppb (1 ppb = 1.913 $\mu g/m^3$ at 20°C).

Regulations

WHO guidelines provide a 1-h limit of 200 $\mu g/m^3$ and an annual limit of 40 $\mu g/m^3$. The US EPA[11] only provides for an annual standard of 100 $\mu g/m^3$. The EU legislation on NO_2 provides one short-term limit value at 200 $\mu g/m^3$ for 1 h and an annual level at 40 $\mu g/m^3$, not to be exceeded[10] after 2010.

In European cities, the median of 24-h NO_2 concentrations currently ranges from about 30 to about 90 µg/m³ and the 90th percentile from about 40 to about 140 µg/m³. The concentrations tend to be higher in cities with higher traffic density and in southern European cities[13].

Health effects

Healthy subjects experience reductions in pulmonary function and increased airway reactivity only at levels of NO_2 exposure much higher (>1500 mg/m³) than those measured outdoors[2,4]. Some people, however, are susceptible to effects at much lower concentrations.

It seems that the most sensitive subjects are asthmatics, though individual asthmatic subjects differ in their sensitivity to NO_2 exposure. In experimental studies with mild asthmatics, hyper-responsiveness and lung function decrements have been reported at NO_2 concentrations as low as 550 µg/m³, though responders cannot be defined *a priori*[2,4]. Subjects with more severe asthma may respond differently.

Epidemiological studies have mainly focused on indoor exposures[43] (see also Chapter 11). From these, there is evidence of increased respiratory symptoms and illness with increased long-term average indoor NO_2 concentration[43,44]. The few studies which were conducted evaluating outdoor exposures[45] showed increasing illness incidence with increasing NO_2 concentrations, but the causal association with NO_2 was not entirely clear. It should be noted that even small changes in susceptibility to respiratory viruses may have important public health significance, mainly because of widespread indoor exposure.

Some epidemiological studies have investigated short-term effects of NO_2 on mortality and hospital admissions[32,46,47]. In some of these studies, significant (but weak) effects of NO_2 have been found. The effects tend to be reduced, however, in multi-pollutant models with inclusion of particles or carbon monoxide in the model, so it is not completely clear whether effects can reliably be attributed to NO_2. NO_2 has also been found to modify the effect of particles: in the APHEA study, the increase in mortality due to particles was higher in cities where the long-term NO_2 concentrations were higher[13]. This was interpreted as an indication that a greater proportion of particles originated from traffic in places with higher NO_2 levels. One cohort study has evaluated the long-term effects of NO_2, though only as a general indicator of traffic pollution[28].

There are also experimental studies investigating changes of host defence against infection. Animal studies have suggested that effects on alveolar macrophage antimicrobial function can occur at NO_2 concentrations of 1000 µg/m³ or higher. This may explain the increased infectivity found in some epidemiological studies. A few clinical studies of controlled exposure and subsequent infection *in vitro* of alveolar macrophage cells have indicated effects on host defence mechanisms of some healthy individuals[4].

The importance of NO_2 for health and the need for regulation thus comes less from its direct effects on health than its role as an O_3 precursor and a contributor to the formation of secondary particles.

Sulphur dioxide (SO_2)

Measurement, sources, and distribution

In the earlier part of the 20th century very high concentrations of SO_2, together with particles, were measured in many urban areas. Because of their close interdependence (both were derived primarily from coal combustion) the effects of SO_2 and particles were often considered together[1,4]. Since the 1970s, SO_2 concentrations in both Europe and the USA have declined as a result of changing fuel quality and fuel use[2]. However, in large cities outside those areas (*e.g.* in China), where coal is still used for domestic cooking and heating, high concentrations are still observed. Because of its historic importance, monitoring of SO_2 has been extensive and there is a large and long-term database of 24-h SO_2 measurements in Europe. SO_2 concentrations are expressed in ppb or $\mu g/m^3$ (1 ppb = 2.704 $\mu g/m^3$ at 20°C).

Regulations

WHO provides a guideline of 125 $\mu g/m^3$ for 24-h SO_2 exposure, 500 $\mu g/m^3$ for 10 min and an annual average of 50 $\mu g/m^3$, independent of the presence of particles. The US EPA gives a 3-h average standard of 1300 $\mu g/m^3$, a 24-h average of 365 $\mu g/m^3$ and an annual standard of 80 $\mu g/m^3$. The EU has limit values for 1 h of 350 $\mu g/m^3$ not to be exceeded by 2005, 125 $\mu g/m^3$ for 24 h and an annual average of 20 $\mu g/m^3$ for the protection of ecosystems with no margin of tolerance[10].

The median levels of 24-h SO_2 are typically below 50 $\mu g/m^3$ in European cities but there are occasional values of 125 $\mu g/m^3$ on a 24-h basis, mainly in the cities of central-eastern Europe[13].

Health effects

Older experimental studies established very short-term responses to high levels of SO_2 which included decreases in lung function, and increases in specific airway resistance and respiratory symptoms[2]. Asthmatics are the most sensitive group, although individuals vary in their responsiveness.

Because of their close association, short-term epidemiological studies in the past were unable to distinguish between the effects of SO_2 and particles. Recent studies, however, consistently demonstrate effects on mortality (total, respiratory and cardiovascular)[46,48] and hospital respiratory and cardiovascular admissions[49,50], in cities with SO_2 levels below the WHO guidelines. This finding is considered by some investigators to be inexplicable at such low levels of SO_2 and merits further investigation. Although the effect of SO_2 appears to be independent of particles in multipollutant models, it may in reality be associated with sulphates and be an indicator of specific particle characteristics.

As far as long-term exposures are concerned, results from cohort studies[24,25] which evaluated SO_2 indicate that health effects are predominantly a result of exposure to ambient particles.

Carbon monoxide (CO)

Measurement, sources, and distribution

Carbon monoxide is mainly produced by incomplete combustion of carbonaceous fuels such as gasoline and natural gas. Outdoors it is mainly emitted from vehicles. Its concentration is relatively high in traffic canyons and may be very high in road tunnels, multi-storey car parks and other such microenvironments. Also CO concentrations inside vehicles may be higher than outdoors, while a range of indoor sources exist, such as ETS and gas appliances[2]. It has been shown that individual exposure to CO in non-smokers mainly happens during motor vehicle travel[4].

CO is routinely measured by monitoring networks and is usually expressed in mg/m^3 or ppm (1 ppm = 1.165 mg/m^3 at 20°C).

Regulations

WHO air quality guidelines give a guideline of 100 mg/m^3 for 15-min exposure, 60 mg/m^3 for 30-min, 30 mg/m^3 for 1-h and 10 mg/m^3 for 8-h exposure. There is no long-term average guideline. The US EPA have adopted a standard of 10 mg/m^3 as an 8-h and 40 mg/m^3 as a 1-h average[11], while the EU[51] proposes an 8-h limit value of 10 mg/m^3, not to be exceeded by 2005. The WHO air quality guidelines are set to prevent levels of COHb (carboxyhaemoglobin) in the blood exceeding 2.5%.

Health effects

The toxic effects of CO are largely attributed to its high affinity with haemoglobin and myoglobin. Its affinity to haemoglobin is 200–250 times that for oxygen. Approximately 80–90% of absorbed CO binds with haemoglobin to form carboxyhaemoglobin (COHb). High exposures to CO cause acute poisoning, but such exposures are not encountered in outdoor urban settings. Unlike other gaseous pollutants presented above, CO appears to have no toxic effect on the lung but its health effects are manifested through the interference with oxygen transport[2,4]. Continuous exposure to levels less than 10 mg/m^3 should not cause COHb levels >2% in normal non-smokers. For continuous exposures to CO concentrations up to 200 ppm at sea level, the COHb% at equilibrium can be approximated as COHb% = COppm × 0.16. In practice, it is difficult to predict the percentage COHb because of the large spatial and temporal variation in CO exposure.

In controlled human exposure studies in patients with coronary artery disease, COHb levels between 2 and 6% have been associated with cardiovascular endpoints such as shortening of time to onset of angina. A limited number of recent epidemiological studies have provided evidence on the association of CO exposure to cardiac arrhythmia[52], hospital admissions for heart disease[53] and mortality[15,54].

Conclusions

In summary, current levels of air pollution in Europe have considerable adverse health effects. These have been better studied for ambient particle concentrations, which appear to have both long- and short-term effects including an increase in mortality. It appears that the health effects of particles mainly concern sensitive population subgroups such as the elderly or those with chronic respiratory illness.

The short-term effects of ozone exposure on health are also well documented. With regard to other pollutants, there is some evidence of NO$_2$ and CO effects. NO$_2$, in addition, is important as a precursor to other pollutants and as a traffic pollution indicator. The levels of SO$_2$ have decreased in Europe and, although this pollutant is consistently associated with health endpoints, it may be acting as a surrogate for a specific mixture of other pollutants.

References

1. Davis DL. *When Smoke Ran Like Water*. New York: Basic Books, 2002
2. WHO. *Air Quality Guidelines for Europe*, 2nd edn. WHO Regional Publications, European Series No 91. Copenhagen: WHO, 2000

3 Holland WW, Bennett AE, Cameron IR, Florey C duV, Leeder SR, Schilling RSE, Swan AV, Waller RE. Health effects of particulate air pollution: reappraising the evidence. *Am J Epidemiol* 1979; **110**: 527–659

4 Committee of the Environmental and Occupational Health Assembly of the American Thoracic Society. Health effects of outdoor air pollution. *Am J Respir Crit Care Med* 1996; **153**: 3–50 and 477–98

5 HEALTH21. *The Health For All Policy Framework For the WHO European Region*. European Health for all Series No 6. Copenhagen: WHO Regional Office for Europe, 1999

6 Department of Health. Committee on the Medical Effects of Air Pollution. *Non-biological Particles and Health*. London: HMSO, 1995

7 US Environmental Protection Agency (EPA). *Review of the National Ambient Air Quality Standards for Particulate Matter. OAQPS Staff Paper*. Office for Air Quality Planning and Standards. EPA-452/R-96-013, 1996

8 Penitten P, Timonen P, Tittanen A, Mirme J, Ruuskanen J, Pekkanen J. Ultrafine particles in urban air and respiratory health among adult asthmatics. *Eur Respir J* 2001; **17**: 428–35

9 Commission of the European Communities. Council Directive 96/62/EC on ambient air quality assessment and management. *Official J Eur Communities* 1996; L296/5521.11.1996

10 Commission of the European Communities. Council Directive 1999/30/EC relating to limit values for sulphur dioxide, oxides of nitrogen, particulate matter and lead in ambient air. *Official J Eur Communities* 1999; L163/41.29.6.1999

11 US EPA. National Ambient Air Quality Standards (NAAQS) www.epa.gov/airs/criteria.html

12 Air Pollution and Health: a European Information System (APHEIS). *Health Impact Assessment of Air Pollution in 26 European cities. Second-Year Report*. Paris: Institut de Veille Sanitaire, 2002

13 Katsouyanni K, Touloumi G, Samoli E *et al*. Confounding and effect modification in the short-term effects of ambient particles on total mortality: Results from 29 European cities within the APHEA2 Project. *Epidemiology* 2001; **12**: 521–31

14 Samet JM, Dominici F, Curriero C, Coursac I, Zeger SL. Fine particulate air pollution and mortality in 20 U.S. cities, 1987–1994. *N Engl J Med* 2000; **343**: 1742–9

15 Samet J, Zeger SL, Dominici F. *The National Morbidity, Mortality and Air Pollution Study. Part II Results. Health Effects Institute Report no 94*. Cambridge MA: HEI, 2000

16 Dominici F, McDermott A, Zeger SL, Samet JM. On the use of generalized additive models in time-series studies of air pollution and health. *Am J Epidemiol* 2002; **156**: 193–203

17 Health Effects Institute (HEI). *Revised Analyses of Time-Series Studies of Air Pollution and Health. Special Report*. Boston: HEI, 2003

18 Atkinson R, Anderson HR, Sunyer J *et al*. Acute effects of particulate air pollution on respiratory admissions. Results from APHEA2 project. *Am J Respir Crit Care Med* 2001; **164**: 1860–6

19 Gouveia N, Fletcher T. Time-series analysis of air pollution and mortality: effects by cause, age and socioeconomic status. *J Epidemiol Community Health* 2000; **54**: 750–5

20 Dockery DW. Epidemiologic evidence of cardiovascular effects of particulate air pollution. *Environ Health Perspect* 2001; **109** (**Suppl 4**): 483–6

21 Goldberg MS, Burnett RT, Bailar 3rd JC *et al*. Identification of persons with cardiorespiratory conditions who are at risk of dying from the acute effects of ambient air particles. *Environ Health Perspect* 2001; **109** (**Suppl 4**): 487–94

22 Zanobetti A, Schwartz J, Samoli E *et al*. The temporal pattern of mortality responses to air pollution: a multicity assessment of mortality displacement. *Epidemiology* 2002; **13**: 87–93

23 Schwartz J. Harvesting and long-term exposure effects in the relation between air pollution and mortality. *Am J Epidemiol* 2000; **151**: 440–8

24 Dockery DW, Pope CA, Xu X, Spengler JD, Ware JII, Fay ME, Ferris BG, Speizer FE. An association between air pollution and mortality in six U.S. cities. *N Engl J Med*, 1993; **329**: 1753–9

25 Pope CA, Thun MJ, Namboodiri MM, Dockery DW, Evans JS, Speizer FE, Heath CW. Particulate air pollution as a predictor of mortality in a prospective study of U.S. adults. *Am J Respir Crit Care Med* 1995; **151**: 669–74

26 Brunekreef B. Air pollution and life expectancy: Is there a relation? *Occup Environ Med* 1997; **54**: 781–4

27 Pope 3rd CA, Burnett RT, Thun MJ, Calle EE, Krewski D, Ito K, Thurston GD. Lung cancer, cardiopulmonary mortality, and long-term exposure to fine particulate air pollution. *JAMA* 2002; **287**: 1132–41

28 Hoek G, Brunekreef B, Goldbohm S, Fischer P, Van Den Brandt PA. Association between mortality and indicators of traffic-related air pollution in the Netherlands: a cohort study. *Lancet* 2002; **360**: 1203–9

29 Kuenzli N, Kaiser R, Medina S *et al*. Public health impact of outdoor and traffic related air pollution: a European assessment. *Lancet* 2000; **356**: 795–801

30 Ezzati M, Lopez AD, Rodgers A, Hoorn SV, Murray CJ. Selected major risk factors and global and regional burden of disease. *Lancet* 2002; **360**: 1347–60

31 Commission of the European Communities. Council Directive 2002/3/EC of 12 February 2002

32 Touloumi G, Katsouyanni K, Zmirou D *et al*. Short-term effects of ambient oxidant exposure on mortality: a combined analysis within the APHEA project. *Am J Epidemiol* 1997; **146**: 177–85

33 Goldberg MS, Burnett RT, Brook J, Bailar 3rd JC, Valois MF, Vincent R. Associations between daily cause-specific mortality and concentrations of ground level ozone in Montreal, Quebec. *Am J Epidemiol* 2001; **154**: 817–26

34 Jorres RA, Holz O, Zachgo W *et al*. The effect of repeated ozone exposures on inflammatory markers in bronchoalveolar lavage fluid and mucosal biopsies. *Am J Respir Crit Care Med* 2000; **161**: 1855–61

35 Broeckaert F, Arsalane K, Hermans C, Bergamaschi E, Brustolin A, Mutti A, Bernard A. Serum claracell protein: a sensitive biomarker of increased lung epithelium permeability caused by ambient ozone. *Environ Health Perspect* 2000; **108**: 1533–7

36 McConnell R, Berhane K, Gilliland F, London SJ, Islam T, Gauderman WJ, Avol E, Margolis HG, Peters JM. Asthma in exercising children exposed to ozone: a cohort study. *Lancet* 2002; **359**: 386–91

37 McDonnell WF, Abbey DE, Nishino N, Lebowitz MD. Long-term ambient ozone concentration and the incidence of asthma in nonsmoking adults: the AHSMOG Study. *Environ Res* 1999; **80**: 110–21

38 Horak F, Studnicka M, Gartner C, Spengler JD, Tauber E, Urbanek R, Veiter A, Frischer T. Particulate matter and lung function growth in children: a 3-yr follow-up study in Austrian schoolchildren. *Eur Respir J* 2002; **19**: 838–45

39 Beeson WL, Abbey DE, Knutsen SF. Long-term concentrations of ambient air pollutants and incidence lung cancer in California adults: results from the AHSMOG study. *Environ Health Perspect* 1998; **106**: 813–23

40 Abbey DE, Nishino N, McDonnell WF, Burchette RJ, Knutsen SF, Lawrence Beeson W, Yang JX. Long-term inhalable particles and other air pollutants related to mortality in nonsmokers. *Am J Respir Crit Care Med* 1999; **159**: 373–82

41 Rijnders E, Janssen NAH, Van Vliet PHN, Brunekreef B. Personal and outdoor nitrogen dioxide concentrations in relation to degree of urbanization and traffic density. *Environ Health Perspect* 2001; **109**: 411–7

42 Spengler J, Brauer M, Koutrakis P. Acid air and health. *Environ Sci Technol* 1990; **24**: 946–56

43 Samet JM, Marbury MC, Spengler JD. Health effects and sources of indoor air pollution. Part I. *Am Rev Respir Dis* 1987; **136**: 1486–508

44 Neas LM, Dockery DW, Ware JH, Spengler JD, Speizer FE, Ferris BG. Association of indoor nitrogen dioxide with respiratory symptoms and pulmonary function in children. *Am J Epidemiol* 1991; **134**: 204–9

45 Braun-Fahrlander C, Ackermann-Liebrich CU, Schwartz J, Gnehn HP, Rutishauser M, Wanner HU. Air pollution and respiratory symptoms in preschool children. *Am Rev Respir Dis* 1992; **145**: 42–7

46 Zmirou D, Schwartz J, Saez M *et al*. Time-series analysis of air pollution and cause-specific mortality. *Epidemiology* 1998; **9**: 495–503

47 Saez M, Ballester F, Barcelo MA, Perez-Hoyos S, Bellido J, Tenias JM, Ocana R, Fidueiras A, Arribas F, Aragones N, Tobias A, Cirera L, Canada A on behalf of the EMECAM group. A combined analysis of the short-term effects of photochemical air pollutants on mortality within the EMECAM project. *Environ Health Perspect* 2002; **110**: 221–8

48 Katsouyanni K, Touloumi G, Spix C *et al*. Short-term effects of ambient sulphur dioxide and particulate matter on mortality in 12 European cities: results from time series data from the APHEA project. *BMJ* 1997; **314**: 1658–63

49 Sunyer J, Anto JM, Murillo C, Saez M. Air pollution and emergency room admissions for chronic obstructive pulmonary diseases. *Am J Epidemiol* 1991; **134**: 277–86

50 Schwartz J and Morris R. Air pollution and hospital admissions for cardiovascular disease in Detroit, Michigan. *Am J Epidemiol* 1995; **142**: 23–35
51 European Commission. Directive 200/69/EC of 16 November 2000 relating to limit values for benzene and carbon monoxide in ambient air. *Official J Eur Communities* 13.12.2000
52 Peters A, Liu E, Verrier RL, Schwartz J, Gold DR, Mittleman M, Baliff J, Oh JA, Allen G, Monahan K, Dockery DW. Air pollution and incidence of cardiac arrhythmia. *Epidemiology* 2000; **11**: 11–7
53 Schwartz J. Air pollution and hospital admissions for heart disease in eight U.S. counties. *Epidemiology* 1999; **10**: 17–22
54 Touloumi G, Samoli E, Katsouyanni K. Daily mortality and 'winter type' air pollution in Athens, Greece—a time-series analysis within the APHEA project. *J Epidemiol Common Health* 1996; **50 (Suppl 1)**: 47–51

Electromagnetic radiation

Anders Ahlbom and **Maria Feychting**

Department of Epidemiology, Institute of Environmental Medicine, Karolinska Institute, Stockholm, Sweden

Electromagnetic fields (EMF) are ubiquitous in modern society. It is well known that exposure to strong fields can result in acute effects, such as burns; the mechanisms behind such effects are well established. There is, however, also a concern that long-term exposure to weak fields might have health effects due to an as-yet unknown mechanism. Because of the already widespread exposure, even small health effects could have profound public health implications. Comprehensive research efforts are therefore warranted, and are indeed ongoing. The strongest evidence for health risks is from exposure to fields generated in connection with use of electric power. As for fields used by telecommunications technology, there is still considerably fewer data available and for the time being there is only very weak support for the existence of health effects. However, extensive research activities are ongoing and much more data will be available in the near future. This situation of scientific uncertainty and considerable public concern creates dilemmas for decision makers.

Introduction

Electromagnetic fields are ubiquitous in modern society. They occur in connection with use of electric power, electronic surveillance systems and various types of wireless communications. While these fields differ with respect to strengths and physical characteristics, they all give rise to concern among those exposed about the possibility of health risks. It is well established that strong fields can give rise to acute health effects, such as burns, but exposure guidelines and regulations protect effectively against such effects. Current concerns are instead directed towards the possibility that long-term exposure to weak fields might have detrimental health effects due to some, to date unknown, biological mechanism.

Due to the widespread use of these techniques and the very prevalent exposure to some of the types of field involved, even a weak association with disease risk could have strong impacts on public health. Although the likelihood of such a scenario is debatable, it is the opinion of many that close monitoring of health risks among exposed subjects is a high

Correspondence to:
Anders Ahlbom, Professor of Epidemiology, Department of Epidemiology, Institute of Environmental Medicine, Karolinska Institute, Box 210, 171 77 Stockholm, Sweden. E-mail: anders.ahlbom@ imm.ki.se

priority. Indeed, extensive research is ongoing and has been so for several years. A report linking childhood cancer mortality to the presence of power lines in the near proximity to the children's homes about 25 years ago spurred the interest in power frequency fields[1]. This interest has remained high ever since, although it has gradually shifted towards other types of field, and in particular towards those used in connection with telecommunications.

The objective of this paper is to discuss the evidence pertaining to the possibility that long-term exposure to weak fields of the types discussed above may be associated with health risks.

The electromagnetic spectrum and low frequency fields

The electromagnetic spectrum encompasses a wide variety of electromagnetic fields, including static fields, radio frequency fields, UV radiation, visible light and X-ray radiation. Electromagnetic fields are characterized by their frequency or wavelength; the wavelength is inversely proportional to the frequency. At the lower end of the electromagnetic spectrum, it is customary to refer to the frequency, while at the upper end one usually uses wavelength. The radiation energy is directly proportional to the frequency, as described by Planck's law. The electromagnetic spectrum can be divided into an ionizing and a non-ionizing segment. In the non-ionizing part, the radiation energy is too weak to break chemical bonds and, thus, to form ions. In contrast, ionizing radiation carries enough energy to break chemical bonds. The border between non-ionizing and ionizing radiation is located roughly at the upper end of the UV-band. This paper deals with potential health risks from a subset of the non-ionizing part of the spectrum, namely fields of frequencies up to 300 GHz; these fields are often called low frequency fields.

This frequency band includes fields that are generated in connection with production, transmission, distribution and use of electric power. Such fields usually have a frequency of 50 or 60 Hz. We will refer to these frequencies as ELF (extremely low frequency). The low frequency band also includes fields that are used for communication with mobile telephony. This technology typically uses frequencies from 450 up 2500 MHz, although new technology will extend this band upwards. We refer to this as RF (radiofrequency). Use of electricity and telecommunications are the technologies that have caused most of the current concern regarding possible health risks and for which most research is available. However, quite a number of other applications also employ fields in the range below 300 GHz. Television and radio transmitters use frequencies similar to mobile phones, as do microwave ovens. Surveillances system of the type found in stores, as well as by cashier machines, often use so-called intermediate frequencies, up to about 40 kHz. Static fields are used for example by magnetic resonance imaging (MRI).

ELF magnetic fields in the environment are usually characterized by their flux density, which is measured in units of Tesla (T) or microTesla (µT). Environmental RF fields are characterized by their power density, measured as Watt per square metre (W/m^2).

Established mechanisms of interaction

The ELF fields have a long wavelength; indeed 50 Hz corresponds to a wavelength of 3500 km, which is similar to the Earth's radius. As a consequence, such fields essentially pass through the body without depositing any energy. The established mechanism of interaction between such fields and the human body is induction of electric currents. The current density induced in the body is a direct consequence of the external magnetic field flux density; this is the reason why such fields are typically characterized by their µT-level. It is well established in laboratory studies and in theoretical calculations that high internal current densities cause acute biological effects. The biophysical quantity that is used to characterize the induced currents is Ampere per square meter. Existing exposure guidelines are expressed in terms of so-called basic restrictions and primarily aim to prevent neurological effects by restricting the internal current density[2]. The environmental flux densities required to produce internal current densities that exceed the basic restrictions are orders of magnitude above what one normally encounters in the general environment. The exception is certain work places where high electric fields can momentarily result in current densities near the guideline reference values.

The RF fields have wavelengths in the order of a few centimetres or less, depending on the actual frequency. Depending on the field strength, some energy is deposited in the body, mainly within one or two centimetres of its surface. The only known consequence of this is heating. The purpose of the basic restrictions in existing guidelines is to prevent excessive heating, locally or in the whole body[2]. The physical quantity that is used in this context is the specific absorption rate (SAR), measured in W/kg. The internal SAR cannot be measured directly inside the body, but is established based on models and theoretical calculations. The SAR values that are indicated on some mobile phones are derived in this way and reflect the maximum SAR value that a particular phone is assumed to be able to produce. The actual field levels that people encounter in connection with mobile phone use produce SAR values below the basic restrictions, although of the same order of magnitude. The fields from base stations that people experience, on the other hand, are much weaker; they are in fact several orders of magnitude below the guideline levels.

Potential health risks from weak long-term exposure: ELF

Given the small amount of energy that is deposited in connection with exposure to ELF fields, any health effects due to weak long-term exposure would have to be produced by a to-date unknown biophysical mechanism. Despite this, researchers have been intrigued by this possibility. Ahlbom et al[3] give a comprehensive overview of the relevant epidemiology. The interest originates to no small extent from the early epidemiological study by Wertheimer and Leeper[1]. The results from this suggested that childhood cancer mortality was associated with the existence of power lines near the children's homes, and particularly with such power lines that were indicative of high magnetic field exposure. Although the implications by many were considered implausible, and despite several methodological problems in the study, that work has been followed by several attempts to replicate the findings. To date, close to 20 studies on childhood cancer and residential exposure to ELF fields have been published[3]. The studies have generally been of increasing methodological strengths, particularly with respect to assessment of magnetic field exposure but also with respect to selection bias and other methodological aspects. The first few of these replications addressed a variety of cancer types and also total cancer, but later studies have increasingly focused on childhood leukaemia. Perhaps to the surprise of the scientific community, the later studies to a great extent confirmed the original finding, although the results have certainly not been identical and indeed some studies failed to find any association. To assess the overall evidence, a pooled analysis was carried out[4] based on primary data from the sub group of nine studies fulfilling certain quality criteria. The principal finding of the pooled analysis was that residential magnetic field exposure in excess of 0.4 µT was associated with about a doubling in the relative risk of childhood leukaemia. It was concluded that chance was an unlikely explanation, but that systematic error could explain some of the observed excess risk.

In parallel with the childhood cancer research, possible associations between other implicated diseases and ELF fields have been explored. Most of this research was directed towards other forms of cancer: brain tumours, leukaemia in adults, and male and female breast cancer are the forms that have attracted the greatest interest. Despite these efforts, however, results have been inconclusive. Interest has prevailed in relation to breast cancer, which remains a focus of attention. However, a recently completed large study with refined occupational exposure information and oestrogen status data for the cases was entirely negative and this is likely to affect the interest in this research track[5].

Outside the cancer field, cardiovascular disease may be the area that has attracted most of the interest. This was based on physiological experiments which noted that ELF magnetic fields appeared to affect heart rate variability[6]. These experiments were followed by a utility worker study showing that chronic heart disease mortality was not associated with ELF exposure but that arrhythmia and myocardial infarction mortality was[7]. However, several later studies that addressed the issue from different perspectives have failed to replicate these results[8,9].

In parallel with the epidemiological research, extensive *in vivo* and *in vitro* research has also been carried out. Despite intense efforts, this has not resulted in the detection of any new mechanisms of interaction between ELF fields and the human body beyond the induction of electric current, nor a strong candidate for such a mechanism. As a consequence, the epidemiological evidence stands alone.

In an evaluation of carcinogenicity, IARC (The International Agency for Research on Cancer) classified ELF magnetic fields as 2B, which translates to *possible carcinogen*; the basis for this classification was the childhood leukaemia results. In essence, over the years, the childhood leukaemia results have increased in strength. At the same time, the exposure level above which effects are seen has been pushed upwards, implying that only a small proportion of homes are exposed at those levels. Based on the combined control groups in the pooled analysis, this percentage was estimated at less than 1%, and considerably less in the European subset. The evidence for other diseases seems instead to have decreased in strength over the years.

Potential health risks from weak long-term exposure: RF

The situation for RF fields is very different from that for ELF. Whereas early research has looked at people with occupational RF exposure (*e.g.* military personnel), studies that specifically address mobile telephony are few and recent. Although the results of the early occupational studies are mainly negative, they are of limited value in an overall assessment because of restrictions in their designs, particularly with regard to exposure assessment. Thus, this is in essence a new research field. To date, half a dozen epidemiological studies on mobile phone users have been published—with predominantly negative findings. Unlike the ELF area there is no seminal study with clear positive results that drives the research. Instead, the driving force appears to be a concern that this new technology penetrates the world population at high speed and therefore warrants close monitoring. The so-called Stewart report has a comprehensive review of the RF research[10].

To date, there are basically two types of studies on mobile phone users, which differ with respect to how exposure information is obtained. One group of studies uses records from the network operators[11–14]. The operators can provide data on number of years of contract, frequency of calls, duration of calls and also, under certain circumstances, more detailed data about individual subscriptions and calls. The studies that have used these data so far have limited themselves to basic data. Some fundamental problems with this approach are that the contract holder is not always the user of the phone and that additional information such as use of hands-free device is required for a more sophisticated assessment of exposure. The next generation of studies may attempt to combine operator data and data obtained directly from the users. Using operator data is, however, not a trivial approach from the logistic point of view. Indeed the study by Rothman and others was aborted for legal reasons relating to privacy issues and can still only report a 1-year follow up. This renders the study uninformative from the substantive point of view, although it has been important for methodological reasons[15,16].

The other group of studies asks subjects in case-control studies about their phone use[17–20]. More detailed data can in principle be obtained by using this approach. However, recall bias is always a concern in such studies. These case-control studies may also be affected by selection bias. In particular, the studies by Hardell and co-workers have been criticized for the possibility of both selection bias and recall bias[21,22]. Later methodological studies have compared operator data and questionnaire data and showed that subjects systematically over-estimate the amount of phone use, which speaks in favour of using a combined approach[23]. An inherent difficulty in this research is the still limited length of the exposure period for most users and also the recent change from analogue to digital systems with different exposure characteristics; obviously the significance of this change is unknown at present. All these studies have focused on brain tumours although some have looked at other intracranial tumours as well. Overall the results are negative but because of the difficulties mentioned above, and because of some glimmers of associations in two studies, the issue cannot be settled yet[14,24].

RF exposure from base stations has also attracted attention, but this is to a great extent initiated by the public rather than by research interests. Scientists normally observe that the exposure levels from base stations are exceeded by about a thousand times by exposure levels from the phones themselves. Thus, from the scientific view it makes more sense to study exposure from phones. Yet, it is true that base stations give rise to whole body exposure for 24 h a day, for those who stay in the neighbourhood. An inherent problem in these studies that remains to be solved is exposure assessment around base stations[25]. For a variety of reasons, distance from the mast is a virtually meaningless proxy for exposure; thus, such studies cannot be designed in a meaningful way before meters or new exposure assessment models adapted to epidemiological purposes are constructed and made available. To date, several studies on radio and TV transmitters have been published in the scientific literature, but they all share the problems discussed above[26–33]. This research area is still in a very premature state and the results of the published studies are of limited interest.

Just as for ELF fields, the *in vivo* and *in vitro* research has been unable so far to come up with results that convincingly show the presence of some biological effect from exposure to RF fields other than heating. There are, however, some as yet unconfirmed results that warrant further follow up. In particular, for the *in vitro* research, however, there is a problem to separate thermal effects from other effects because of difficulties to design the dosimetry such that temperature is unaffected. This experimental research is currently very intense and results are to be expected in the near future.

Balancing risk

Electromagnetic fields are associated with several of the factors that are known to produce public concern. The fields are invisible; they represent new technology; power line, base station and other sources of exposure are uncontrollable by the exposed individual; for many of the exposure sources, the exposed subject has no direct use of the exposure source. The current scientific situation is one of uncertainty and it is often pointed out that the existence of risks cannot be excluded. Indeed ELF fields were classified by IARC as a possible carcinogen; from the public health point of view, this classification often results in the agent being considered carcinogenic. The evidence for the presence of health risks from RF fields is of course very weak, yet the existence of a risk cannot be excluded. At the same time everyone would agree that modern society is inconceivable without electricity. While some people might disagree about the necessity of mobile phones in particular, most people would still conclude that telecommunications are essential.

This presents decision makers with several dilemmas. Even if the risks from ELF field exposure were taken for granted, it would not follow automatically what actions should be taken. The dilemma is that very few people are exposed at high levels and that the disease for which there is the hardest evidence is very rare. So the decision maker would have to balance the public health benefits and the costs and technical and practical consequences of various schemes that could be considered in order to reduce exposure to the population.

For RF fields, the public health consequences would probably be large if a risk were to be detected. On the other hand, the evidence for a risk is at present very weak—bordering on non-existent. Yet, there are some risk management actions that could be invoked at low cost. Examples are recommendations to use hands-free equipment and to limit calls as much as possible. A particular issue is whether to recommend that children in particular should restrict their use. Such a recommendation would not be based on scientific data specifically pointing towards children being at risk, but rather on some common understanding that children are more sensitive because they are still developing, perhaps combined with the moral concept that one should be more careful with children. However, whether, and by whom, even very low cost actions should be recommended in situations with such weak scientific support for the existence of a risk is currently the topic of intense discussions.

References

1 Wertheimer N, Leeper E. Electrical wiring configurations and childhood cancer. *Am J Epidemiol* 1979; **109**: 273–84
2 ICNIRP. Guidelines for limiting exposure to time-varying electric, magnetic and electromagnetic fields (up to 300 GHz). *Health Phys* 1998; **74**: 494
3 Ahlbom A, Day N, Feychting M, Roman E, Skinner J, Dockerty J, Linet M, McBride M, Michaelis J, Olsen JH, Tynes T, Verkasalo PK. A pooled analysis of magnetic fields and childhood leukaemia. *Br J Cancer* 2000; **83**: 692–8
4 Ahlbom A, Cardis E, Green A, Linet M, Savitz D, Swerdlow A. Review of the epidemiological literature on EMF and health. *Environ Health Perspect* 2001; **109 (Suppl 6)**: 911–33
5 Forssen U. Electromagnetic Fields and Breast Cancer. Unpublished PhD Thesis. Stockholm: Karolinska Institute, 2003
6 Sastre A, Cook MR, Graham C. Nocturnal exposure to intermittent 60-Hz magnetic fields alters human cardiac rhythm. *Bioelectromagnetics* 1998; **19**: 98–106
7 Savitz DA, Liao D, Sastre A *et al*. Magnetic field exposure and cardiovascular disease mortality among electric utility workers. *Am J Epidemiol* 1999; **149**: 135–42
8 Sahl J, Mezei G, Kavet R *et al*. Occupational magnetic field exposure and cardiovascular mortality in a cohort of electric utility workers. *Am J Epidemiol* 2002; **156**: 913–8
9 Johansen C, Feychting M, Möller M *et al*. Risk of severe cardiac arrhythmia in male utility workers: nationwide Danish cohort study. *Am J Epidemiol* 2002; **156**: 857–61
10 IEGMP. Independent Expert Group On Mobile Phones (Chairman: Sir William Stewart). *Mobile Phones and Health*. Chilton, Didcot: Independent Expert Group On Mobile Phones, 2000
11 Rothman KJ, Loughlin JE, Funch DP, Dreyer NA. Overall mortality of cellular telephone customers. *Epidemiology* 1996; **7**: 303–5
12 Dreyer NA, Loughlin JE, Rothman KJ. Cause-specific mortality in cellular telephone users. *JAMA* 1999; **282**: 1814–6
13 Johansen C, Boice JD, McLaughlin JK, Olsen JH. Cellular telephones and cancer—a nationwide cohort study in Denmark. *J Natl Cancer Inst* 2001; **93**: 203–7
14 Auvinen A, Hietanen M, Luukkonen R, Riitta-Sisko K. Brain tumors and salivary gland cancers among cellular telephone users. *Epidemiology* 2002; **13**: 356–9
15 Rothman JK, Chou C-K, Morgan R, Balzano Q, Guy AW, Funch DP, Preston-Martin S, Mandel J, Steffens R, Carlo G. Assessment of cellular telephone and other radio frequency exposure for epidemiologic research. *Epidemiology* 1996; **7**: 291–8
16 Funch DP, Rothman KJ, Loughlin JE, Dreyer NA. Utility of telephone company records for epidemiologic studies of cellular telephones. *Epidemiology* 1996; **7**: 299–302
17 Hardell L, Näsman Å, Påhlson A, Hallquist A, Hansson Mild K. Use of cellular telephones and the risk for brain tumours: A case-control study. *Int J Oncol* 1999; **15**: 113–6
18 Hardell L, Hansson Mild K, Carlberg M. Case-control study on the use of cellular and cordless phones and the risk for malignant brain tumours. *Int J Radiat Biol* 2002; **78**: 931–6
19 Muscat JE, Malkin MG, Thompson S, Shore RE, Stellman SD, McRee D, Neugut AI, Wynder EL. Handheld cellular telephone use and the risk of brain cancer. *JAMA* 2000; **284**: 3001–7

20 Inskip P, Tarone RE, Hatch EE, Wilcosky TC, Shapiro WR, Selker RG, Fine HA, Black PM, Loeffler JS, Linet MS. Cellular-telephone use and brain tumors. *N Engl J Med* 2001; **344**: 79–86
21 Ahlbom A, Feychting M. Re: Use of cellular phones and the risk of brain tumours: a case-control study [letter]. *Int J Oncol* 1999; **15**: 1045 (and reply by Hardell *et al*)
22 Rothman KJ. Epidemiologic evidence on health risks of cellular telephones. *Lancet* 2000; **356**: 1837–40
23 Hillert L, Ahlbom A, Feychting M, Järup L, Larsson A, Neasham D, Elliott P. Mobile phone use: validation of exposure assessment. *Bioelectromagnetic Society Conference, Maui*, June 2003 (Abstract)
24 Hardell L, Hallquist A, Hansson Mild K, Carlberg M, Påhlsson A, Lilja A. Cellular and cordless telephones and the risk for brain tumors. *Eur J Cancer Prevent* 2002; **17**: 377–86
25 Mann S, Cooper TG, Allen SG. *Exposure to Radio Waves Near Mobile Phone Base Stations*. NRPB-R321. Didcot, UK: National Radiological Protection Board, 2000
26 Selvin S, Schulman J, Merrill DW. Distance and risk measures for the analysis of spatial data: a study of childhood cancers. *Soc Sci Med* 1992; **34**: 769–77
27 Maskarinec G, Cooper J, Swygert L. Investigation of increased incidence in childhood leukemia near radio towers in Hawaii: preliminary observations. *J Environ Pathol Toxicol Oncol* 1994: **13**: 33–7
28 Hocking B, Gordon IR, Grain HL, Hatfield GE. Cancer incidence and mortality and proximity to TV towers. *Med J Aust* 1996; **165**: 601–5
29 Dolk H, Shaddick G, Walls P, Grundy C, Thakrar B, Kleinschmidt I, Elliott P. Cancer incidence near radio and television transmitters in Great Britain. 1. Sutton Coldfield transmitter. *Am J Epidemiol* 1997; **145**: 1–9
30 Dolk H, Elliott P, Shaddick G, Walls P, Thakrar B. Cancer incidence near radio and television transmitters in Great Britain. 2. All high power transmitters. *Am J Epidemiol* 1997; **145**: 10–7
31 McKenzie DR, Yin Y, Morrell S. Childhood incidence of acute lymphoblastic leukaemia and exposure to broadcast radiation in Sydney—a second look. *Aust NZ J Public Health* 1998; **22**: 360–7
32 Cooper D, Hemmings K, Saunders P. Re: Cancer incidence near radio and television transmitters in Great Britain. I. Sutton Coldfield transmitter; II. All high power transmitters. *Am J Epidemiol* 2001; **153**: 202–4
33 Michelozzi P, Capon A, Kirchmayer U, Forastiere F, Biggeri A, Barca A, Perucci CA. Adult and childhood leukemia near a high-power radio station in Rome, Italy. *Am J Epidemiol* 2002; **155**: 1096–103

Hazards of heavy metal contamination

Lars Järup

Department of Epidemiology and Public Health, Imperial College, London, UK

The main threats to human health from heavy metals are associated with exposure to lead, cadmium, mercury and arsenic. These metals have been extensively studied and their effects on human health regularly reviewed by international bodies such as the WHO. Heavy metals have been used by humans for thousands of years. Although several adverse health effects of heavy metals have been known for a long time, exposure to heavy metals continues, and is even increasing in some parts of the world, in particular in less developed countries, though emissions have declined in most developed countries over the last 100 years. Cadmium compounds are currently mainly used in re-chargeable nickel–cadmium batteries. Cadmium emissions have increased dramatically during the 20th century, one reason being that cadmium-containing products are rarely re-cycled, but often dumped together with household waste. Cigarette smoking is a major source of cadmium exposure. In non-smokers, food is the most important source of cadmium exposure. Recent data indicate that adverse health effects of cadmium exposure may occur at lower exposure levels than previously anticipated, primarily in the form of kidney damage but possibly also bone effects and fractures. Many individuals in Europe already exceed these exposure levels and the margin is very narrow for large groups. Therefore, measures should be taken to reduce cadmium exposure in the general population in order to minimize the risk of adverse health effects. The general population is primarily exposed to mercury *via* food, fish being a major source of methyl mercury exposure, and dental amalgam. The general population does not face a significant health risk from methyl mercury, although certain groups with high fish consumption may attain blood levels associated with a low risk of neurological damage to adults. Since there is a risk to the fetus in particular, pregnant women should avoid a high intake of certain fish, such as shark, swordfish and tuna; fish (such as pike, walleye and bass) taken from polluted fresh waters should especially be avoided. There has been a debate on the safety of dental amalgams and claims have been made that mercury from amalgam may cause a variety of diseases. However, there are no studies so far that have been able to show any associations between amalgam fillings and ill health. The general population is exposed to lead from air and food in roughly equal proportions. During the last century, lead emissions to ambient air have caused considerable pollution, mainly due to lead emissions from petrol. Children are particularly susceptible to lead exposure due to high gastrointestinal uptake and the permeable blood–brain barrier. Blood levels in children should be

Correspondence to: Lars Järup, Department of Epidemiology and Public Health, Imperial College, London, UK. E-mail: l.jarup@imperial.ac.uk reduced below the levels so far considered acceptable, recent data indicating that there may be neurotoxic effects of lead at lower levels of exposure than previously anticipated. Although lead in petrol has dramatically decreased over the last decades, thereby reducing environmental exposure, phasing out any remaining uses of lead additives in motor fuels should be encouraged. The use of lead-based paints should be abandoned, and lead should not be used in food containers. In particular, the public should be aware of glazed food containers, which may leach lead into food. Exposure to arsenic is mainly *via* intake of food and drinking water, food being the most important source in most populations. Long-term exposure to arsenic in drinking-water is mainly related to increased risks of skin cancer, but also some other cancers, as well as other skin lesions such as hyperkeratosis and pigmentation changes. Occupational exposure to arsenic, primarily by inhalation, is causally associated with lung cancer. Clear exposure–response relationships and high risks have been observed.

Introduction

Although there is no clear definition of what a heavy metal is, density is in most cases taken to be the defining factor. Heavy metals are thus commonly defined as those having a specific density of more than 5 g/cm^3. The main threats to human health from heavy metals are associated with exposure to lead, cadmium, mercury and arsenic (arsenic is a metalloid, but is usually classified as a heavy metal).

Heavy metals have been used in many different areas for thousands of years. Lead has been used for at least 5000 years, early applications including building materials, pigments for glazing ceramics, and pipes for transporting water. In ancient Rome, lead acetate was used to sweeten old wine, and some Romans might have consumed as much as a gram of lead a day. Mercury was allegedly used by the Romans as a salve to alleviate teething pain in infants, and was later (from the 1300s to the late 1800s) employed as a remedy for syphilis. Claude Monet used cadmium pigments extensively in the mid 1800s, but the scarcity of the metal limited the use in artists' materials until the early 1900s.

Although adverse health effects of heavy metals have been known for a long time, exposure to heavy metals continues and is even increasing in some areas. For example, mercury is still used in gold mining in many parts of Latin America. Arsenic is still common in wood preservatives, and tetraethyl lead remains a common additive to petrol, although this use has decreased dramatically in the developed countries. Since the middle of the 19th century, production of heavy metals increased steeply for more than 100 years, with concomitant emissions to the environment (Fig. 1).

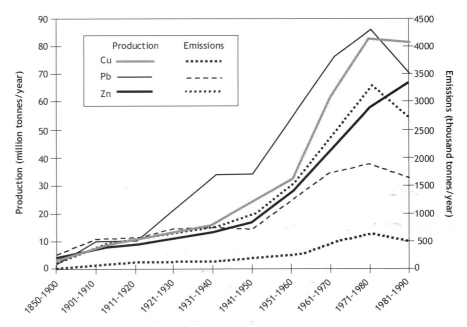

Fig. 1 Global production and consumption of selected toxic metals, 1850–1990. Source: Ref. 43.

At the end of the 20th century, however, emissions of heavy metals started to decrease in developed countries: in the UK, emissions of heavy metals fell by over 50% between 1990 and 2000[1].

Emissions of heavy metals to the environment occur *via* a wide range of processes and pathways, including to the air (*e.g.* during combustion, extraction and processing), to surface waters (*via* runoff and releases from storage and transport) and to the soil (and hence into groundwaters and crops) (see Chapter 1). Atmospheric emissions tend to be of greatest concern in terms of human health, both because of the quantities involved and the widespread dispersion and potential for exposure that often ensues. The spatial distributions of cadmium, lead and mercury emissions to the atmosphere in Europe can be found in the Meteorological Synthesizing Centre-East (MSC-E) website (http://www.msceast.org/hms/emission.html#Spatial). Lead emissions are mainly related to road transport and thus most uniformly distributed over space. Cadmium emissions are primarily associated with non-ferrous metallurgy and fuel combustion, whereas the spatial distribution of anthropogenic mercury emissions reflects mainly the level of coal consumption in different regions.

People may be exposed to potentially harmful chemical, physical and biological agents in air, food, water or soil. However, exposure does not result only from the presence of a harmful agent in the environment. The key word in the definition of exposure is contact[2]. There must be contact between the agent and the outer boundary of the human body, such as the airways, the skin or the mouth. Exposure is often defined as a function of concentration and time: "an event that occurs when there is contact at a boundary between a human and the environment with a contaminant of a specific concentration for an interval of time"[3]. For exposure to happen, therefore, co-existence of heavy metals and people has to occur (see Chapter 1).

Cadmium

Occurrence, exposure and dose

Cadmium occurs naturally in ores together with zinc, lead and copper. Cadmium compounds are used as stabilizers in PVC products, colour pigment, several alloys and, now most commonly, in re-chargeable nickel–cadmium batteries. Metallic cadmium has mostly been used as an anti-corrosion agent (cadmiation). Cadmium is also present as a pollutant in phosphate fertilizers. EU cadmium usage has decreased considerably during the 1990s, mainly due to the gradual phase-out of cadmium products other than Ni-Cd batteries and the implementation of more stringent EU environmental legislation (Directive 91/338/ECC). Notwithstanding these reductions in Europe, however, cadmium production, consumption and emissions to the environment worldwide have increased dramatically during the 20th century. Cadmium containing products are rarely re-cycled, but frequently dumped together with household waste, thereby contaminating the environment, especially if the waste is incinerated.

Natural as well as anthropogenic sources of cadmium, including industrial emissions and the application of fertilizer and sewage sludge to farm land, may lead to contamination of soils, and to increased cadmium uptake by crops and vegetables, grown for human consumption. The uptake process of soil cadmium by plants is enhanced at low pH[4].

Cigarette smoking is a major source of cadmium exposure. Biological monitoring of cadmium in the general population has shown that cigarette smoking may cause significant increases in blood cadmium (B-Cd) levels, the concentrations in smokers being on average 4–5 times higher than those in non-smokers[4]. Despite evidence of exposure from environmental tobacco smoke[5], however, this is probably contributing little to total cadmium body burden.

Food is the most important source of cadmium exposure in the general non-smoking population in most countries[6]. Cadmium is present in most foodstuffs, but concentrations vary greatly, and individual intake also varies considerably due to differences in dietary habits[4]. Women usually have lower daily cadmium intakes, because of lower energy consumption than men. Gastrointestinal absorption of cadmium may be influenced by nutritional factors, such as iron status[7].

B-Cd generally reflects current exposure, but partly also lifetime body burden[8]. The cadmium concentration in urine (U-Cd) is mainly influenced by the body burden, U-Cd being proportional to the kidney concentration. Smokers and people living in contaminated areas have higher urinary cadmium concentrations, smokers having about twice as high concentrations as non-smokers[4].

Health effects

Inhalation of cadmium fumes or particles can be life threatening, and although acute pulmonary effects and deaths are uncommon, sporadic cases still occur[9,10]. Cadmium exposure may cause kidney damage. The first sign of the renal lesion is usually a tubular dysfunction, evidenced by an increased excretion of low molecular weight proteins [such as β_2-microglobulin and α_1-microglobulin (protein HC)] or enzymes [such as N-Acetyl-β-D-glucosaminidase (NAG)][4,6]. It has been suggested that the tubular damage is reversible[11], but there is overwhelming evidence that the cadmium induced tubular damage is indeed irreversible[4].

WHO[6] estimated that a urinary excretion of 10 nmol/mmol creatinine (corresponding to *circa* 200 mg Cd/kg kidney cortex) would constitute a 'critical limit' below which kidney damage would not occur. However, WHO calculated that *circa* 10% of individuals with this kidney concentration would be affected by tubular damage. Several reports have since shown that kidney damage and/or bone effects are likely to occur at lower kidney cadmium levels. European studies have shown signs of cadmium induced kidney damage in the general population at urinary cadmium levels around 2–3 µg Cd/g creatinine[12,13].

The initial tubular damage may progress to more severe kidney damage, and already in 1950 it was reported that some cadmium exposed workers had developed decreased glomerular filtration rate (GFR)[14]. This has been confirmed in later studies of occupationally exposed workers[15,16]. An excess risk of kidney stones, possibly related to an increased excretion of calcium in urine following the tubular damage, has been shown in several studies[4].

Recently, an association between cadmium exposure and chronic renal failure [end stage renal disease (ESRD)] was shown[17]. Using a registry of patients, who had been treated for uraemia, the investigators found a double risk of ESRD in persons living close to (<2 km) industrial cadmium emitting plants as well as in occupationally exposed workers.

Long-term high cadmium exposure may cause skeletal damage, first reported from Japan, where the itai-itai (ouch-ouch) disease (a combination of osteomalacia and osteoporosis) was discovered in the 1950s. The exposure was caused by cadmium-contaminated water used for irrigation of local rice fields. A few studies outside Japan have reported similar findings[4]. During recent years, new data have emerged suggesting that also relatively low cadmium exposure may give rise to skeletal damage, evidenced by low bone mineral density (osteoporosis) and fractures[18–20].

Animal experiments have suggested that cadmium may be a risk factor for cardiovascular disease, but studies of humans have not been able to confirm this[4]. However, a Japanese study showed an excess risk of cardiovascular mortality in cadmium-exposed persons with signs of tubular kidney damage compared to individuals without kidney damage[21].

Cancer

The IARC has classified cadmium as a human carcinogen (group I) on the basis of sufficient evidence in both humans and experimental animals[22]. IARC, however, noted that the assessment was based on few studies of lung cancer in occupationally exposed populations, often with imperfect exposure data, and without the capability to consider possible confounding by smoking and other associated exposures (such as nickel and arsenic). Cadmium has been associated with prostate cancer, but both positive and negative studies have been published. Early data indicated an association between cadmium exposure and kidney cancer[23]. Later studies have not been able clearly to confirm this, but a large multi-centre study showed a (borderline) significant over-all excess risk of renal-cell cancer, although a negative dose–response relationship did not support a causal relation[24]. Furthermore, a population-based multicentre-study of renal cell carcinoma found an excess risk in occupationally exposed persons[25]. In summary, the evidence for cadmium as a human carcinogen is rather weak, in particular after oral exposure. Therefore, a classification of cadmium as 'probably carcinogenic to humans' (IARC group 2A) would be more appropriate. This conclusion also complies with the EC classification of some cadmium compounds (Carcinogen Category 2; Annex 1 to the directive 67/548/EEC).

Mercury

Occurrence, exposure and dose

The mercury compound cinnabar (HgS), was used in pre-historic cave paintings for red colours, and metallic mercury was known in ancient Greece where it (as well as white lead) was used as a cosmetic to lighten the skin. In medicine, apart from the previously mentioned use of mercury as a cure for syphilis, mercury compounds have also been used as diuretics [calomel (Hg$_2$Cl$_2$)], and mercury amalgam is still used for filling teeth in many countries[26].

Metallic mercury is used in thermometers, barometers and instruments for measuring blood pressure. A major use of mercury is in the chlor-alkali industry, in the electrochemical process of manufacturing chlorine, where mercury is used as an electrode.

The largest occupational group exposed to mercury is dental care staff. During the 1970s, air concentrations in some dental surgeries reached 20 μg/m^3, but since then levels have generally fallen to about one-tenth of those concentrations.

Inorganic mercury is converted to organic compounds, such as methyl mercury, which is very stable and accumulates in the food chain. Until the 1970s, methyl mercury was commonly used for control of fungi on seed grain.

The general population is primarily exposed to mercury *via* food, fish being a major source of methyl mercury exposure[27], and dental amalgam. Several experimental studies have shown that mercury vapour is released from amalgam fillings, and that the release rate may increase by chewing[28].

Mercury in urine is primarily related to (relatively recent) exposure to inorganic compounds, whereas blood mercury may be used to identify exposure to methyl mercury. A number of studies have correlated the number of dental amalgam fillings or amalgam surfaces with the mercury content in tissues from human autopsy, as well as in samples of blood, urine and plasma[26]. Mercury in hair may be used to estimate long-term exposure, but potential contamination may make interpretation difficult.

Health effects

Inorganic mercury

Acute mercury exposure may give rise to lung damage. Chronic poisoning is characterized by neurological and psychological symptoms, such as tremor, changes in personality, restlessness, anxiety, sleep disturbance and depression. The symptoms are reversible after cessation of exposure. Because of the blood–brain barrier there is no central nervous involvement related to inorganic mercury exposure. Metallic mercury may cause kidney damage, which is reversible after exposure has stopped. It has also been possible to detect proteinuria at relatively low levels of occupational exposure.

Metallic mercury is an allergen, which may cause contact eczema, and mercury from amalgam fillings may give rise to oral lichen. It has been feared that mercury in amalgam may cause a variety of symptoms. This so-called 'amalgam disease' is, however, controversial, and although some authors claim proof of symptom relief after removal of dental amalgam fillings[29], there is no scientific evidence of this[30].

Organic mercury

Methyl mercury poisoning has a latency of 1 month or longer after acute exposure, and the main symptoms relate to nervous system damage[31]. The earliest symptoms are parestesias and numbness in the hands and feet. Later, coordination difficulties and concentric constriction of the visual field may develop as well as auditory symptoms. High doses may lead to death, usually 2–4 weeks after onset of symptoms. The Minamata catastrophe in Japan in the 1950s was caused by methyl mercury poisoning from fish contaminated by mercury discharges to the surrounding sea. In the early 1970s, more than 10,000 persons in Iraq were poisoned by eating bread baked from mercury-polluted grain, and several thousand people died as a consequence of the poisoning. However, the general population does not face significant health risks from methyl mercury exposure with the exception of certain groups with high fish consumption.

A high dietary intake of mercury from consumption of fish has been hypothesized to increase the risk of coronary heart disease[32]. In a recent case-control study, the joint association of mercury levels in toenail clippings and docosahexaenoic acid levels in adipose tissue with the risk of a first myocardial infarction in men was evaluated[33]. Mercury levels in the patients were 15% higher than those in controls (95% CI, 5–25%), and the adjusted odds ratio for myocardial infarction associated with the highest compared with the lowest quintile of mercury was 2.16 (95% CI, 1.09–4.29; P for trend = 0.006).

Another recent case-control study investigated the association between mercury levels in toenails and the risk of coronary heart disease among male health professionals with no previous history of cardiovascular disease. Mercury levels were significantly correlated with fish consumption, and the mean mercury level was higher in dentists than in non-dentists. When other risk factors for coronary heart disease had been controlled for, mercury levels were not significantly associated with the risk of coronary heart disease[34].

These intriguing contradictory findings need to be followed up by more studies of other similarly exposed populations.

Lead

Occurrence, exposure and dose

The general population is exposed to lead from air and food in roughly equal proportions. Earlier, lead in foodstuff originated from pots used

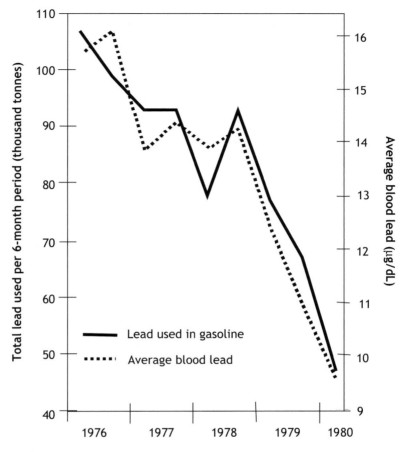

Fig. 2 Lead concentrations in petrol and children's blood (USA).
Source: redrawn from Annest (1983), as reproduced in National Academy of Sciences/National Research Council. Measuring Lead Exposure in Infants, Children, and Other Sensitive Populations. Washington, DC, USA: National Academy Press, 1993.

for cooking and storage, and lead acetate was previously used to sweeten port wine. During the last century, lead emissions to ambient air have further polluted our environment, over 50% of lead emissions originating from petrol. Over the last few decades, however, lead emissions in developed countries have decreased markedly due to the introduction of unleaded petrol. Subsequently blood lead levels in the general population have decreased (Fig. 2).

Occupational exposure to inorganic lead occurs in mines and smelters as well as welding of lead painted metal, and in battery plants. Low or moderate exposure may take place in the glass industry. High levels of air emissions may pollute areas near lead mines and smelters. Airborne lead can be deposited on soil and water, thus reaching humans *via* the food chain.

Up to 50% of inhaled inorganic lead may be absorbed in the lungs. Adults take up 10–15% of lead in food, whereas children may absorb up to 50% *via* the gastrointestinal tract. Lead in blood is bound to erythrocytes, and elimination is slow and principally *via* urine. Lead is accumulated in the skeleton, and is only slowly released from this body compartment. Half-life of lead in blood is about 1 month and in the skeleton 20–30 years[35].

In adults, inorganic lead does not penetrate the blood–brain barrier, whereas this barrier is less developed in children. The high gastrointestinal uptake and the permeable blood–brain barrier make children especially susceptible to lead exposure and subsequent brain damage. Organic lead compounds penetrate body and cell membranes. Tetramethyl lead and tetraethyl lead penetrate the skin easily. These compounds may also cross the blood–brain barrier in adults, and thus adults may suffer from lead encephalopathy related to acute poisoning by organic lead compounds.

Health effects

The symptoms of acute lead poisoning are headache, irritability, abdominal pain and various symptoms related to the nervous system. Lead encephalopathy is characterized by sleeplessness and restlessness. Children may be affected by behavioural disturbances, learning and concentration difficulties. In severe cases of lead encephalopathy, the affected person may suffer from acute psychosis, confusion and reduced consciousness. People who have been exposed to lead for a long time may suffer from memory deterioration, prolonged reaction time and reduced ability to understand. Individuals with average blood lead levels under 3 µmol/l may show signs of peripheral nerve symptoms with reduced nerve conduction velocity and reduced dermal sensibility. If the neuropathy is severe the lesion may be permanent. The classical picture includes a dark blue lead sulphide line at the gingival margin. In less serious cases, the most obvious sign of lead poisoning is disturbance of haemoglobin synthesis, and long-term lead exposure may lead to anaemia.

Recent research has shown that long-term low-level lead exposure in children may also lead to diminished intellectual capacity. Figure 3 shows a meta-analysis of four prospective studies using mean blood lead level over a number of years. The combined evidence suggests a weighted mean decrease in IQ of 2 points for a 0.48 µmol/l (10 µg/dl) increase in blood lead level (95% confidence interval from –0.3 points to –3.6 points)[35].

Acute exposure to lead is known to cause proximal renal tubular damage[35]. Long-term lead exposure may also give rise to kidney damage and, in a recent study of Egyptian policemen, urinary excretion of NAG

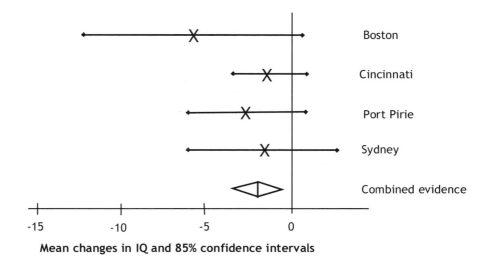

Fig. 3 Estimated mean change in IQ for an increase in blood lead level from 0.48 to 0.96 µmol/l (10–20 µg/dl) from a meta-analysis of four prospective studies[35].

was positively correlated with duration of exposure to lead from automobile exhaust, blood lead and nail lead[36].

Despite intensive efforts to define the relationship between body burden of lead and blood pressure or other effects on the cardiovascular system, no causal relationship has been demonstrated in humans[35].

Using routinely collected data on mortality (1981–96), hospital episode statistics data 1992–1995 and statutory returns to the Health and Safety Executive (RIDDOR), one death and 83 hospital cases were identified[37]. The authors found that mortality and hospital admission ascribed to lead poisoning in England were rare, but that cases continue to occur and that some seem to be associated with considerable morbidity.

Blood lead levels in children below 10 µg/dl have so far been considered acceptable, but recent data indicate that there may be toxicological effects of lead at lower levels of exposure than previously anticipated. There is also evidence that certain genetic and environmental factors can increase the detrimental effects of lead on neural development, thereby rendering certain children more vulnerable to lead neurotoxicity[38].

IARC classified lead as a 'possible human carcinogen' based on sufficient animal data and insufficient human data in 1987. Since then a few studies have been published, the overall evidence for lead as a carcinogen being only weak, the most likely candidates are lung cancer, stomach cancer and gliomas[39].

Arsenic

Occurrence, exposure and dose

Arsenic is a widely distributed metalloid, occurring in rock, soil, water and air. Inorganic arsenic is present in groundwater used for drinking in several countries all over the world (*e.g.* Bangladesh, Chile and China), whereas organic arsenic compounds (such as arsenobetaine) are primarily found in fish, which thus may give rise to human exposure[40].

Smelting of non-ferrous metals and the production of energy from fossil fuel are the two major industrial processes that lead to arsenic contamination of air, water and soil, smelting activities being the largest single anthropogenic source of atmospheric pollution[41]. Other sources of contamination are the manufacture and use of arsenical pesticides and wood preservatives.

The working group of the EU DG Environment concluded that there were large reductions in the emissions of arsenic to air in several member countries of the European Union in the 1980s. In 1990, the total emissions of arsenic to the air in the member states were estimated to be 575 tonnes. In 1996, the estimated total releases of arsenic to the air in the UK were 50 tonnes[42].

Concentrations in air in rural areas range from <1 to 4 ng/m^3, whereas concentrations in cities may be as high as 200 ng/m^3. Much higher concentrations (>1000 ng/m^3) have been measured near industrial sources. Water concentrations are usually <10 μg/l, although higher concentrations may occur near anthropogenic sources. Levels in soils usually range from 1 to 40 mg/kg, but pesticide application and waste disposal can result in much higher concentrations[40].

General population exposure to arsenic is mainly *via* intake of food and drinking water. Food is the most important source, but in some areas, arsenic in drinking water is a significant source of exposure to inorganic arsenic. Contaminated soils such as mine-tailings are also a potential source of arsenic exposure[40].

Absorption of arsenic in inhaled airborne particles is highly dependent on the solubility and the size of particles. Soluble arsenic compounds are easily absorbed from the gastrointestinal tract. However, inorganic arsenic is extensively methylated in humans and the metabolites are excreted in the urine[40].

Arsenic (or metabolites) concentrations in blood, hair, nails and urine have been used as biomarkers of exposure. Arsenic in hair and nails can be useful indicators of past arsenic exposure, if care is taken to avoid external arsenic contamination of the samples. Speciated metabolites in urine expressed as either inorganic arsenic or the sum of metabolites (inorganic arsenic + MMA + DMA) is generally the best estimate of recent arsenic dose. However, consumption of certain seafood may confound estimation of inorganic arsenic exposure, and should thus be avoided before urine sampling[40].

Health effects

Inorganic arsenic is acutely toxic and intake of large quantities leads to gastrointestinal symptoms, severe disturbances of the cardiovascular and central nervous systems, and eventually death. In survivors, bone marrow depression, haemolysis, hepatomegaly, melanosis, polyneuropathy and encephalopathy may be observed. Ingestion of inorganic arsenic may induce peripheral vascular disease, which in its extreme form leads to gangrenous changes (black foot disease, only reported in Taiwan).

Populations exposed to arsenic *via* drinking water show excess risk of mortality from lung, bladder and kidney cancer, the risk increasing with increasing exposure. There is also an increased risk of skin cancer and other skin lesions, such as hyperkeratosis and pigmentation changes.

Studies on various populations exposed to arsenic by inhalation, such as smelter workers, pesticide manufacturers and miners in many different countries consistently demonstrate an excess lung cancer. Although all these groups are exposed to other chemicals in addition to arsenic, there is no other common factor that could explain the findings. The lung cancer risk increases with increasing arsenic exposure in all relevant studies, and confounding by smoking does not explain the findings.

The latest WHO evaluation[40] concludes that arsenic exposure *via* drinking water is causally related to cancer in the lungs, kidney, bladder and skin, the last of which is preceded by directly observable precancerous lesions. Uncertainties in the estimation of past exposures are important when assessing the exposure–response relationships, but it would seem that drinking water arsenic concentrations of approximately 100 µg/l have led to cancer at these sites, and that precursors of skin cancer have been associated with levels of 50–100 µg/l.

The relationships between arsenic exposure and other health effects are less clear. There is relatively strong evidence for hypertension and cardiovascular disease, but the evidence is only suggestive for diabetes and reproductive effects and weak for cerebrovascular disease, long-term neurological effects, and cancer at sites other than lung, bladder, kidney and skin[40].

Conclusions

Recent data indicate that adverse health effects of cadmium exposure, primarily in the form of renal tubular damage but possibly also effects on bone and fractures, may occur at lower exposure levels than previously anticipated. Many individuals in Europe already exceed these exposure levels and the margin is very narrow for large groups. Therefore, measures should be taken to reduce cadmium exposure in the general population in order to minimize the risk of adverse health effects.

The general population does not face a significant health risk from methylmercury, although certain groups with high fish consumption may attain blood levels associated with a low risk of neurological damage to adults. Since there is a risk to the fetus in particular, pregnant women should avoid a high intake of certain fish, such as shark, swordfish and tuna. Fish, such as pike, walleye and bass, taken from polluted fresh waters should especially be avoided.

There has been a debate on the safety of dental amalgams and claims have been made that mercury from amalgam may cause a variety of diseases, but to date no studies have been able to show any associations between amalgam fillings and ill health.

Children are particularly vulnerable to lead exposure. Blood levels in children should be reduced below the levels so far considered acceptable, recent data indicating that there may be neurotoxic effects of lead at lower levels of exposure than previously anticipated. Although lead in petrol has dramatically declined over the last decades, thereby reducing environmental exposure, there is a need to phase out any remaining uses of lead additives in motor fuels. The use of lead-based paints should also be abandoned, and lead should not be used in food containers. In particular, the public should be aware of glazed food containers, which may leach lead into food.

Long-term exposure to arsenic in drinking water is mainly related to increased risks of skin cancer, but also some other cancers, and other skin lesions such as hyperkeratosis and pigmentation changes. Occupational exposure to arsenic, primarily by inhalation, is causally associated with lung cancer. Clear exposure–response relationships and high risks have been observed.

References

1. Department of the Environment, Transport and the Regions. *Statistics Release 184 1999 UK Air Emissions Estimates* (28 March 2001)
2. Berglund M, Elinder CG, Järup L. *Humans Exposure Assessment. An Introduction.* WHO/SDE/OEH/01.3, 2001
3. NRC. *Human Exposure Assessment for Airborne Pollutants. Advances and Opportunities.* Washington, DC: National Research Council, National Academy Press, 1991
4. Jarup L, Berglund M, Elinder CG, Nordberg G, Vahter M. Health effects of cadmium exposure—a review of the literature and a risk estimate. *Scand J Work Environ Health* 1998; **24** (**Suppl 1**): 1–51
5. Hossn E, Mokhtar G, El-Awady M, Ali I, Morsy M, Dawood A. Environmental exposure of the pediatric age groups in Cairo City and its suburbs to cadmium pollution. *Sci Total Environ* 2001; **273**: 135–46

6 WHO. *Cadmium*. Environmental Health Criteria, vol. 134. Geneva: World Health Organization, 1992
7 Flanagan PR, McLellan JS, Haist J, Cherian MG, Chamberlain MJ, Valberg LS. Increased dietary cadmium absorption in mice and human subjects with iron deficiency. *Gastroenterology* 1978; **74**: 841–6
8 Järup L, Rogenfelt A, Elinder CG, Nogawa K, Kjellström T. Biological half-time of cadmium in the blood of workers after cessation of exposure. *Scand J Work Environ Health* 1983; **9**: 327–31
9 Seidal K, Jorgensen N, Elinder CG, Sjogren B, Vahter M. Fatal cadmium-induced pneumonitis. *Scand J Work Environ Health* 1993; **19**: 429–31
10 Barbee Jr JY, Prince TS. Acute respiratory distress syndrome in a welder exposed to metal fumes. *South Med J* 1999; **92**: 510–2
11 Hotz P, Buchet JP, Bernard A, Lison D, Lauwerys R. Renal effects of low-level environmental cadmium exposure: 5-year follow-up of a subcohort from the Cadmibel study. *Lancet* 1999; **354**: 1508–13
12 Buchet JP, Lauwerys R, Roels H, Bernard A, Bruaux P, Claeys F, Ducoffre G, DePlaen P, Staessen J, Amery A, Lijnen P, Thijs L, Rondia D, Sartor F, Saint Remy A, Nick L. Renal effects of cadmium body burden of the general population. *Lancet* 1990; **336**: 699–702
13 Jarup L, Hellstrom L, Alfven T, Carlsson MD, Grubb A, Persson B *et al*. Low level exposure to cadmium and early kidney damage: the OSCAR study. *Occup Environ Med* 2000; **57**: 668–72
14 Friberg L. Health hazards in the manufacture of alkaline accumulators with special reference to chronic cadmium poisoning. *Acta Med Scand* 1950; **Suppl 240**: 1–124
15 Bernard A, Roels H, Buchet JP, Cardenas A, Lauwerys R. Cadmium and health: the Belgian experience. *IARC Scientific Publications* 1992; **118**: 15–33
16 Järup L, Persson B, Elinder C-G. Decreased glomerular filtration rate in cadmium exposed solderers. *Occup Environ Med* 1995; **52**: 818–22
17 Hellström L, Elinder CG, Dahlberg B, Lundberg M, Järup L, Persson B, Axelson O. Cadmium exposure and end-stage renal disease. *Am J Kidney Dis* 2001; **38**: 1001–8
18 Staessen JA, Roels HA, Emelianov D, Kuznetsova T, Thijs L, Vangronsveld J *et al*. Environmental exposure to cadmium, forearm bone density, and risk of fractures: prospective population study. Public Health and Environmental Exposure to Cadmium (PheeCad) Study Group. *Lancet* 1999; **353**: 1140–4
19 Alfven T, Elinder CG, Carlsson MD, Grubb A, Hellstrom L, Persson B *et al*. Low-level cadmium exposure and osteoporosis. *J Bone Miner Res* 2000; **15**: 1579–86
20 Nordberg G, Jin T, Bernard A, Fierens S, Buchet JP, Ye T, Kong Q, Wang H. Low bone density and renal dysfunction following environmental cadmium exposure in China. *Ambio* 2002; **6**: 478–81
21 Nishijo M, Nakagawa H, Morikawa Y, Tabata M, Senma M, Miura K *et al*. Mortality of inhabitants in an area polluted by cadmium: 15 year follow up. *Occup Environ Med* 1995; **52**: 181–4
22 IARC. Cadmium and cadmium compounds. In: *Beryllium, Cadmium, Mercury and Exposure in the Glass Manufacturing Industry*. IARC Monographs on the Evaluation of Carcinogenic Risks to Humans, vol. **58**. Lyon: International Agency for Research on Cancer, 1993; 119–237
23 Kolonel LN. Association of cadmium with renal cancer. *Cancer* 1976; **37**: 1782–7
24 Mandel JS, McLaughlin JK, Schlehofer B, Mellemgaard A, Helmert U, Lindblad P, McCredie M, Adami HO. International renal-cell cancer study. IV. Occupation. *Int J Cancer* 1995; **61**: 601–5
25 Pesch B, Haerting J, Ranft U, Klimpel A, Oelschlagel B, Schill W. Occupational risk factors for renal cell carcinoma: agent-specific results from a case-control study in Germany. MURC Study Group. Multicentre urothelial and renal cancer study. *Int J Epidemiol* 2000; **29**: 1014–24
26 WHO. *Inorganic Mercury*. Environmental Health Criteria, vol. **118**. Geneva: World Health Organization, 1991
27 WHO. *Methyl Mercury*. Environmental Health Criteria, vol. **101**. Geneva: World Health Organization, 1990
28 Sallsten G, Thoren J, Barregard L, Schutz A, Skarping G. Long-term use of nicotine chewing gum and mercury exposure from dental amalgam fillings. *J Dent Res* 1996; **75**: 594–8
29 Lindh U, Hudecek R, Danersund A, Eriksson S, Lindvall A. Removal of dental amalgam and other metal alloys supported by antioxidant therapy alleviates symptoms and improves quality of life in patients with amalgam-associated ill health. *Neuroendocrinol Lett* 2002; **23**: 459–82

30 Langworth S, Bjorkman L, Elinder CG, Jarup L, Savlin P. Multidisciplinary examination of patients with illness attributed to dental fillings. *J Oral Rehabil* 2002; **29**: 705–13
31 Weiss B, Clarkson TW, Simon W. Silent latency periods in methylmercury poisoning and in neurodegenerative disease. *Environ Health Perspect* 2002; **110 (Suppl 5)**: 851–4
32 Salonen JT, Seppanen K, Nyyssonen K, Korpela H, Kauhanen J, Kantola M, Tuomilehto J, Esterbauer H, Tatzber F, Salonen R. Intake of mercury from fish, lipid peroxidation, and the risk of myocardial infarction and coronary, cardiovascular, and any death in eastern Finnish men. *Circulation* 1995; **91**: 645–55
33 Guallar E, Sanz-Gallardo MI, van't Veer P, Bode P, Aro A, Gomez-Aracena J, Kark JD, Riemersma RA, Martin-Moreno JM, Kok FJ; Heavy Metals and Myocardial Infarction Study Group. Mercury, fish oils, and the risk of myocardial infarction. *N Engl J Med* 2002; **347**: 1747–54
34 Yoshizawa K, Rimm EB, Morris JS, Spate VL, Hsieh CC, Spiegelman D, Stampfer MJ, Willett WC. Mercury and the risk of coronary heart disease in men. *N Engl J Med* 2002; **347**: 1755–60
35 WHO. *Lead*. Environmental Health Criteria, vol. **165**. Geneva: World Health Organization, 1995
36 Mortada WI, Sobh MA, El-Defrawy MM, Farahat SE. Study of lead exposure from automobile exhaust as a risk for nephrotoxicity among traffic policemen. *Am J Nephrol* 2001; **21**: 274–9
37 Elliott P, Arnold R, Barltrop D, Thornton I, House IM, Henry JA. Clinical lead poisoning in England: an analysis of routine sources of data. *Occup Environ Med* 1999; **56**: 820–4
38 Lidsky TI, Schneider JS. Lead neurotoxicity in children: basic mechanisms and clinical correlates. *Brain* 2003; **126**: 5–19
39 Steenland K, Boffetta P. Lead and cancer in humans: where are we now? *Am J Ind Med* 2000; **38**: 295–9
40 WHO. *Arsenic and Arsenic Compounds*. Environmental Health Criteria, vol. **224**. Geneva: World Health Organization, 2001
41 Chilvers DC, Peterson PJ. Global cycling of arsenic. In: Hutchinson TC, Meema KM (eds) *Lead, Mercury, Cadmium and Arsenic in the Environment*. Chichester: John Wiley & Sons, 1987; 279–303
42 DG Environment. Ambient air pollution by As, Cd and Ni compounds. Position paper, Final version, October 2000. Brussels: European Commission DG Environment
43 Nriagu JO. History of global metal pollution. *Science* 1996; **272**: 223–4
44 Annest JL. Trends in the blood level leads of the US population: The Second National Health and Nutrition Examination Survey(NHANES II) 1976–1980. In: Rutter M, Jones RR (eds) Lead Versus Health, John Wiley & Sons, New York, 1934; 33–58

Health hazards and waste management

Lesley Rushton

MRC Institute for Environment and Health, Leicester, UK

Different methods of waste management emit a large number of substances, most in small quantities and at extremely low levels. Raised incidence of low birth weight births has been related to residence near landfill sites, as has the occurrence of various congenital malformations. There is little evidence for an association with reproductive or developmental effects with proximity to incinerators. Studies of cancer incidence and mortality in populations around landfill sites or incinerators have been equivocal, with varying results for different cancer sites. Many of these studies lack good individual exposure information and data on potential confounders, such as socio-economic status. The inherent latency of diseases and migration of populations are often ignored. Waste management workers have been shown to have increased incidence of accidents and musculoskeletal problems. The health impacts of new waste management technologies and the increasing use of recycling and composting will require assessment and monitoring.

Introduction

The generation of waste and the collection, processing, transport and disposal of waste—the process of 'waste management'—is important for both the health of the public and aesthetic and environmental reasons. Waste is anything discarded by an individual, household or organization. As a result waste is a complex mixture of different substances, only some of which are intrinsically hazardous to health. The potential health effects of both waste itself and the consequences of managing it have been the subject of a vast body of research. This chapter gives an overview of waste, waste management processes, and the research into health hazards associated with these, discusses the limitations of studies to date and outlines some future developments and challenges.

Correspondence to: Lesley Rushton, MRC Institute for Environment and Health, 94 Regent Road, Leicester LE1 7DD, UK. E-mail: lr24@le.ac.uk

What is meant by waste?

The UK Environment Agency classifies waste as either controlled waste or non-controlled waste[1]. Controlled waste includes waste generated from households (municipal solid waste), commercial and industrial organizations and from construction and demolition. Non-controlled waste includes waste generated from agriculture, mines and quarries and from dredging operations. In 1998–99 over 470 million tonnes of waste were generated in the UK (Fig. 1). The mean production[2] of daily household and commercial waste in EU Member States in 1993–96 was approximately 370 kg/capita/annum, ranging from 350 to 430 kg. Municipal solid waste (MSW) consists of many different things including food and garden waste, paper and cardboard, glass, metals, plastics and textiles. These are also generated by commercial and industrial organizations although large volumes of chemical and mineral waste are produced in addition, depending on the sector. Agricultural waste comprises mainly slurry and farmyard manure with significant quantities of straw, silage effluent, and vegetable and cereal residues. Most of this is spread on land. Certain types of waste are defined as hazardous because of the inherent characteristics (*e.g.* toxic, explosive). The three largest waste streams in this category are oils and oily wastes, construction and demolition waste and asbestos, and wastes from organic chemical processes.

Methods of waste management

Waste management is now tightly regulated in most developed countries and includes the generation, collection, processing, transport and disposal of waste. In addition the remediation of waste sites is an important issue, both to reduce hazards whilst operational and to prepare the site for a change of use (*e.g.* for building).

The major methods of waste management are:

- Recycling—the recovery of materials from products after they have been used by consumers.

- Composting—an aerobic, biological process of degradation of biodegradable organic matter.

- Sewage treatment—a process of treating raw sewage to produce a non-toxic liquid effluent which is discharged to rivers or sea and a semi-solid sludge, which is used as a soil amendment on land, incinerated or disposed of in land fill.

- Incineration—a process of combustion designed to recover energy and reduce the volume of waste going to disposal.

- Landfill—the deposition of waste in a specially designated area, which in modern sites consists of a pre-constructed 'cell' lined with an impermeable layer (man-made or natural) and with controls to minimize emissions.

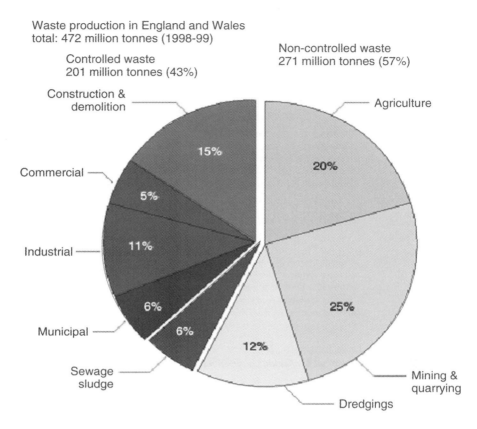

Fig. 1 Waste production in England and Wales, 1998–99. Source: Environment Agency website: http://www.environment-agency.gov.uk

Table 1 (adapted from Pheby et al[3]) outlines some of the advantages and disadvantages of different methods of waste disposal.

Hazardous substances associated with waste management

Environmental monitoring of all potential sources of pollution from different waste management options has been, and is being continuously, carried out and thus a great deal is known about the types and amount of substances emanating from them. Whatever the waste management option, it is generally the case that: (a) there are usually a large number of different substances; and (b) only a few of these are produced in any quantity with many being at extremely low levels[4]. Gases emitted from landfill sites, for example, consist principally of methane and carbon dioxide, with other gases, such as hydrogen sulphide and mercury vapour being emitted at low concentrations, and a mixture of volatile organic compounds

Table 1 Waste management options—key advantages and disadvantages

Option	Advantages	Disadvantages
Recycling	Conservation of resources	Diverse range of processes
	Supply of raw materials to industry	Emissions from recycling process
	Reduction of waste disposed to landfill and incineration	May be more energy used for processes than original manufacture
		Currently low demand for products
		Requires co-operation from individuals
Composting	Reduction of waste to dispose to landfill and incineration	Odours, noise, vermin nuisance
	Recovery of useful organic matter for use as soil amendment	Bio-aerosols—organic dust containing bacteria or fungal spores
	Employment opportunities	Emits volatile organic compounds
		Potential pathway from use on land for contaminants to enter food chain
Sewage treatment	Safe disposal of human waste	Discharges may contain organic compounds, endocrine disrupting compounds, heavy metals, pathogenic microorganisms
	Protects sources of potable water supply	Odour nuisance
Incineration	Reduces weight and volume of waste, about 30% is left as ash which can be used for materials recovery	Produces hazardous solid waste
	Reduces potential infectivity of clinical waste	Discharges contaminated waste water
	Produces energy for electricity generation	Emits toxic pollutants, heavy metals, and combustion products
Landfill	Cheap disposal method	Water pollution from leachate and run off
	Waste used to back fill quarries before reclamation	Air pollution from anaerobic decomposition of organic matter to produce methane, carbon dioxide, nitrogen, sulphur and volatile organic compounds
	Landfill gas contributes to renewable energy supply	Emission of known or suspected carcinogens or teratogens (e.g. arsenic, nickel, chromium, benzene, vinyl chloride, dioxins, polycyclic aromatic hydrocarbons)
		Animal vectors (seagulls, flies, rats) for some diseases
		Odour, dust, road traffic problems

(VOCs) comprising approximately 0.5%[5]. A WHO exposure assessment expert group suggested that priority pollutants should be defined on the basis of toxicity, environmental persistence and mobility, bioaccumulation and other hazards such as explositivity[6]. In addition to the substances above, they suggested that landfill site investigations should consider metals, polycyclic aromatic hydrocarbons (PAH), polychlorinated biphenyls (PCB), chlorinated hydrocarbons, pesticides, dioxins, asbestos, pharmaceuticals and pathogens. Waste incineration also produces a large number of pollutants from the combustion of sewage sludge, chemical, clinical and municipal waste, which can be grouped as particles and gases, metals, and organic compounds[7]. Ten pollutants considered[8] to have the greatest potential impact on human health based on environmental persistence, bioaccumulation and amount emitted and/or on inherent toxicity were cadmium, mercury, arsenic, chromium, nickel, dioxins, PCBs, PAHs,

PM_{10} and SO_2. Microbial pathogens are a potential source of hazard, particularly in composting and sewage treatment but also in landfill. Dust and the production of particulate matter are produced in landfill, incineration and composting processes and by road traffic involved in all waste management options.

Less easily quantifiable hazards, which might nevertheless impact on the population near a waste disposal site include odours, litter, noise, heavy traffic, flies and birds.

Impact of waste management practices on health

Introduction

There is a large body of literature on the potential adverse health effects of different waste management options, particularly from landfill and incineration. There is little on potential problems resulting from environmental exposures from composting and very little on recycling. Although much research has focused on the health of the general population, particularly those living near a waste disposal site, occupational health problems of the workforce involved in waste management are also important to consider.

Much of the health literature on the toxicity of the individual substances highlighted above relates to occupational or accidental exposure and thus generally to higher levels of exposure than those expected from waste disposal methods. Many of the substances, such as cadmium, arsenic, chromium, nickel, dioxins and PAHs are considered to be carcinogenic, based on animal studies or studies of people exposed to high levels. Evidence that these substances cause cancer at environmental levels, however, is often absent or equivocal. In addition to carcinogenicity, many of these substances can produce other toxic effects (depending on exposure level and duration) on the central nervous system, liver, kidneys, heart, lungs, skin, reproduction, *etc*. For other pollutants such as SO_2 and PM_{10}, air pollution studies have indicated that there may be effects on morbidity and mortality at background levels of exposure, particularly in susceptible groups such as the elderly. Chemicals such as dioxins and organochlorines may be lipophilic and accumulate in fat-rich tissues and have been associated with reproductive or endocrine-disrupting endpoints.

Landfill sites

One of the mostly widely known and publicized landfill sites is that of Love Canal in New York State. Large quantities of toxic materials, including residues from pesticides production, were deposited in the 1930s and

1940s, followed by the building of houses and a school on and around the landfill in the 1950s. By the mid 1970s, chemicals leaking from the site were detected in local streams, sewers, soil and indoor air of houses. This site and the subsequent studies of the health of the population in the vicinity fuelled public opinion on the problems of waste disposal practices and raised public concern more generally.

Since then there have been many studies of populations living near landfill sites, frequently carried out near one specific site in response to public concern. These studies have varied in design and include cross-sectional, case-control, retrospective follow-up and ecological (geographical comparison) studies (see Chapter 2). The last of these have often been initiated after apparent clusters of specific diseases have been reported near a site. In addition, several large studies have been carried out investigating health outcomes near hundreds of sites.

There have been several comprehensive reviews of epidemiological studies[9–12].

Birth defects and reproductive disorders

Reproductive effects associated with landfill sites have been extensively researched and include low birth weight (less than 2500 g), fetal and infant mortality, spontaneous abortion, and the occurrence of birth defects. Vianna and Polan[13] and Goldman et al[14] both found increased incidence of low birth weight in the populations around the Love Canal site, the former during the period of active dumping (1940–1953) and the latter among house owners (although not among those renting) from 1965 to 1978. A similar increase in the proportion of low birth weight babies was found in those living within a radius of 1 km of the Lipari Landfill in New Jersey, particularly in 1971–75 following a period of heavy pollution of streams and a nearby lake from leachate from the site[15]. Trends in low birth weight and neonatal deaths were found to correspond closely with time and quantities of dumping at a large hazardous waste disposal site in California, with significantly lower birth weights in exposed areas than control areas during the periods of heaviest dumping[16]. It should be noted that exposed areas were defined according to the number of odour complaints rather than any more objective measure.

The results from these single site studies for low birth weight contrast with results from two large multiple site case-control studies in the USA[17,18]. These used residence as an exposure measure and found no association with low birth weight. However, a geographical study of adverse birth outcomes associated with living within 2 km of a landfill site between 1982 and 1997 in Great Britain found a significantly excess risk, which increased during operation or after closure compared with the risk before opening[19].

An interesting finding from this study was that 80% of the population in Great Britain live within 2 km of an operating or closed landfill site.

The results of studies of congenital malformations are less convincing than those of low birth weight. In the two US multiple site studies, one[17] found a small increase (1.5-fold) in heart and circulatory malformations but no increased risk for other malformations. The other[18] found no association, although the response to the questionnaire used to collect data was relatively poor (63%) and it is unclear how congenital malformations were defined. The UK study[19] found significantly elevated risks for several defects, including neural tube defects, hypospadias and epispadias, abdominal wall defects and surgical correction of gastroschisis and exomphalos, although there was a tendency for there to be a higher risk in the period before opening compared with after opening of a landfill site, for several anomalies. A similar finding was also reported in the analysis of congenital malformation rates among the population living near the Welsh landfill of Nant-y-Gwyddon where nearly double the risk was found in exposed areas both before and after the site opened. However four cases, a nine-fold excess, of gastroschisis, were observed after the site opened[20]. A study of 21 European hazardous waste sites found that residence within 3 km of a site was associated with a significantly raised risk of congenital anomaly, with a fairly consistent decrease in risk with distance away from the sites. Risk was raised for neural-tube defects, malformations of the cardiac septa and anomalies of great arteries and veins[21]. A study by the same group showed similar increases in chromosomal anomalies, even after adjustment for maternal age[22].

The studies of congenital malformations described above have generally used residential proximity as a measure of exposure. A similar study was carried out in New York State but also attempted to investigate associations with off-site migration of chemicals and certain categories of chemicals present at the sites[23]. A small (12%) statistically significant risk of congenital malformations was associated with maternal proximity to a site which increased with off-site chemical leaks. Significant associations were found for pesticides with musculoskeletal system defects, metals and solvents with nervous system defects, and plastics with chromosomal anomalies. However, a case-control study to follow-up these findings which established the probability of low, medium or high exposure for four potential pathways of exposure (groundwater ingestion and inhalation, air, vapour, particulates) found no increased risk for mothers assigned a medium or high exposure[24].

Cancer

Several geographical comparison studies have investigated cancer mortality and incidence around waste sites. Increased frequency of cancers in counties containing hazardous waste sites was found in two US studies[25,26], particularly for gastrointestinal, oesophageal, stomach, colon and rectal cancer. These studies are, however, limited by a lack of chemical release data. No increase in cancer rates or the frequency of chromosome changes was found in relation to the Love Canal site[27,28]. Two reports[29,30] of cancer incidence among persons living near the Miron Quarry site, the third largest in North America found increased incidence of cancers of the liver, kidney, pancreas and non-Hodgkin's lymphomas. Once again no measurements of exposure were available, and there was a relatively short period from first exposure (1968) to cancer onset (1979–1985).

Studies of self-reported health symptoms

Many of the studies investigating health outcomes other than birth defects and reproductive orders and cancers have been community health surveys and have relied on the self-reporting of symptoms through interviews or questionnaires. These are comprehensively reviewed by Vrijheid[10]. The health problems investigated include respiratory symptoms, irritation of the skin, nose and eyes, gastrointestinal problems, fatigue, headaches, psychological problems and allergies. It has been suggested that evaluation of a relationship between these symptoms is complicated by confounding by stress, public perception of risk, odours and nuisance related to the site, and recall bias. For example, a survey[31] found that residents who indicated they were worried about pollution reported more symptoms than those who were not worried, both in the exposed and control areas.

Incineration

Evaluation of the potential health effects of the large number of pollutants which can be produced by waste incineration can be approached by assessing the effects of individual pollutants[8,32] or through more general studies of community residents[33] and incinerator workers (see below).

Individual pollutants

From the health aspect, the most important pollutants associated with incineration are particles, acidic gases and aerosols, metals and organic compounds. There is an extensive research literature on both the acute and chronic effects of particles (see Chapters 10 and 11). Despite methodological limitations, epidemiological studies worldwide have demonstrated considerable consistency of findings with regard to the association of particle exposure and acute health effects such as increased overall mortality and emergency hospital admissions, particularly cardiovascular and respiratory mortality and morbidity[34,35]. Effects appear to be more severe in susceptible groups such as children, the elderly, or those with chronic conditions such as asthma or pre-existing cardiovascular disease[36].

Although less well established, results from large US cohort studies suggest that long-term exposure to low concentrations is associated with chronic health effects such as increased rates of bronchitis and reduced lung function[37], shortened life span, elevated rates of respiratory symptoms and lung cancer[38].

Studies of outdoor levels of NO_2 and health effects may be hampered by difficulties in separating the effects of the various components of ambient pollution. Results for mortality and morbidity have been inconsistent between cities and studies. Less equivocal are results relating to acidic gases, in particular sulphur dioxide (SO_2), for which there is also an extensive literature[36]. Asthmatics appear to be particularly sensitive to the effects of SO_2 on lung function, although the concentrations at which these occur vary between studies. Environmental exposure has been shown, like particulates, to be associated with increased cardiovascular and respiratory mortality and morbidity.

Metals associated with incinerator emissions include lead, cadmium, mercury, chromium, arsenic and beryllium. Different forms of these at various levels and *via* various media and exposure pathways have all been shown to cause a range of carcinogenic and non-carcinogenic health effects. In general, however, epidemiological evidence for increased risk at environment levels of exposure is scarce or equivocal and it is therefore extremely difficult to assess what impact, if any, a relatively small additional exposure from incinerators would have.

The organic compounds which have received the most attention relating to incineration are dioxins and PCBs, partly because of their ability to accumulate in the body. High levels of dioxin exposure found in workplaces and after accidents such as that at Seveso have caused chloracne and an increase in cardiovascular disease. Some studies have also found increased risk from some cancers, although the results vary depending on the specific substance. Extrapolation of these results to the low levels of exposure generally experienced environmentally remains problematical.

Health effects in communities

Most of the studies of communities living near incinerators have assessed exposure using some measure of distance from the site or an estimate of areas at most risk from emissions. Little evidence has been found for an association between modern waste incinerators and reproductive or developmental effects. In addition, there is little evidence of increased prevalence of respiratory illness near incinerators, using either self-reported symptoms or physiological measures.

Studies focusing on a single waste incinerator suggested some relationship between distance from the site and mortality or incidence from some cancers, for example laryngeal and lung cancers, childhood cancers and leukaemias and soft-tissue sarcoma and non-Hodgkin's lymphoma. A series of studies in the UK[39–41] of multiple sites compared observed cancer incidence rates in bands of increasing distance from each incinerator with rates based on national data. Adjustment was made for age, sex and deprivation and lagged analyses were also carried out. No evidence of an increasing risk of lung or laryngeal cancer was found with proximity to incinerators used for the disposal of solvents and oils. However, a study of residence near MSW incinerators found statistically increasing risk with increasing proximity for all cancers and for colorectal, lung, liver and stomach cancers, although there was evidence of residual confounding for all cancers, stomach and lung. Because of the substantial level of misdiagnosis which can occur among registrations and death certificates for liver cancer, the authors carried out a histological review of the cases and reanalysis for this disease. Reduced estimates of excess incidence of primary liver cancer were reported[41].

Worker populations

There is a large workforce employed in waste collection, sorting and disposal. Workers may be exposed to the same potential hazards as the general population, although the amount of exposure and risk may differ. The type of work varies between waste management options with some, such as landfill and incineration, being more automated than others, such as waste collection, sorting and recycling. The incidence of occupational accidents in waste collection workers has been found to be higher than the general workforce[42]. The work of waste collectors involves considerable heavy lifting as well as other manual handling of containers, increasing the risk of musculoskeletal problems. It has been suggested that increased exposure to bio-aerosols and volatile compounds may lead to elevated incidence of work-related respiratory gastrointestinal and skin problems in waste collections compared to the general workforce. Cross-sectional studies of workers in the waste sorting and recycling industries and in landfill sites, have observed similar work-related problems to those of waste collectors[43].

In addition to VOCs and bioaerosols, dust levels have been found to be high at refuse transfer stations and incinerators. An excess of deaths [Standardized Mortality Ratio (SMR) = 355, 95% confidence interval (CI) 162–165] due to lung cancer was observed in the workforce of a large

Swedish incinerator[44]. However, reduced mortality from lung cancer was found for Italian incinerator workers[45]. In the Italian study, a non-significant increased risk was found for gastric cancer (SMR 2.79, 90% CI 0.94–6.35).

There is little published information on the health risks of compost workers. A small cross-sectional study of 58 compost workers[46] found significantly increased antibody concentrations against fungi and actinomycetes compared to a control group of 40 newly employed compost workers and biowaste collectors. This was associated with significantly more symptoms and diseases of the airways ($P = 0.003$) and skin ($P = 0.02$) diagnosed by occupational health physicians.

Remediation of waste and waste sites is an expanding activity, particularly the mediation of hazardous or toxic waste. The health of the workforce involved in this is an important issue. Although no data to date indicate any adverse health effects in remediation workers, countries like the USA have introduced a surveillance programme[47]. An assessment of the risk of occupational fatalities associated with hazardous waste site remediation estimated that the fatality risks to remediation workers were orders of magnitude greater than human cancer risk, and that truck drivers and labourers were particularly at risk[48].

Discussion

There is no doubt that, given the diversity of material coming under the heading of waste, there is considerable potential for hazardous exposure to occur through waste management. High levels of contamination of air, soil and water in a few well publicized situations have led to widespread unease about the potential health effects of waste management processes, particularly within communities living in the proximity to relevant sites. Overall, however, the vast body of literature does not generally support these concerns, particularly for the two most common methods, incineration and landfill disposal. There is also a lack of evidence as to the precise substance(s) implicated. Any emissions from waste management processes are likely to be a mixture of many substances for which a toxicological profile is unknown.

Many of the studies are hampered by a lack of good exposure information and use surrogate indirect measures perhaps leading to exposure misclassification. The levels of most of the potential substances would also be expected to be extremely low, even if all sources of exposure were taken into account. Lack of specificity can also occur in defining health outcomes, particularly if these are self-reported. Many outcomes, such as cancers, would not be expected to occur until several years after exposure, requiring analysis for latency which is lacking in many studies. Migration into and out of relevant areas is also often ignored.

The greatest challenge, however, is to eliminate the effects of factors which might relate to both health outcome and environmental exposure, such as age, ethnicity, gender, socio-economic or deprivation status, smoking, access to health care and occupational history. Lack of complete adjustment for such confounders probably exists in many of the studies relating to waste management particularly those using geographical designs. Studies have shown that socio-economically disadvantaged populations and minority groups may be disproportionately located in areas around waste disposal sites[47].

Studies based on individuals rather than communities are thus perhaps the way forward for future evaluations of potential health effects relating to waste management. However, all of the limitations described above would need to be addressed. Epidemiology is increasingly making use of developing biomarker technology both for estimating internal dose (exposure) and the biological response (effect). This would be particularly relevant in situations where one or two specific substances are of concern (either because of high levels of exposure or because of the health effects the substance may cause), but may be less appropriate for investigating the more general exposures emitted from waste management processes, which tend to be heterogeneous in nature. In addition to a potential reduction in misclassification, biomarkers offer the possibility of identification of lower level exposures and the total burden of exposure, the identification of health events earlier in the natural history of clinical disease and insight into the mechanisms relating exposure and disease. However care is also needed to ensure that the chosen biomarkers are appropriate for the epidemiological design. For example, the use of urinary cotine levels can confirm whether someone is currently smoking or exposed to environmental tobacco smoke but would not aid the assessment of long-term exposure. In contrast, biomarkers of genetic susceptibility can be valuable for use in studies of chronic disease. The field of molecular epidemiology offers the opportunity to combine epidemiology with molecular toxicology to investigate interactions between genetic factors and environmental factors in the cause of disease and identify susceptible groups.

The need to keep the emission of pollutants and exposure to other nuisances arising from waste management operations is widely acknowledged and increasingly stringent regulations have resulted in the development of waste management technologies to achieve this. It is also likely that the proportion of waste managed by different processes will change and that these proportions will vary between communities depending on the characteristics of the waste generated, the facilities already available, economic considerations, and public opinion. The general trend at the moment is towards an increasing proportion of waste being recycled. However, this may generate new challenges, not only a likely considerable financial investment, but a need for a larger workforce for waste sorting and recycling, increasing the need for the issues previously highlighted relating to worker health to be addressed. Wider use of alternative technologies is likely, including advanced thermal treatment, such as gasification and pyrolysis, and bio-mechanical waste treatment which refers to a number of mechanical and biological processes to treat waste before disposal. The health impacts of these technologies will need to be assessed and monitored.

Although the possible physical health effects arising from waste management processes have been addressed, there has been little research into socio-economic impacts of waste-management options. Public perceptions of the relative health risks reflect not only differences in understanding but underlying social values. The development of effective participatory programmes is essential to ensure the public right and responsibility to be involved in the assessment and management of hazards in their communities is addressed, leading hopefully to improved assessments and management strategies.

References

1 Environment Agency. *Waste Statistics for England and Wales 1998–99*. Environment Agency
2 Fischer C, Crowe M. *Household and Municipal Waste: Comparability of Data in EEA Member Countries*. Copenhagen: European Environment Agency, 2000
3 Pheby D, Grey M, Giusti L, Saffron L. *Waste Management and Public Health: The State of the Evidence*. Bristol: South West Public Health Observatory, 2002
4 Johnson BL, DeRosa CT. The toxicological hazards of superfund hazardous waste sites. *Rev Environ Health* 1997; **12**: 235–51
5 Zmirou D, Deloraine A, Savinc P, Tillier C, Bouchanlat A, Maury N. Short term health effects of an industrial toxic waste landfill: a retrospective follow-up study in Montchanin, France. *Arch Environ Health* 1994; **49**: 228–38
6 WHO European Centre for Environment and Health. *Methods of Assessing Risk to Health from Exposure to Hazards Released from Waste Landfills*. Lodz, Poland: WHO Regional Office for Europe, 2000
7 Harrad SJ, Harrison RM. *The Health Effects of the Products of Waste Combustion*. Birmingham, UK: Institute of Public and Environmental Health, University of Birmingham, 1996
8 IEH. *Health Effects of Waste Combustion Products*. Leicester: Institute for Environment and Health, 1997
9 Agency for Toxic Substances and Disease Register. *Hazardous Substances Emergency Events Surveillance*. Annual report. Atlanta: US Department of Health and Human Services, Public Health Service, 1995
10 Vrijheid M. Health effects of residence near hazardous waste landfill sites: a review of epidemiologic literature. *Environ Health Perspect* 2000; **108**: 101–12
11 Upton AC. Public health aspects of toxic chemical disposal sites. *Annu Rev Public Health* 1989; **10**: 1–25
12 National Research Council. *Environmental Epidemiology: Public Health and Hazardous Wastes*, vol. **1**. Washington, DC: National Academy Press, 1991
13 Vianna NJ, Polan AK. Incidence of low birthweight among Love Canal residents. *Science* 1984; **226**: 1217–9

14 Goldman LR, Paigen B, Magnant MM, Highland JH. Low birthweight, prematurity and birth defects in children living near the hazardous waste site, Love Canal. *Hazard Waste Hazard Mater* 1985; **2**: 209–23

15 Berry M, Bove F. Birthweight reduction association with residence near a hazardous waste landfill. *Environ Health Perspect* 1997; **105**: 856–61

16 Kharazi M, Von Behren J, Smith M, Lomas T, Armstrong M, Broadwin R *et al*. A community based study of adverse pregnancy outcomes near a large hazardous waste landfill in California. *Toxicol Ind Health* 1997; **13**: 299–310

17 Shaw GM, Schulman J, Frisch JD, Cummins SK, Harris JA. Congenital malformations and birthweight in areas with potential environmental contamination. *Arch Environ Health* 1992; **47**: 147–54

18 Sosniak WA, Kaye WE, Gomez TM. Data linkage to explore the risk of low birthweight associated with maternal proximity to hazardous waste sites from the National Priorities list. *Arch Environ Health* 1994; **49**: 251–5

19 Elliott P, Briggs D, Morris S, de Hoogh C, Hurt C, Jensen TK *et al*. Risk of adverse outcomes in populations living near landfill sites. *BMJ* 2001; 363–8

20 Fielder HMP, Poon-King C, Palmer SR, Coleman G. Assessment of the impact on health of residents living near the Nant-y-Gwyddon landfill site: Retrospective analysis. *BMJ* 2000; **320**: 19–23

21 Dolk H, Vrijheid M, Armstrong B, Abramsky L, Bianchi F, Garne E *et al*. Risk of congenital anomalies near hazardous-waste landfill sites in Europe: the EUROHAZCON study. *Lancet* 1998; **352**: 423–7

22 Vrijheid M, Dolk H, Armstrong B, Abramsky L, Bianchi F, Fazarinc I *et al*. Chromosomal congenital anomalies and residence near hazardous waste landfill sites. *Lancet* 2002; **359**: 320–2

23 Geshwind SA, Stowijk JAJ, Bracken M *et al*. Risk of congenital malformations associated with proximity to hazardous waste sites. *Am J Epidemiol* 1992; **135**: 1197–207

24 Marshall EG, Gensburg LJ, Deres DA, Geary NS, Cayo MR. Maternal residential exposure to hazardous wastes and risk of central nervous system and musculoskeletal birth defects. *Arch Environ Health* 1997; **52**: 416–25

25 Najem GR, Louria DB, Lavenhar MA, Feuerman M. Clusters of cancer mortality in New Jersey municipalities: with special reference to chemical toxic waste disposal sites and per capita income. *Int J Epidemiol* 1983; **12**: 276–88

26 Griffith J, Duncan R, Riggan W, Peloma A. Cancer mortality in US counties with hazardous waste sites and ground water pollution. *Arch Environ Health* 1989; **44**: 69–74

27 Janerich DT, Burnett WS, Feck G, Hoff M, Nasca P, Polednak AP *et al*. Cancer incidence in the Love Canal area. *Science* 1981; **212**: 1404–7

28 Heath CW, Nadel MR, Zack MM, Chen ATL, Beuder MA *et al*. Cytogenetic findings in persons living near the Love Canal. *JAMA* 1984; **251**: 1217–9

29 Goldberg MS, Al-Homsi N, Goulet L, Riberdy H. Incidence of cancer among persons living near a municipal solid waste landfill site in Montreal, Quebec. *Arch Environ Health* 1995; **50**: 416–24

30 Goldberg MS, DeWar R, Desy M, Riberdy H. Risk of developing cancer relative to living near a municipal solid waste landfill site in Montreal, Quebec, Canada. *Arch Environ Health* 1999; **54**: 291–6

31 Ozonoff D, Colten ME, Cupples A, Heeren T, Schatzin A, Mangione T *et al*. Health problems reported by residents of a neighbourhood contaminated by a hazardous waste facility. *Am J Ind Med* 1987; **11**: 581–97

32 Committee on Health Effects of Waste Incineration. *Waste Incineration and Public Health*. Washington, DC: National Academy Press, 2000

33 Hu S, Shy CM. Health effects of waste incineration: a review of epidemiologic studies. *J Air Waste Manage Assoc* 2001; **51**: 1100–9

34 Dockery DW, Pope III CA. Acute respiratory effects of particulate air pollution. *Annu Rev Public Health* 1994; **15**: 107–32

35 Katsouyanni K, Touloumi G, Spix C, Schwartz J, Balducci F, Medina S *et al*. Short term effects of ambient sulphur dioxide and particulate matter on mortality in 12 European cities: Results from time series data from the APHEA project. *BMJ* 1997; **314**: 1658–63

36 Zanobetti A, Schwartz J, Gold D. Are there sensitive subgroups for the effects of airborne particles? *Environ Health Perspect* 2000; **108**: 841–5

37 WHO. *Air Quality Guidelines for Europe* (European Series No 91). Copenhagen: World Health Organization Regional Office for Europe, 2000
38 EPAQPS. *Airborne Particles*. London: The Stationery Office, 2001
39 Elliott P, Hills M, Beresford J, Kleinschmidt I, Jolley D, Pattenden S et al. Incidence of cancers of the larynx and lung near incinerators of waste solvents and oils in Great Britain. *Lancet* 1992; **339**: 854–8
40 Elliott P, Shaddick G, Kleinschmidt I, Jolley D, Walls P, Beresford J et al. Cancer incidence near municipal solid waste incinerators in Great Britain. *Br J Cancer* 1996; **73**: 702–10
41 Elliott P, Eaton N, Shaddick G, Carter R. Cancer incidence near municipal solid waste incinerators in Great Britain. Part 2: Histopathological and case-note review of primary liver cancer cases. *Br J Cancer* 2000; **82**: 1103–6
42 Poulsen OM, Breum NO, Ebbehøj N, Hansen AM, Ivens UI, van Lelieveld D et al. Collection of domestic waste. Review of occupational health problems and their possible causes. *Sci Total Environ* 1994; **170**: 1–19
43 Poulsen OM, Breum NO, Ebbehøj N, Hansen AM, Ivens UI, van Lelieveld D et al. Sorting and recycling of domestic waste. Review of occupational health problems and their possible causes. *Sci Total Environ* 1995; **168**: 33–56
44 Gustavsson P. Mortality among workers at a municipal waste incinerator. *Am J Ind Med* 1989; **15**: 245–53
45 Rapiti E, Sperati A, Fano V, Dell'Orco V, Forastiere F. Mortality amongst workers at municipal waste incinerators in Rome: a retrospective cohort study. *Am J Ind Med* 1997; **31**: 659–61
46 Bunger J, Antlauf-Lammers M, Schulz TG, Westphal GA, Muller MM, Ruhnan P et al. Health complaints and immunological makers of exposure to bioaerosols among biowaste collectors and compost workers. *Occup Environ Med* 2000; **57**: 458–64
47 Johnson BL. Hazardous waste: human health effects. *Toxicol Ind Health* 1997; **13**: 121–43
48 Hoskin A, Leigh J, Planek T. Estimated risk of occupational fatalities associated with hazardous waste site remediation. *Risk Anal* 1994; **14**: 1011–7

Contaminants in drinking water

John Fawell* and **Mark J Nieuwenhuijsen**[†]

Flackwell Heath, Bucks, and †Department of Environmental Science and Technology, Imperial College, Royal School of Mines, Prince Consort Road, London, UK

An adequate supply of safe drinking water is one of the major prerequisites for a healthy life, but waterborne disease is still a major cause of death in many parts of the world, particularly in children, and it is also a significant economic constraint in many subsistence economies. The basis on which drinking water safety is judged is national standards or international guidelines. The most important of these are the WHO Guidelines for Drinking Water Quality. The quality of drinking water and possible associated health risks vary throughout the world with some regions showing, for example, high levels of arsenic, fluoride or contamination of drinking water by pathogens, whereas elsewhere these are very low and no problem. Marked variations also occur on a more local level within countries due, for example, to agricultural and industrial activities. These and others are discussed in this chapter.

Introduction

An adequate supply of safe drinking water is one of the major prerequisites for a healthy life, but waterborne disease is still a major cause of death in many parts of the world, particularly in children, and it is also a significant economic constraint in many subsistence economies.

Drinking water is derived from two basic sources: surface waters, such as rivers and reservoirs, and groundwater. All water contains natural contaminants, particularly inorganic contaminants that arise from the geological strata through which the water flows and, to a varying extent, anthropogenic pollution by both microorganisms and chemicals. In general, groundwater is less vulnerable to pollution than surface waters. There are a number of possible sources of man-made contaminants, some of which are more important than others. These fall into the categories of point and diffuse sources. Discharges from industrial premises and sewage treatment works are point sources and as such are more readily identifiable and controlled; run off from agricultural land and from hard surfaces, such as roads, are not so obvious, or easily controlled. Such sources can give rise to a significant variation in the contaminant

Correspondence to: John Fawell, Independent Consultant on Drinking Water and Environment, 89 Heath End Road, Flackwell Heath, Bucks HP10 9EW, UK. E-mail: John.Fawell@ johnfawell.co.uk

load over time. There is also the possibility of spills of chemicals from industry and agriculture and slurries from intensive farm units that can contain pathogens. In some countries, badly sited latrines and septic tanks are a significant source of contamination, especially of wells. Local industries can also give rise to contamination of water sources, particularly when chemicals are handled and disposed of without proper care. The run-off or leaching of nutrients into slow flowing or still surface waters can result in excessive growth of cyanobacteria or blue-green algae[1]. Many species give rise to nuisance chemicals that can cause taste and odour and interfere with drinking water treatment. However, they frequently produce toxins, which are of concern for health, particularly if there is only limited treatment.

If treatment is not optimized, unwanted residues of chemicals used in water treatment can also cause contamination, and give rise to sediments in water pipes. Contamination during water distribution may arise from materials such as iron, which can corrode to release iron oxides, or from ingress of pollutants into the distribution system. Diffusion through plastic pipes can occur, for example when oil is spilt on the surrounding soil, giving rise to taste and odour problems. Contamination can also take place in consumers' premises from materials used in plumbing, such as lead or copper, or from the back-flow of liquids into the distribution system as a consequence of improper connections. Such contaminants can be either chemical or microbiological.

Drinking water treatment as applied to public water supplies consists of a series of barriers in a treatment train that will vary according to the requirements of the supply and the nature and vulnerability of the source. Broadly these comprise systems for coagulation and flocculation, filtration and oxidation. The most common oxidative disinfectant used is chlorine. This provides an effective and robust barrier to pathogens and provides an easily measured residual that can act as a marker to show that disinfection has been carried out, and as a preservative in water distribution.

The basis on which drinking water safety is judged is national standards or international guidelines. The most important of these are the WHO Guidelines for Drinking-Water Quality[2]. These are revised on a regular basis and are supported by a range of detailed documents describing many of the aspects of water safety. The Guidelines are now based on Water Safety Plans that encompass a much more proactive approach to safety from source-to-tap.

Microbial contamination

The contamination of drinking water by pathogens causing diarrhoeal disease is the most important aspect of drinking water quality. The problem arises as a consequence of contamination of water by faecal matter, particularly human faecal matter, containing pathogenic organisms. One of the great scourges of cities in Europe and North America in the 19th century was outbreaks of waterborne diseases such as cholera and typhoid. In many parts of the developing world it remains a major cause of disease. It is therefore essential to break the faecal–oral cycle by preventing faecal matter from entering water sources and/or by treating drinking water to kill the pathogens. However, these approaches need to operate alongside hygiene practices such as hand washing, which reduce the level of person-to-person infection.

Detection and enumeration of pathogens in water are not appropriate under most circumstances in view of the difficulties and resources required so *Escherichia coli* and faecal streptococci are used as indicators of faecal contamination. The assumption is that if the indicators are detected, pathogens, including viruses, could also be present and therefore appropriate action is required. However, the time taken to carry out the analysis means that if contamination is detected, the contaminated water will be well on the way to the consumer and probably drunk by the time the result has been obtained. In addition the small volume of water sampled (typically 100 ml) means that such check monitoring on its own is not an adequate means of assuring drinking water safety. It is also essential to ensure that the multiple barriers are not only in place but working efficiently at all times, whatever the size of the supply. Drinking water is not, however, sterile and bacteria can be found in the distribution system and at the tap. Most of these organisms are harmless, but some opportunist pathogens such as *Pseudomonas aeruginosa* and *Aeromonas* spp. may multiply during distribution given suitable conditions[3]. Currently there is some debate as to whether these organisms are responsible for any waterborne, gastrointestinal disease in the community but *P. aeruginosa* is known to cause infections in immunocompromised patients and weakened patients in hospitals.

A number of organisms are emerging as potential waterborne pathogens and some are recognized as significant pathogens that do give rise to detectable waterborne outbreaks of infection. The most important of these is *Cryptosporidium parvum*, a protozoan, gastrointestinal parasite which gives rise to severe, self limiting diarrhoea and for which there is, currently, no specific treatment. *Cryptosporidium* is excreted as oocysts from infected animals, including humans, which enables the organism to survive in the environment until ingested by a new host[3]. This organism has given rise to a number of waterborne or water associated outbreaks in the UK, and an outbreak of cryptosporidiosis in Milwaukee in the USA resulted in many thousands of cases[4], and probably a number of deaths among the portion of the population which were immunocompromised[5]. The most important barriers to infection are those that remove particles, including coagulation, sedimentation and filtration. However, water is not the only source of infection. It is probable that person-to-person spread following contact with faecal matter from infected animals is more important and there have been outbreaks involving milk and swimming pools[6]. Currently, there is no scientifically based standard for *Cryptosporidium* in drinking water. A similar parasite, *Giardi*, has been responsible for a number of cases of gastrointestinal illness and in the USA, illness was referred to as beaver fever because beavers were shown to be a source in some areas. As with *Cryptosporidium*, water is not the only source but, unlike *Cryptosporidium*, it is reasonably susceptible to chlorine and because of its larger size can be more easily removed by particle removal processes[3].

Although the common waterborne diseases of the 19th century are now almost unknown in developed countries, it is vital that vigilance is maintained at a high level because these diseases are still common in many parts of the world. The seventh cholera pandemic, which started in 1961, arrived in South America in 1991 and caused 4700 deaths in 1 year[7]. According to the WHO World Health Report 1998, over 1 billion people do not have an adequate and safe water supply of which 800 million are in rural areas. WHO also estimate that there are 2.5 million deaths and 4 billion cases due to diarrhoeal disease, including dysentery, to which waterborne pathogens are a major contributor. There are still an estimated 12.5 million cases of *Salmonella typhi* per year and waterborne disease is endemic in many developing countries. In this age of rapid global travel, the potential for the reintroduction of waterborne pathogens in developed countries still remains. In addition, as our knowledge of microbial pathogens improves, we are able to identify other organisms that cause waterborne disease. The Norwalk-like viruses are named after a major waterborne outbreak in North America, and there is a range of emerging pathogens including *Campylobacter*, a major cause of food poisoning, and *E. coli* O157, which has caused deaths in North America where chlorination was not present, or failed, and other barriers were inadequate[3].

Microbial contamination of drinking water thus remains a significant threat and constant vigilance is essential, even in the most developed countries.

Chemical contaminants

As indicated above, there are many sources of chemical contaminants in drinking water. However, the most important contaminants from a health standpoint are naturally occurring chemicals that are usually found in groundwater.

Arsenic

Waterborne arsenic is a major cause of disease in many parts of the world including the Indian sub-continent—particularly Bangladesh and Bengal—South America, and the Far East. It is the only contaminant that has been shown to be the cause of human cancers following exposure through drinking water. Besides cancer of the skin, lung and bladder and probably liver, arsenic is responsible for a range of adverse effects, including hyperkeratosis and peripheral vascular disease[8,9]. However, the epidemiological data also demonstrate that many local factors are important, including nutritional status. There are considerable difficulties in assessing arsenic exposure. In Bangladesh, where millions of tube wells were sunk, the concentration of arsenic can vary significantly between wells only a short distance apart. WHO have set a provisional guideline value of 10 µg/l based on the practical limit of achievability, but there is an ongoing discussion on the scientific basis for this guideline, including whether the available data would allow distinction between a standard of say 5, 10, or 15 µg/l and whether exposure to 50 µg/l, the old guideline, will result in illness.

Fluoride

Waterborne fluoride is another major cause of morbidity in a number of parts of the world, including the Indian sub-continent, Africa and the Far East, where concentrations of fluoride can exceed 10 mg/l. High intakes of fluoride can give rise to dental fluorosis, an unsightly brown mottling of teeth, but higher intakes result in skeletal fluorosis, a condition arising from increasing bone density and which can eventually lead to fractures and crippling skeletal deformity. A WHO working group concluded that skeletal fluorosis and an increased risk of bone fractures occur at a total intake of 14 mg fluoride per day, and there is evidence suggestive of an increased risk of bone effects at intakes above about 6 mg fluoride per day[10,11].

This is a major cause of morbidity and can manifest itself at a relatively early age with the result that affected individuals cannot work properly and may be economically as well as physically disadvantaged for life. Many factors appear to influence the risk of such adverse effects, including volume of drinking water, nutritional status and, particularly, fluoride intake from other sources.

Selenium and uranium

Selenium and uranium have also both been shown to cause adverse effects in humans through drinking water. In seleniferous areas, drinking water can contribute to high selenium intakes, which can give rise to loss of hair, weakened nails and skin lesions, and more seriously, changes in peripheral nerves and decreased prothrombin time[12]. Uranium is found in groundwater associated with granitic rocks and other mineral deposits. It is a kidney toxin and has been associated with an increase in fractional calcium excretion and increased microglobulinurea, although within the normal range found in the population. Uranium is a current topic of research with regard to exposure through drinking water[13].

Iron and manganese

Both iron and manganese can occur at high concentrations in some source waters that are anaerobic[14]. When the water is aerated they are oxidized to oxides that are of low solubility. These will cause significant discolouration and turbidity at concentrations well below those of any concern for health. They may, however, cause consumers to turn to alternative supplies which may be more aesthetically acceptable but which are microbiologically unsafe.

Agricultural chemicals

Agriculture is another source of chemical contamination. In this case the most important contaminant is nitrate, which can cause methaemoglobinaemia, or blue-baby syndrome, in bottle-fed infants under 3 months of age[15]. There remains uncertainty about the precise levels at which clinically apparent effects occur and it also seems that the simultaneous presence of microbial contamination, causing infection, is an important risk factor[16]. WHO have proposed a guideline value of 50 mg/l nitrate based on studies in which the condition was rarely seen below that concentration, but was increasingly seen above 50–100 mg/l. However, when nitrite is also present this must also be taken into account, since it is about 10 times as potent a methaemoglobinaemic agent as nitrate.

Concern is often expressed about pesticides in drinking water but there is little evidence that this is a cause of illness, except perhaps following a spill with very high concentrations[17]. In Cambodia, media and public concern regarding pesticides in drinking water resulted in an expensive analytical exercise being carried out that found none of the pesticides of concern (Steven Iddings, personal communication, 2001). Of greater concern is the run-off of nutrients to surface waters, often combined with sewage discharges, that lead to significant growths of cyanobacteria referred to above[1]. There is a wide range of toxins produced by these organisms and it is probable that not all the toxins have been identified to date. Where drinking water treatment is limited, there is a potential for undesirable concentrations to be present in drinking water. Concerns are particularly directed at hepatotoxins such as the microcystins and cylindrospermopsin, and the neurotoxins such as saxitoxin.

Urban pollution

Industry and human dwellings are also a source of potential contaminants. The most common are heavy metals, and solvents, such as tri and tetrachloroethene, which are sometimes found in groundwater and hydrocarbons, particularly from petroleum oils[2]. There is little good evidence that these pollutants occur at concentrations in drinking water that are sufficient to cause health effects, but some of the low molecular weight aromatic hydrocarbons can give rise to severe odour problems in drinking water at concentrations of less than 30 µg/l.

By-products of water treatment

Drinking water treatment is intended to remove microorganisms and, increasingly in many cases, chemical contaminants. Nevertheless, the process can in itself result in the formation of other contaminants such as the trihalomethanes and haloacetic acids from the reaction of chemical oxidants with naturally occurring organic matter. This requires a balance to be struck between the benefits of the chemical oxidants in destroying microorganisms and the potential risks from the by-products. Of these by-products, only trihalomethanes tend to be routinely monitored in drinking water and the standard for total THMs in the UK is 100 µg/l. Water treatment, however, can take many forms and can use different chemicals including chlorine, chloramines, chlorine dioxide and ozone. Each treatment methodology has certain advantages and disadvantages, but all of them form by-products of some sort. The type and quantities of by-products formed depend on a number of factors. The formation of by-products during chlorination (one of the most common treatments), for example, depends on the amount and content of organic matter, bromine levels, temperature, pH and residence time.

Uptake of trihalomethanes, generally the most common volatile DBP, can occur not only through ingestion, but also by inhalation and skin absorption during activities such as swimming, showering and bathing. For most other DBPs, ingestion is the main route for uptake[18]. DBPs have been associated with cancers of the bladder, colon and rectum and adverse birth outcomes such as spontaneous abortion, (low) birth weight, stillbirth and congenital malformations in epidemiological studies and to a much lesser extent at high levels in toxicological studies. Overall, however, the evidence is inconsistent and inconclusive[9,19,20].

Endocrine disrupters

Endocrine disrupters are chemicals that interfere with the endocrine system, for example by mimicking the natural hormones. They may be associated with a range of adverse reproductive health effects, including sperm count decline, hypospadias and cryptorchidism, and cancer of the breast and testes, although the current human evidence is weak[21]. Phthalates, bisphenols, alkyl phenols, alkyl phenol ethoxylates, polyethoxylates, pesticides, human hormones and pharmaceuticals have all been implicated and sewage effluent discharged to surface water has been shown to contain many of these substances[17,22]. Since many surface waters which receive sewage effluent are subsequently used as drinking water sources (*i.e.* re-use of water), it is important that the water is properly treated, which will remove these substances. Effects on wildlife, such as fish exposed to sewage effluent, have been reported but there is currently little if any evidence that humans drinking tap water are affected.

Discussion

The quality of drinking water and possible associated health risks vary throughout the world. Whilst some regions show high levels of arsenic, fluoride or contamination of drinking water by pathogens, for example, elsewhere these are very low and present no problem for human health. Marked variations in levels of contamination also occur more locally, often as a result of agricultural and industrial activities. The differences in health risks that these variations represent lead to different priorities for the treatment and provision of drinking water. Microbial contamination of drinking water remains a significant threat and constant vigilance is essential, even in the most developed countries. More recent research has suggested a possible association between disinfection by-products and cancer and adverse reproductive outcomes, but potential risks are largely outweighed by the benefits of drinking water with a low microbial load. Where possible, however, further efforts should be made to reduce levels of disinfection by-products without compromising the disinfection process and at a reasonable cost to the consumer. To be able to set priorities, good quality data on the levels of contaminants in water and related morbidity and mortality are needed, although the interpretation may be complicated by the multi-factorial nature of many diseases. Well-designed epidemiological studies are also needed for some of the contaminants such as chlorination by-products, arsenic, fluoride and uranium where information on exposure–response relationships is missing or of insufficient quality. In other cases, toxicological studies are also required to help to determine the potential risk.

There is evidence from a number of countries of consumers rejecting microbially safe public supplies, because of problems with discolouration and chlorine tastes, in favour of more expensive and microbiologically less satisfactory local supplies or bottled water. There is little point in making a considerable investment in providing safe public supplies if the water is not accepted by consumers. In particular, this can lead to poorer consumers, who are more likely to receive unacceptable water supplies, paying more for their water than better off consumers. Delivering safe and acceptable water, therefore, is a key target in improving public health in many developing countries. Even in developed countries, however, the same priority remains, as shown by waterborne outbreaks such as that at Walkerton in Canada that resulted in several deaths.

There also remains a need for high quality research in a number of areas, though this must be set in the appropriate context for the countries in which the problems occur. Increased knowledge has shown the complexity of many of the issues that are related to drinking water and health. Overall, however, it is evident that the supply and maintenance of safe drinking water remain key requirements for public health.

References

1. Chorus I, Bartram J. Toxic cyanobacteria in water. *A Guide to their Public Health Consequences, Monitoring and Management.* Published on behalf of WHO by E &FN Spon, London and New York, 1999
2. World Health Organization. *Guidelines for Drinking-Water Quality,* 3rd edn. www.who.int/water_sanitation_health/GDWQ/draftchemicals/list.htm. Last accessed June 2003. Geneva: WHO, 2003
3. Hunter P. *Waterborne Disease. Epidemiology and Ecology.* Chichester: Wiley, 1997
4. MacKenzie WR, Hoxie NJ, Proctor ME, Gradus MS, Blair KA, Peterson DE, Kazmierczak JJ, Addiss DG, Fox KR, Rose JB *et al.* A massive outbreak in Milwaukee of cryptosporidium infection transmitted through the public water supply. *N Engl J Med* 1994; **331**: 161–7
5. Hoxie NJ, Davis JP, Vergeront JM, Nashold RD, Blair KA. Cryptosporidiosis-associated mortality following a massive waterborne outbreak in Milwaukee, Wisconsin. *Am J Public Health* 1997; **87**: 2032–5
6. MacKenzie WR, Kazmierczak JJ, Davis JP. An outbreak of cryptosporidiosis associated with a resort swimming pool. *Epidemiol Infect* 1995; **115**: 545–53
7. Reeves PR, Lan R. Cholera in the 1990s. *Br Med Bull* 1998; **54**: 611–23
8. IPCS. *Arsenic and Arsenic Compound,* 2nd edn. Environmental Health Criteria 224. Geneva: World Health Organization, 2001
9. IARC. *Some Drinking Water Disinfectants and Contaminants, Including Arsenic.* IARC Monographs on the evaluation of carcinogenic risks to humans. Volume 84. Lyon: IARC, 2003
10. IPCS. *Fluorides.* Environmental Health Criteria 227. Geneva: World Health Organization, 2002
11. World Health Organization. *Fluoride.* http://www.who.int/water_sanitation_health/GDWQ/draftchemicals/fluoride2003.pdf. Accessed June 2003. Geneva: WHO, 2003
12. Barceloux DG. Selenium. *J Toxicol Clin Toxicol* 1999; **37**: 145–72
13. World Health Organization. *Uranium.* http://www.who.int/water_sanitation_health/GDWQ/draftchemicals/uranium2003.pdf. Accessed June 2003. Geneva: WHO, 2003

14 World Health Organization. *Manganese.* http://www.who.int/water_sanitation_health/GDWQ/draftchemicals/manganese2003.pdf. Accessed June 2003. Geneva: WHO, 2003
15 Fan AM, Steinberg VE. Health implications of nitrate and nitrite in drinking water: an update on methemoglobinemia occurrence and reproductive and developmental toxicity. *Regul Toxicol Pharmacol* 1996; **23**: 35–43
16 Avery AA. Infantile methemoglobinemia: reexamining the role of drinking water nitrates. *Environ Health Perspect* 1999; **107**: 583–6
17 Fawell JK, Standfield G. Drinking water quality and health. In: Harrison RM (ed.) *Pollution: Causes, Effects and Control,* 4th edn. London: Royal Society of Chemistry, 2001
18 Nieuwenhuijsen MJ, Toledano MB, Elliott P. Uptake of chlorination disinfection by-products; a review and a discussion of its implications for epidemiological studies. *J Expos Anal Environ Epidemiol* 2000; **10**: 586–99
19 Nieuwenhuijsen MJ, Toledano MB, Eaton NE, Elliott P, Fawell J. Chlorination disinfection by-products in water and their association with adverse reproductive outcomes: a review. *Occup Environ Med* 2000; **57**: 73–85
20 IPCS *Disinfectants and Disinfectant By-products.* Environmental Health Criteria 216. Geneva: World Health Organization, 2000
21 Joffe M. Are problems with male reproductive health caused by endocrine disruption? *Occup Environ Med* 2001; **58**: 281–7
22 Fawell J, Chipman K. Endocrine disrupters, drinking water and public reassurance. *Water Environ Manage* 2000; **5**: 4–5

Indoor air pollution: a global health concern

Junfeng (Jim) Zhang* and Kirk R Smith[†]

*Environmental and Occupational Health Sciences Institute & School of Public Health, University of Medicine and Dentistry of New Jersey, NJ and [†]Environmental Health Sciences, School of Public Health, University of California at Berkeley, Berkeley, CA, USA

Indoor air pollution is ubiquitous, and takes many forms, ranging from smoke emitted from solid fuel combustion, especially in households in developing countries, to complex mixtures of volatile and semi-volatile organic compounds present in modern buildings. This paper reviews sources of, and health risks associated with, various indoor chemical pollutants, from a historical and global perspective. Health effects are presented for individual compounds or pollutant mixtures based on real-world exposure situations. Health risks from indoor air pollution are likely to be greatest in cities in developing countries, especially where risks associated with solid fuel combustion coincide with risk associated with modern buildings. Everyday exposure to multiple chemicals, most of which are present indoors, may contribute to increasing prevalence of asthma, autism, childhood cancer, medically unexplained symptoms, and perhaps other illnesses. Given that tobacco consumption and synthetic chemical usage will not be declining at least in the near future, concerns about indoor air pollution may be expected to remain.

Long history

Correspondence to: Dr Junfeng (Jim) Zhang, Environmental and Occupational Health Sciences Institute & School of Public Health, University of Medicine and Dentistry of New Jersey, 170 Frelinghuysen Road, Piscataway, New Jersey 08854, USA. E-mail: jjzhang@eohsi.rutgers.edu

One of the basic human needs is shelter. Throughout human history, people have depended upon rock shelters, caves and rude huts to protect themselves from the vagaries of weather and climate. Today, some of these crude shelters can still be found in parts of the world, mainly in less developed countries. On the other hand, civilization has brought a significant fraction of the world's population to live in modern single and multifamily dwellings, work in modern office buildings, and carry out various activities in public facilities and other built environments that provide amenities and convenience far beyond the basic needs of sheltering.

People living in crude shelters have relied on crude fuels for their cooking and heating needs. The combustion of these crude biomass fuels—such as crop residues, animal dung and wood—generates smoke that adversely affects the health of occupants. Industrialization resulted in extensive use of coal, a significant fraction of which was burnt indoors for space heating during winter months. The coal smoke emitted from household chimneys in London, for example, was the main cause of the Great London Smog in the winter of 1952—an event that led to thousands of excess deaths during a week-long episode. This and other similar events triggered the introduction of legislation banning domestic use of coal in many cities of developed countries. Fifty years later, however, coal combustion is still taking place in households worldwide, and indoor smoke from solid fuels (biomass and coal) is affecting something like half the world's households, today[1].

In modern residences, cooking and space-heating needs are usually met by fossil fuels such as natural gas, liquefied petroleum gas, heating oil (petroleum product) and electricity. Occasional carbon monoxide (CO) poisoning cases are reported, mainly as a consequence of the improper use or inadequate ventilation of appliances. Concerns have also been reported about exposure to nitrogen dioxide (NO_2) emitted from gas stoves[2]. Compared to combustion of solid fuels, however, gaseous fuels in simple devices emit substantially smaller amounts of pollution, including particulate matter (PM), CO, eye irritating volatile organic compounds (*e.g.* aldehydes), and carcinogenic compounds such as benzene and 1,3-butidiene and polycyclic aromatic hydrocarbons[3–5] (Table 1).

Although indoor air pollution from fuel combustion is therefore generally of lesser concern in modern homes and buildings, than in traditional homes in the developing world, there may nevertheless be important sources of exposure, related to materials used in construction, furnishing,

Table 1 Major health-damaging pollutants generated from indoor sources

Pollutant	Major indoor sources
Fine particles	Fuel/tobacco combustion, cleaning, cooking
Carbon monoxide	Fuel/tobacco combustion
Polycyclic aromatic hydrocarbons	Fuel/tobacco combustion, cooking
Nitrogen oxides	Fuel combustion
Sulphur oxides	Coal combustion
Arsenic and fluorine	Coal combustion
Volatile and semi-volatile organic compounds	Fuel/tobacco combustion, consumer products, furnishings, construction materials, cooking
Aldehydes	Furnishing, construction materials, cooking
Pesticides	Consumer products, dust from outside
Asbestos	Remodelling/demolition of construction materials
Lead[a]	Remodelling/demolition of painted surfaces
Biological pollutants	Moist areas, ventilation systems, furnishings
Radon	Soil under building, construction materials
Free radicals and other short-lived, highly reactive compounds	Indoor chemistry

[a]Pb-containing dust from deteriorating paint is an important indoor pollutant for occupants in many households, but the most critical exposure pathways are not usually through air.

furniture and consumer products. For example, asbestos had been widely used as insulating materials in buildings built before mid-1970s, when its use was banned in the USA and many other countries. These buildings may also contain lead-based paints. It is therefore important to protect workers and occupants from exposure to asbestos fibres and lead due to dust re-suspension during renovation or demolition of old buildings.

Since the 1970s, many energy-conserving buildings have been built in North America and Europe. Improved energy conservation was mainly achieved through reducing exchanges between outdoor fresh air and indoor air. Meanwhile, synthetic materials and chemical products have been extensively used in these airtight buildings. The combination of low ventilation rate and the presence of numerous sources of synthetic chemicals has resulted in elevated concentrations of volatile organic compounds (VOCs) (*e.g.* benzene, toluene, formaldehyde), semivolatile organic compounds (SVOCs) (*e.g.* phthalate plasticizers and pesticides) and human bioeffluents. This has been suggested as a major contributing factor to occupant complaints of illness symptoms, or so-called 'sick building syndrome', in the last three decades[6]. Although the aetiology is still not clear, many cases of respiratory diseases, allergies and asthma, medically unexplained symptoms including sick building syndrome, and cancer, are believed to be attributable to poor indoor air quality in both developing and developed countries[7,8].

Indoor smoke from burning solid fuels

Combustion of solid fuels in households often takes place in simple, poorly designed and maintained stoves. This kind of combustion contributes directly to low energy efficiency, adding pressure on fuel resources. Low combustion efficiency means that a large fraction of fuel carbon is converted to compounds other than carbon dioxide (CO_2)—*i.e.* products of incomplete combustion. These incomplete combustion products mainly comprise CO and fine (respirable) particles, as well as a large suite of VOCs and SVOCs[3–5,9]. Daily exposure to products of incomplete combustion poses both acute and chronic health risks[4,5]. A recent World Health Organization (WHO) report estimated that indoor smoke from solid fuels ranked as one of the top ten risk factors for the global burden of disease, accounting for an estimated 1.6 million premature deaths each year[10] (see also Chapter 1). Among all environmental risk factors, it ranked second only to poor water/sanitation/hygiene (Fig. 1)[10,11]. In this estimate, the burden of disease is defined as lost healthy life years, which includes those lost to premature death and those lost to illness as weighted by a disability factor (severity)[12]. It needs to be recognized,

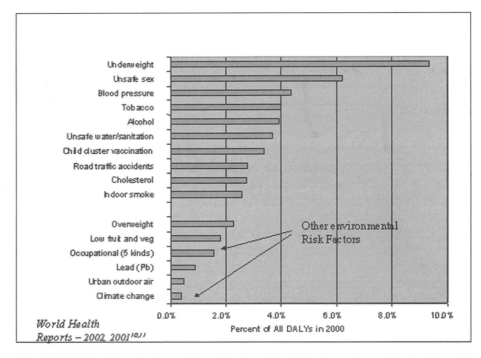

Fig. 1 Global burden of disease from the top 10 risk factors plus selected other risk factors[10,11]. Note: Indoor smoke category here includes only solid fuel use in households and not smoke from other fuels or tobacco.

however, that such estimates are associated with relatively large uncertainties, because the data available on exposure and on exposure-effects relationships are limited, despite the apparently large risks and populations involved.

The existing literature provides strong evidence that smoke from solid fuels is a risk factor for acute respiratory infections (ARI), chronic obstructive pulmonary disease (COPD) and lung cancer (from coal smoke)[8]. Evidence from 13 studies in developing countries indicates that young children living in solid-fuel using households have two to three times more risk of serious ARI than unexposed children after adjustment for potential confounders including socio-economic status[13]. An evaluation of eight studies in developing countries indicates that women cooking over biomass fires for many years have two to four times more risk of COPD than those unexposed after adjustment for potential confounding factors[14]. Excess lung cancer mortality rates are reported in Chinese women who had been exposed to the smoke from household use of so-called 'smoky coal' which has high sulphur content and emits a large quantity of smoke and PAHs compared to other types of coals[15]. The existing epidemiological literature provides moderate evidence that solid fuel smoke is a risk factor for cataracts, tuberculosis, asthma attacks and adverse pregnancy outcomes. Current literature in developed countries would also suggest that exposure to smoke from solid fuels produces cardiovascular disease, though no studies have been done to date in households in developing countries. More studies involving developing country households are needed to evaluate further the suggestive evidence that has so far been found[8].

Household coal combustion produces the same kinds of pollutants as biomass fuels, but the amounts vary according to such parameters as the content of volatiles and fixed carbon. Depending on the geological conditions of coal formation, household use of coal can also produce differing levels of sulphur dioxide (SO_2) and certain toxic elements[3,16]. Chronic fluorine and arsenic poisoning are, for example, particular problems in parts of China relying on local dirty coal deposits used for household fuel[17].

Environmental tobacco smoke (ETS)

Tobacco smoke has been an accepted and well-documented cause of ill health for more than half a century. As shown in Figure 1, tobacco accounted for an estimated 4% of the global burden of disease in 2000, mainly as a result of active smoking exposure. However, there has been growing concern about the health effects of exposure to ETS, also called passive smoking, involuntary smoking, or second-hand smoking. ETS refers to the mixture of primary smoke exhaled by smokers and the secondary smoke produced by the burning tobacco between puffs. ETS exposure is lower than that experienced by active smokers, but the smoke is generally similar and contains the same gases and particles including a wide range of irritating compounds and carcinogens. Available data show that ETS exposure is a significant health risk factor in adults, children and infants (Table 2)[18–26]. In addition to the health outcomes shown in Table 2, breast cancer[27] and pulmonary tuberculosis[28,29] have been suggested, with limited evidence, to be associated with ETS exposure. In a recent cross-sectional study of 1718 school-age children whose mothers were never smokers, a monotonic exposure–response relationship was observed between paternal smoking and decline of pulmonary function[30]. Estimates indicate that 3000 lung cancer deaths each year can be attributed to passive smoking in the USA along with hundreds of thousands of childhood respiratory disease cases[31].

ETS exposure affects large numbers of people living in both developing and developed countries. Globally, ETS is perhaps the largest modern source of indoor air pollution, reflecting the alarmingly high levels of smoking prevalence (29% in adults worldwide)[32]. In each of the four

Table 2 Diseases and risks associated with ETS exposure

Illness	Population	Exposure assessment	Number of studies in meta-analysis	Relative risk			Reference
				Point estimate	95% CI$_{low}$	95% CI$_{high}$	
LRI	Children <3 years of age	Smoking by either parent	24 community and hospital based studies	1.57	1.4	1.77	18
Asthma	Children >1 year of age	Smoking by either parent	14 case-control studies	1.37	1.15	1.64	19
Otitis media (recurrent)	Children <7 years of age	Smoking by either parent	9 case-control, survey, and cohort studies	1.48	1.08	2.04	20
Ischaemic heart disease	Adults	Lifelong non-smokers married to smokers	17 studies	1.25	1.17	1.33	21
Lung cancer	Adults	Lifelong non-smokers with spouses who currently smoke	37 studies	1.24	1.13	1.36	22
Nasal-sinus cancer	Adults	Household exposure to passive smoking	3 studies (no meta-analysis)	2.3[a]	1.7[a]	3.0[a]	23
				–	Low	High	
LBW or SGA	Infants	Prenatal maternal smoking		–	1.5	3.5	24
Sudden Infant Death Syndrome	Infants <1 year	Prenatal maternal smoking		–	1.8	2.4	25
LBW or SGA	Infants	ETS exposure of non-smoking mothers		–	1.1	1.3	26
Sudden Infant Death Syndrome	Infants <1 year	Postnatal maternal smoking		–	1.6	2.4	25

LRI = lower respiratory tract infection; LBW = low birth weight; SGA = small for gestational age.
[a]Relative risk estimates for nasal-sinus cancer adapted from Ref. 23. See Health outcomes section for more details.

Chinese cities selected for an air pollution epidemiological study, more than 59% of fathers of school-age children were regular smokers, based on a survey[33] of >8000 households during 1993–1996. According to the US Centre for Disease Control and Prevention (CDC), nearly 3000 children under 18 become regular smokers every day in the USA. The CDC also estimates that in the USA active tobacco smoking causes more than 400,000 deaths each year and results in more than $50 billion in direct medical costs annually. It is true that the direct health impacts of passive smoking are much smaller than those of active smoking. Efforts to control ETS, however, can have a much larger health benefit than just the reduction in ETS exposure itself, because they are also one of the best avenues by which to change the social acceptance of smoking itself and to encourage smokers to quit.

Indoor inorganic contaminants

Inorganic gases found in contaminated indoor air include CO_2, CO, SO_2, NO_2, ozone (O_3), hydrogen chloride (HCl), nitrous acid (NHO_2), nitric acid vapour (HNO_3) and radon.

In residences where combustion appliances are present, major indoor sources of CO_2 are fuel combustion and occupants' expired air. Exposure to CO_2 itself is normally not a health concern. However, indoor CO_2 levels can be used as an indicator of the presence of other human bioeffluents in occupied facilities (*e.g.* office building, workshops, theatres, commercial buildings) and also are often employed as an indicator of whether the facilities have adequate ventilation. It is generally accepted that CO_2 levels inside an occupied building, when no combustion source is present, should be no more than 650 ppm above outdoor levels.

Household combustion is mainly responsible for elevated indoor levels of CO, SO_2 and NO_2. If properly operated and maintained, appliances that burn gaseous fossil fuels have high combustion efficiencies and thus generate insignificant amounts of CO. However, the high combustion temperature associated with gas combustion favours the formation of NO_2. Epidemiological studies suggest that long-term exposure to NO_2 (through the use of gas stoves) is a modest risk factor for respiratory illnesses compared to the use of electric stoves[2]. The concern about CO, on the other hand, is primarily for its acute poisoning—*i.e.* its ability to bind strongly to haemoglobins (see Chapter 10). Acute exposure to high levels of CO from improperly operated and maintained appliances is the leading cause of poisoning death in USA and claims many lives worldwide[34].

Ozone (O_3) is a strong oxidizing agent. For this reason, some so-called air purifiers are intended to produce ozone indoors in order to 'purify' the air, on the largely erroneous belief that O_3 may 'kill' odorous contaminants[35]. Besides this intentional indoor O_3 source, use of photocopiers and laser printers may generate O_3. However, the amount of O_3 created through typical office activities will normally not significantly elevate indoor O_3 concentrations. Exposure to O_3 may cause breathing problems, reduce lung function, exacerbate asthma, irritate eyes and nose, reduce resistance to colds and other infections, and speed up ageing of lung tissue. Importantly, indoor O_3, whether penetrating from the outdoors or derived indoors, can drive chemical reactions among chemical species present indoors, generating secondary pollutants that may be of greater health concern compared to primary pollutants[36]. For example, several terpenes are present indoors at concentrations several orders of magnitude higher than their outdoor concentrations because of the wide use of these terpenes as solvents in consumer products (*e.g. d*-limonene contained in lemon scented detergents, α-pinene contained in pine scented paints). Under typical indoor conditions, these terpenes can react with O_3 at a rate faster than, or comparable to, the air exchange rate, to form ultra-fine and fine particles, aldehydes, hydrogen peroxide, carboxylic acids, reactive intermediates, and free radicals including the hydroxyl radical[37–40]. The hydroxyl radical can, in turn, further react with almost all the organics present in the air. It is clear that some of the secondary products from O_3-initiated indoor reactions are strong airway irritants and that exposure to ultra-fine and fine particles have respiratory and cardiovascular effects[41]. Therefore, it is not desirable, and can be problematic, to use ozone generators indoors because of the risk associated both with O_3 exposure and with secondary pollutants resulting from indoor O_3 chemistry.

Acidic gases such as HCl and HNO_3 are highly corrosive to materials and biological tissues. The main sources of HCl include outdoor-to-indoor transport and thermal decomposition of polyvinyl chloride (PVC). During a fire event, high exposure to HCl from PVC burning can be a real health concern. Major sources of indoor HNO_3 include penetration of outdoor HNO_3 formed in photochemical smog episodes and HNO_3 formed indoors *via* reactions involving O_3, NO_2 and water vapour[42]. Normally, indoor HCl and HNO_3 concentrations are lower than their outdoor levels due to their high reactivity (loss to walls). Compared to HCl and HNO_3, HNO_2 is less corrosive. However, indoor levels of HNO_2 can be substantially higher than outdoor levels and indoor concentrations of the other two acidic gases, due to the presence of a strong indoor source of HNO_2 resulting from heterogeneous reactions involving NO_2 and water films on indoor surfaces[40]. Hence, combustion appliances are sources for both NO_2 exposure and HNO_2 exposure. Epidemiological studies of NO_2 health effects should consequently consider the potential confounding effects of HNO_2 and *vice versa*.

Radon-222, an odourless, colourless, and tasteless noble gas, is an isotope produced as a result of the decay of radium-226, which is found in the Earth's crust as a decay product of uranium. Radon has a half-life of 3.8 days and its decay produces a series of short-lived solid-phase daughter products over a few days until lead-210 is produced. These products tend to become associated with airborne particles. The effect of radon decay indoors, therefore, is to make fine particles slightly radioactive and thus expose lung tissue when they deposit during breathing. During the decay process, three types of radiation (α, β, γ) are emitted, all capable of ionizing atoms in living cells, leading to cell damage (see Chapter 13). However, most of the dose is to the respiratory tract and most of this dose comes from the α radiation. Primary sources of radon in buildings are the soil beneath and adjacent to buildings, domestic water supplies (*e.g.* well water) and building materials. In soil, radon moves through air spaces between soil particles. The fraction of radon that enters soil pores depends on the soil type, pore volume and water content. This is why buildings on sandy or gravelly soils typically have higher radon levels than those on clay soils. Movement of soil radon into buildings is primarily through convection—*i.e.* driven by pressures due to indoor–outdoor temperature differences and pressures associated with winds. Hence, substantial seasonal and diurnal variations are typical in indoor radon concentrations. (Concentrations based on short-term measurements should thus be reported carefully because the measurements may not be representative of long-term cumulative exposure.) Highest radon concentrations in water have been found in drilled wells, particularly in areas with granitic bedrock containing uranium. Radon in groundwater is released when temperature is increased, pressure is increased, and/or water is aerated. (Showering provides optimum conditions for radon release from water.)

The primary concern about radon exposure is its potential to cause lung cancer, which has been shown in uranium miners and others. It has been reported that lung cancer risk is dependent on cumulative dose in a linear dose–response fashion. Although some ecologic studies have suggested links between radon exposure and other types of cancers, these have not been confirmed by data obtained from underground miners[43,44]. The US Environmental Protection Agency (EPA) has set a guideline value for indoor radon level of 4 pCi/l measured as an annual average. It has been estimated that this guideline value is exceeded in approximately 6% of US residences and in approximately 30% of residences in Midwestern states. Estimated to cause 7–30 thousand deaths annually in the USA, radon exposure is the second leading cause of lung cancer, following smoking. Much of its damage actually occurs in smokers, however, because of a synergistic relationship between the two risk factors. Because radon contamination is naturally derived and imperceptible to human senses, its risk typically causes less alarm than other cancer-causing substances that may pose a significantly smaller risk.

In addition to gases and airborne particles, airborne fibres present indoors may pose health risks. Due to the known health effects of asbestos fibres, the use of asbestos in US buildings was banned nearly three decades ago. Avoiding asbestos fibres in old buildings is a top priority in indoor air quality management for those buildings. Although newer buildings do not contain asbestos, synthetic vitreous fibres (also referred to as man-made mineral fibres, glass fibres) can be found in spray-applied fireproofing, ceiling tiles, thermal insulation, sound insulation, fabrics, filtration components, plasters and acoustic surface treatments. Health concerns relating to synthetic vitreous fibres arise when erosion of fibres occurs from the parent material into the air stream of buildings. Vitreous fibres have been suspected as possible causes of certain SBS symptoms and may cause irritation to the eye, skin, mucous membranes and respiratory tract[45]. It is believed that cancer risks associated with typical building levels of vitreous fibres are low. Marked as a 'healthier' and greener alternative of vitreous fibres, the cellulose fibre is a recycled product made from newsprint. It contains boric acid for fire retardation. However, little information is available on the health effects of cellulose fibre exposure, although concerns have been raised about their potential to cause irritation to the mucous membranes and the upper respiratory tract. Concerns also arise when comparing cellulose fibres with sawdust that is composed mainly of cellulose, polyoses and lignin, as sawdust is classified as a known human carcinogen by IARC[45].

Indoor organic contaminants

Indoor organic contaminants are conventionally classified by volatility. VOCs have boiling points from <0°C to 240–260°C and are present in the gas phase at typical indoor concentrations. SVOCs have boiling points from 240–260 to 380–400°C, partitioning between the gas phase and the particulate phase under typical indoor conditions. Particulate organic matter comprises components of airborne/suspended dust, with boiling points >380°C.

Indoor organic compounds are released from a variety of building materials such as vinyl tile and coving: compounds include phthalate esters, 2-ethyl-1-hexanol), carpets (4-PCH, 4-VCH, styrene), linoleum (C_5–C_{11} aldehydes and acids), particleboard (formaldehyde, other aldehydes, ketones) and power cables (acetophenone, dimethylbenzyl alcohol). A large variety of consumer products can contribute to indoor levels of VOCs and SVOCs, including paints (texanols, ethylene glycol, pinene, butoxyethoxyethanol), paint thinners (C_7–C_{12} alkanes), paint strippers (methylene chloride), adhesives (benzene, alkyl benzenes), caulks (ketones, esters, glycols) and cleaners (2-butoxyethanol, limonene, 2-butanone). Other indoor sources of VOCs and SVOCs include frying foods (1,3-butadiene, acrolein, PAHs), smoking (nicotine, aldehydes, benzene, PAHs), dry cleaned clothing (tetrachloroethylene), deodorizers (*p*-dichlorobenzene), showering (chloroform), moulds (sesquiterpenes) and pesticides (chlorpyrifos, diazinon, dichlorvos). Due to the presence of these numerous indoor sources, many organic compounds are present indoors at concentrations substantially higher than outdoors. High indoor concentrations, coupled with the fact that people spend a larger fraction of time indoors, often make the outdoor contribution to total personal VOC exposure insignificant or negligible[46,47].

Some VOCs and SVOCs are mutagenic and/or carcinogenic—for example, benzene, styrene, tetrachloroethylene, 1,1,1-trichloroethane, trichloroethylene, dichlorobenzene, methylene chloride and chloroform. Long-term exposure to these compounds is thus a concern in terms of cancer risks. Many VOCs and SVOCs found indoors have the potential to cause sensory irritation (*e.g.* aldehydes) and central nervous system symptoms (*e.g.* pesticides). Available studies also suggest that paternal exposure to VOCs (*e.g.* in chlorinated solvents, spray paints, dyes/pigments, cutting oils) during work, and maternal VOC exposures during pregnancy, are responsible for increased risk of childhood leukaemia[48]. There was a conventional misconception that residential and office buildings have VOC concentrations typically two or more orders of magnitude lower than occupational standards or guidelines, in which case exposure to VOCs in residential and office settings would not be likely to be responsible for acute symptoms. However, there are several significant differences between workplace (industrial) exposure and residential/office exposure:

1. Personal protection (*e.g.* respirators, safety gargles) is normally used in workplace settings, but not in residences or offices. People usually spend longer time in residences (and offices) than in workplaces where VOC levels are high for a certain period of time. Therefore people may receive higher cumulative VOC doses in residences and offices.

2. Residential/office exposure affects a much larger population, including those individuals more susceptible to chemical exposure (*e.g.* asthmatics, children and the elderly).

3. Workplace exposure involves one or more known chemicals. However, residential/office exposure usually involves a complex mixture.

For measurement convenience, indoor VOC mixture is often characterized as total volatile organic compounds (TVOC). Indoor TVOC has been used as an indicator of building healthiness because the prevalence rate of SBS symptoms or complaints was suggested to correlate with TVOC concentration[48]. The effectiveness of using TVOC as an indicator of indoor air quality, however, has been increasingly questioned recently, given that large differences in health effects exist among different individual VOCs and that different indoor environments may comprise distinct VOC mixtures.

Despite efforts made over the past two decades, the aetiology of SBS is still poorly understood. Other terminologies, such as medically unexplained symptoms and non-specific-building-related illness, have appeared in the literature, describing SBS symptoms or similar symptoms and illnesses[49]. Non-specific-building-related illness (NSBRI) is characterized by the following symptoms: mucous membrane irritation (ocular, nasal), headache, fatigue, shortness of breath, rash and odour complaints[50]. Recent studies suggest that indoor pollutant mixtures, along with psycho-physiological factors, may play an important role in causing NSBRI. Epidemiological investigations of building-related health complaints document multiple factors including VOCs, characteristics of the ventilation system, work-related stressors and gender as contributory to symptoms[51]. Laboratory animal studies suggest that formation of irritating particles from reactions between ozone and terpenes could contribute to the health effects of NSBRI[41]. Controlled human exposure studies have consistently shown that several hours exposure to a VOC mixture representative of typical problem buildings, compared to clean air, increased symptoms and complaints about the odour[52–55]. Objective effects of this VOC exposure have been demonstrated for neurobehavioural performance[52], lung function[53] and nasal inflammation[56], in some but not all studies.

Exposure to SVOCs can occur not only *via* inhalation but, because they may be present in settled dust, also *via* ingestion. Ingestion of house dust is an important exposure route for small children who usually have frequent hand-to-mouth activities. It is suggested that exposure to plasticizer chemicals (*e.g.* diethylhexyl phthalate) may be partly responsible for the significant increase in asthma prevalence in the last two decades because the hydrolysis products of diethylhexyl phthalate cause bronchial hyper-reactivity in rats. It is also known that prostaglandin can mediate inflammatory responses such as those that cause asthmatic attacks, yet diethylhexyl phthalate and other phthalic acid esters have a similar chemical structure to prostaglandin. In addition, Finnish scientists reported an association between plastic interior surfaces and bronchial obstruction in young children[48]. Another group of SVOCs commonly found indoors are pesticides. It has been demonstrated that furniture, stuffed toys and carpeting can serve as reservoirs of pesticides for weeks after application and that ingestion exposure can be the most important exposure route in children[57]. It has been suggested that the extensive use of indoor pesticides may contribute to acute symptoms, cancer, immunological effects and reproductive effects. The magnitude of risk is not well known, however, due to the lack of data.

A global health concern in the future?

Given the wide range of indoor chemical contaminants highlighted above, it is clear that concern about indoor air pollution is nearly ubiquitous, although the pollutants of concern in modern buildings are different from those in solid-fuel-burning households. Nevertheless, indoor air pollution fits within the risk transition framework[58], in which the traditional risks of household fuel combustion subside and modern risks from building materials emerge. As shown in the conceptual diagram in Figure 2, the absolute risk declines dramatically as traditional solid fuels are replaced, but the trend of relative risk (percent of total risk) is less clear because overall health risk declines as well. However, total risk is likely to peak in cities in developing countries where there is an overlap between the traditional and modern risks. In Figure 2, a question mark is placed at the arrow end of each curve, presenting questions on the future trends of traditional and modern risks of indoor air pollution. Whether indoor air pollution is a future health concern, therefore, would depend on the future trends of the two curves. Globally the traditional risk curve will approach zero as the use of solid fuels for household energy continues to decline. The modern risk curve, however, is more difficult to predict, but may be a cause for pessimism at least for the near future given that the consumption of tobacco and use of synthetic chemicals and materials do not seem likely to fall. Although society routinely substitutes new materials and chemicals for existing ones, the health effects of most of these are not fully understood before they have been put on the market. Chronic effects of cumulative exposure to low-level multiple chemicals have been recognized recently to be an important area of research. It is vital to understand whether this type of exposure has contributed to the increase in the prevalence of asthma, autism, childhood cancer and medically unexplained symptoms and to the apparent decline in human reproductive function (*e.g.* sperm counts) in some populations.

Unfortunately, health risks associated with indoor air pollution have not, in general, received adequate attention from the regulatory sector, building designers or even health professionals. Although the health and other benefits brought by switching the use of solid fuels to that of liquid or gas fuels, or just introducing improved stoves with chimneys, are tremendous and can well offset their cost[59], only one large-scale implementation effort has been made—the improved stove programme in China, which has introduced nearly 200 million stoves since the early 1980s[60]. Just as clean water and sanitation at the household level have come to have high priority as primary health measures in all poor parts of the world, however, so should clean household fuels and ventilation.

Not well recognized by many observers in this context is the huge difference in what is called the 'intake fraction'—that is, the fraction of material released that actually passes across the population's body barriers and is thus swallowed or inhaled. (Previously called exposure factor, exposure efficiency, and exposure effectiveness, among other terms, it is now agreed to use the term 'intake fraction' for this concept[61].) The intake fractions for typical air pollution sources vary by several orders

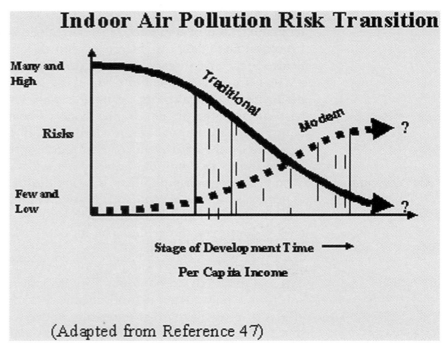

Fig. 2 Conceptualized indoor air pollution risk transition (adapted from Ref. 27).

of magnitude, for example from some 10^{-6} for a large power plant to 5×10^{-2} for a cigarette smoked indoors[62]. In general, one can use the 'rule of one thousand' (*i.e.* a gram of pollution released indoors produces about 1000 times more exposure than one released outdoors), although this obviously varies by situation[63,64]. This means that the place (and time) of release for a pollutant is just as important in determining health effects as its toxicity, which also can vary by several orders of magnitude for different pollutants. Thus, to extend Paracelsus' famous dictum that the 'dose makes the poison', it is also true to say that 'the place makes the poison'[65]. Consequently, small changes in indoor pollutant sources can have the equivalent health benefit as large changes in outdoor sources for the same pollutant.

Nazarof and Weschler state that '[In the USA], health risk assessment constitutes part of the basis for governmental action on environmental matters... Environmental regulations for carcinogens commonly aim to limit the individual lifetime risk of premature death to ~10^{-6}–10^{-5} for contaminants in drinking water and outdoor air. Yet, average lifetime risks of premature death from exposure to indoor air pollutants are at least ~10^{-4}–10^{-3}, and maximum individual risks exceed 10^{-2}. Does it make sense to spend large sums to mitigate environmental risks at a hazardous waste site to 10^{-6} when indoor air quality risks remain unchecked in the range of 10^{-4} to 10^{-2}?' Without the active participation of building designers who appreciate the importance of indoor air quality, it would be impossible to design and construct healthy buildings. Likewise, without the active participation of health scientists and professionals in resolving the puzzles of building-related medically unexplained symptoms and other illnesses, it would be impossible to develop reliable and effective guidelines to prevent excess health risks associated with poor indoor air quality.

Acknowledgement

We are grateful to Drs Charles Weschler, Nancy Fiedler, and Howard Kipen, UMDNJ—Robert Wood Johnson Medical School, for their valuable insights on the chemistry and health effects of indoor air pollutant mixtures. We appreciate Professor Weschler's comment on the manuscript and Robert Harrington's editorial assistance.

References

1. Smith KR, Mehta S, Feuz M. Indoor smoke from household solid fuels. In: Ezzati M, Rodgers AD, Lopez AD, Murray CJL (eds) *Comparative Quantification of Health Risks: Global and Regional Burden of Disease Due to Selected Major Risk Factors*, 2 volumes. Geneva: World Health Organization, 2003; In press
2. Basu R, Samet JM. A review of the epidemiological evidence on health effects of nitrogen dioxide exposure from gas stoves. *J Environ Med* 1999; **1**: 173–7
3. Zhang J, Smith KR, Ma Y et al. Greenhouse gases and other pollutants from household stoves in China: A database for emission factors. *Atmos Environ* 2000; **34**: 4537–49
4. Zhang J, Smith KR. Hydrocarbon emissions and health risks from cookstoves in developing countries. *J Expos Anal Environ Epidemiol* 1996; **6**: 147–61
5. Zhang J, Smith KR. Emissions of carbonyl compounds from various cookstoves in China. *Environ Sci Technol* 1999; **33**: 2311–20
6. Apte MG, Fisk WJ, Daisey JM. Associations between indoor CO2 concentrations and sick building syndrome symptoms in U.S. office buildings: an analysis of the 1994–1996 BASE study data. *Indoor Air* 2000; **10**: 246–57
7. Fisk WJ. Estimates of potential nationwide productivity and health benefits from better indoor environments: An update. In: Spengler JD, Samet JM, McCarthy JF (eds) *Indoor Air Quality Handbook*. New York: McGraw-Hill, 2001; 4.1–4.36
8. Smith KR. Indoor air pollution in developing countries: recommendations for research. *Indoor Air* 2002; **12**: 198–207
9. Tsia SM, Zhang J, Smith KR, Ma Y, Rasmussen RA, Khalil MAK. Characterization of non-methane hydrocarbons emitted from various cookstoves used in China. *Environ Sci Technol* 2003; **37**: 2689–2877
10. WHO. *World Health Report: Reducing Risks, Promoting Healthy Life*. Geneva: WHO, 2002
11. WHO. *World Health Report: Mental Health: New Understandings, New Hope*. Geneva: WHO, 2001
12. Murray CJL, Lopez AD. Global Health Statistics: a compendium of incidence, prevalence, and mortality estimates for over 200 conditions. In: Murray CJL, Lopez AD (eds) *The Global Burden of Disease: A Comprehensive Assessment of Mortality and Disability from Diseases, Injuries, and Risk Factors in 1990 and Projected to 2020*. Cambridge: Harvard University Press, 1996

13 Smith KR, Samet JM, Romieu I, Bruce N. Indoor air pollution in developing countries and acute lower respiratory infections in children. *Thorax* 2000; **55**: 518–32
14 Bruce N, Perez-Padilla R, Albalak R. Indoor air pollution in developing countries: a major environmental and public health challenge for the new millennium. *Bull World Health Organ* 2000; **78**: 1078–92
15 Smith KR, Liu Y. Indoor air pollution in developing countries, Ch 7. In: Samet J (ed.) *The Epidemiology of Lung Cancer*. New York: Marcel Dekker, 1994; 151–84
16 Ge S, Bai Z, Liu W *et al*. The boiler briquette coal versus raw coal: 1. Stack gas emissions. *J Air Waste Manage Assoc* 2001; **51**: 524–33
17 Finkelman RB, Belkin HE, Zheng B. Health impacts of domestic coal use in China. *Proc Natl Acad Sci* 1999; **96**: 3427–31
18 Strachan DP, Cook DG. Health effects of passive smoking. 1. Parental smoking and lower respiratory illness in infancy and early childhood. *Thorax* 1997; **52**: 905–14
19 Strachan DP, Cook DG. Health effects of passive smoking. 4. Parental smoking, middle ear disease and adenotonsillectomy in children. *Thorax* 1998; **53**: 50–6
20 Strachan DP, Cook DG. Health effects of passive smoking. 6. Parental smoking and childhood asthma: longitudinal and case-control studies. *Thorax* 1998; **53**: 204–12
21 Thun M, Henley J, Apicella L. Epidemiologic studies of fatal and nonfatal cardiovascular disease and ETS exposure from spousal smoking. *Environ Health Perspect* 1999; **107** (**Suppl 6**): 841–6
22 Hackshaw AK, Law MR, Wald NJ. The accumulated evidence on lung cancer and environmental tobacco smoke. *BMJ* 1997; **315**: 980–8
23 CalEPA. *Health Effects of Exposure to Environmental Tobacco Smoke*. Smoking and Tobacco Control Monograph 10, Office of Environmental Health Hazard Assessment, California Environmental Protection Agency, 1997
24 Center for Disease Control. The Surgeon General's 1990 Report on the Health Benefits of Smoking Cessation Executive Summary. *MMWR* 1990; **39**: viii–xv
25 Anderson HR, Cook DG. Passive smoking and sudden infant death syndrome: review of the epidemiological evidence [published erratum appears in *Thorax* 1999; **54**: 365–6]. *Thorax* 1997; **52**: 1003–9
26 Windham GC, Eaton A, Hopkins B. Evidence for an association between environmental tobacco smoke exposure and birthweight: a meta-analysis and new data. *Paediatr Perinat Epidemiol* 1999; **13**: 35–57
27 Khuder SA, Simon Jr VJ. Is there an association between passive smoking and breast cancer? *Eur J Epidemiol* 2000; **16**: 1117–21
28 Alcaide J, Altet MN, Plans P *et al*. Cigarette smoking as a risk factor for tuberculosis in young adults: A case-control study. *Tubercle Lung Dis* 1996; **77**: 112–6
29 Altet MN, Alcaide J *et al*. Passive smoking and risk of pulmonary tuberculosis in children immediately following infection. A case-control study. *Tubercle Lung Dis* 1996; **77**: 537–44
30 Venners SA, Xiaobin W, Changzhong C *et al*. Exposure-response relationship between paternal smoking and children's pulmonary function. *Am J Respir Crit Care Med* 2001; **164**: 973–6
31 U.S. Environmental Protection Agency (EPA). Respiratory Health Effects of Passive Smoking: Lung Cancer and Other Disorders. Washington, DC: US Environmental Protection Agency 1992; EPA/600/6–90/006F (NTIS PB 93–134419)
32 Jha P. Curbing the epidemic: governments and the economics of tobacco control. Washington, DC, USA: World Bank, 1999
33 Zhang J, Hu W, Wei F, Wu G, Korn L, Chapman RS. Children's respiratory morbidity prevalence in relation to air pollution in four Chinese cities. *Environ Health Perspect* 2002; **110**: 961–7
34 Girman JR, Chang YL, Hayward SB, Liu KS. Causes of unintentional deaths from carbon monoxide poisonings in California. *West J Med* 1998; **168**: 158–65
35 Weschler CJ. Ozone in indoor environments: concentration and chemistry. *Indoor Environ* 2000; **10**: 269–88
36 Weschler CJ. Reactions among indoor air pollutants. *Sci World* 2001; **1**: 443–57
37 Fan Z, Lioy P, Weschler C, Fiedler N, Kipen H, Zhang, J. Ozone-initiated reactions with volatile organic compounds under a simulated indoor environment. *Environ Sci Technol* 2003; **37**: 1811–1821
38 Li SH, Turpin BJ, Shields HC, Weschler CJ. Indoor hydrogen peroxide derived from ozone/d-limonene reactions. *Environ Sci Technol* 2002; **36**: 3295–302

39 Weschler CJ, Shields HC. Production of the hydroxyl radical in indoor air. *Environ Sci Technol* 1996; **30**: 3250–8
40 Wainman T, Zhang J, Weschler C, Lioy PJ. Ozone and limonene in indoor air: a source of submicron particle exposure. *Environ Health Perspect* 2000; **108**: 1139–45
41 Wolkoff P, Clausen PA, Wilkins CK, Nielsen GD. Formation of strong airway irritants in terpene/ozone mixtures. *Indoor Air* 2000; **10**: 82–91
42 Weschler CJ, Brauer M, Koutrakis P. Indoor ozone and nitrogen dioxide: a potential pathway to the generation of nitrate radicals, dinitrogen pentaoxide, and nitric acid indoors. *Environ Sci Technol* 1992; **26**: 179–84
43 US National Academy of Sciences, Committee on Health Risks of Exposure to Radon (BEIR VI). Health effects of exposure to radon. Washington, DC: National Academy Press, 1999
44 Darby SC, Whitley E, Howe GR *et al.* Radon and cancers other than lung cancer in underground miners: A collaborative analysis of 11 studies. *J Natl Cancer Inst* 1995; **87**: 378–84
45 Vallarino J. Fibers. In: Spengler JD, Samet JM, McCarthy JF (eds) *Indoor Air Quality Handbook*. New York: McGraw-Hill, 2001; 37.1–37.22
46 Ott WR, Roberts JW. Everyday exposure to toxic pollutants: environmental regulations have improved the quality of outdoor air. But problems that persist indoors have received too little attention. *Sci Am* 1998; **278**: 86–91
47 Wallace L. A decade of studies of human exposure: what have we learned? *Risk Anal* 1993; **13**: 135–43
48 Godish T (ed.) Organic contaminants. In: *Indoor Environmental Quality*. Boca Raton, FL: CRC Press, 2000; 95–141
49 Kipen HM, Fiedler N. The role of environmental factors in medically unexplained symptoms and related syndromes: conference summary and recommendations. *Environ Health Perspect* 2002; **110**: 591–5
50 Fiedler N, Zhang J, Fan Z *et al*. Health effects of a volatile organic mixture with and without ozone. *Proceedings of Indoor Air* 2002
51 Ryan CM, Morrow KA. Dysfunctional buildings or dysfunctional people: an examination of the sick building syndrome and allied disorders. *J Consult Clin Psychol* 1992; **60**: 220–4
52 Molhave L, Bach B, Pedersen OF. Human reactions to low concentrations of volatile organic compounds. *Environ Int* 1986; **12**: 167–75
53 Kjaergaard SK, Moelhave L, Pedersen OF. Human reactions to a mixture of indoor air volatile organic compounds. *Atmos Environ* 1991; **25A**: 1417–26
54 Hudnell HK, Otto DA, House DE, Moelhave L. Exposure of humans to a volatile organic mixture. II. Sensory. *Arch Environ Health* 1992; **47**: 31–8
55 Prah JD, Case MW, Goldstein GM. Equivalence of sensory responses to single and mixed volatile organic compounds at equimolar concentrations. *Environ Health Perspect* 1998; **106**: 739–44
56 Koren HS, Devlin RB. Human upper respiratory tract responses to inhaled pollutants with emphasis on nasal lavage. *Ann N Y Acad Sci* 1992; **641**: 215–24
57 Gurunathan S, Robson M, Freeman N, Buckley B, Roy A, Meyer R, Bukowski J, Lioy PJ. Accumulation of chlorpyrifos on residential surfaces and toys accessible to children. *Environ Health Perspect* 1998; **106**: 9–16
58 Smith KR. Development, health, and the environmental risk transition, Ch. 3. In: Shahi G, Levy BS, Binger A, Kjellstrom T, Lawrence R (eds) *International Perspectives in Environment, Development, and Health*. New York: Springer, 1997; 51–62
59 Smith KR. In praise of petroleum? *Science* 2002; **298**: 1847
60 Goldemberg J, Reddy AKN, Smith KR, Williams RH. Rural energy in developing countries, Ch. 10. In: Goldemberg J *et al* (eds) *World Energy Assessment*. New York: United Nations Development Programme, 2000; 367–89
61 Bennett DH, McKone TE, Evans JS *et al*. Defining intake fraction. *Environ Sci Technol* 2002; 207A–11A
62 Smith KR. Fuel combustion, air pollution exposure, and health: the situation in developing countries. *Annu Rev Energy Environ* 1993; **18**: 529–66
63 Smith KR. Air pollution: assessing total exposure in the United States. *Environ* 1988; **30**: 10–5
64 Smith KR. Air pollution: assessing total exposure in developing countries. *Environ* 1988; **30**: 16–20, 28
65 Smith KR. Place makes the poison. *J Expos Anal Environ Epidemiol* 2002; **12**: 167–71

Asthma: environmental and occupational factors

Paul Cullinan and **Anthony Newman Taylor**

Department of Occupational and Environmental Lung Disease, Imperial College, London, UK

Asthma is in several ways a difficult disease to study. Generally arising in childhood, its pattern is often one of remission and relapse; at any point there are difficulties in translating its characteristic, clinical features into an operational definition. Geographical and temporal patterns in its distribution – whereby the disease appears to have increased in frequency in more 'westernised' countries – suggest strong environmental determinants in its causation although there are, too, undoubted and important genetic influences on both its incidence and presentation. Recent aetiological research has concentrated on the function of allergen exposure or on the role of early-life microbial contact that may regulate the development of a range of childhood allergies, including asthma. To date the 'hygiene hypothesis' offers the most efficient explanation for the distribution of the disease in time and place although convincing evidence for it remains elusive.

Definitions of asthma

More easily described than defined, asthma is a disease characterized by reversible airflow obstruction giving rise to intermittent chest tightness, wheeze and breathlessness. Episodes of these symptoms may occur spontaneously or as a result of 'airway hyperresponsiveness', a cardinal feature of the disease in which acute airway narrowing is easily elicited by non-specific stimuli such as exercise, cold air or inhalation of airway irritants including pharmacological agents such as histamine or metacholine. The underlying pathology of asthma is a Th2 lymphocyte-driven eosinophilic airway inflammation frequently accompanied by an IgE-associated allergy to inhaled allergens. When, as in most instances, there is continual exposure to the causative allergen(s), the inflammatory response is persistent.

In western countries, where asthma has been most closely studied, the relevant allergens are those derived from house dust mite species, from domestic pets (particularly cats), from grass or tree pollens or—especially in the USA—cockroaches in the home. In the great majority of asthmatic patients, immune sensitization to one or more of these allergens may be detected by skin prick testing or by the measurement of raised levels of specific IgE antibodies in serum. This state of sensitization is known as

Correspondence to: Paul Cullinan, Department of Occupational and Environmental Lung Disease, Imperial College (NHLI), London SW3 6LR, UK. E-mail: p.cullinan@imperial.ac.uk

'atopy'. Asthma is frequently accompanied by hayfever, an identical immune response to pollens in the nose.

Measuring asthma

In the individual, asthma is diagnosed through a combination of characteristic symptoms and the demonstration of labile or reversible airflow obstruction. Several techniques for the latter are available. Most simply, they include the measurement of airflow (usually forced expiratory volume in the first second of exhalation, or peak expiratory flow) repeatedly over time, or in response to a bronchodilating drug. Other methods, designed to measure airway hyperresponsiveness, involve the measurement of airflow after inhalation of increasing concentrations of non-specific bronchial irritants such as histamine, metacholine, cold air or even exercise.

The extension of these methods to the study of asthma epidemiology is not straightforward. Appropriate tools for making measurements in populations need to combine high diagnostic performance with simplicity and acceptability. Questionnaires are universally used but their value is compromised by different cultural and temporal understandings of specific symptoms and, because asthma is a disease which generally develops in childhood, by the necessity of using proxy (parental) responses. The English word 'wheeze', derived from the old Norse word *hvæsa* ('to hiss'), translates awkwardly into many other languages and not at all, for example, into German or Greek. The doubling in the prevalence of reported wheeze (but not cough) among young children in Leicester over just 8 years[1] may at least in part be attributable to a shift in reporting behaviour in addition to any changes in aetiological factors. The switch in the sex distribution of asthma, from a predominantly male disease in early childhood to one which is equally common in teenage boys and girls, coincides with the age at which questionnaires are no longer completed by parents on behalf of their children; the strongest risk factor for a diagnosis of asthma being missed in childhood is female sex[2]. Despite these misgivings, and beyond the first 2 years of life, there are very few conditions (other than obesity) which commonly give rise to shortness of breath and wheeze on exercise in childhood. The same cannot be claimed for adults, in whom the responses to respiratory questionnaires are less specific.

In population studies, questionnaires are frequently accompanied by other measurement techniques. Strictly speaking these may not be measuring the same process as asthma. Although most asthmatics have airway hyperresponsiveness, this is not always the case all the time, and a considerable proportion of the population with demonstrable hyperresponsiveness are asymptomatic[3]. Similarly with atopy, which is commonly studied. It should not be assumed that the epidemiologies of these conditions are identical to that of asthma although they may show many similarities. For example, in parts of eastern and southern Europe, and in China, there appears to be a much greater divergence between atopy and asthma than there is in western Europe and Australasia[4,5]. In these latter countries, the proportion of asthma which can, epidemiologically speaking, be 'attributed' to atopy is no more than about 45%[6].

Distributions

Age

In the great majority of cases, asthma begins in childhood[7]. This observation should not obscure the fact that a high proportion of children wheeze before the age of two, commonly in response to a respiratory tract infection. Most however will not progress to typical asthma; indeed, they may be less likely than otherwise to do so (Martinez *et al*[8]— and see below). Asthma arising in adulthood is far less frequent with an annual incidence estimated to be around 1–2/1000[9]. A proportion of this may not strictly be incident disease but the failure to recollect childhood symptoms. A particular exception is occupational asthma, which is caused by an agent inhaled at work. This subtype of asthma, not uncommon in several industries, is discussed in more detail below.

Cross-sectional population surveys reflect the age of incidence of most asthma with the highest frequencies in childhood; and also the tendency for the disease to remit, often during teenage years, and perhaps an additional cohort effect (Fig. 1).

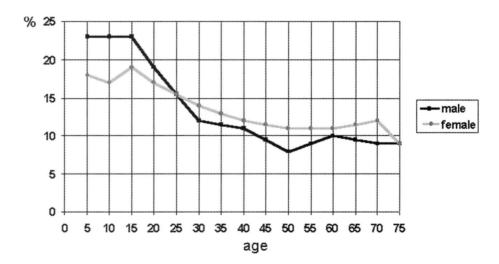

Fig. 1 Prevalence of current asthma in England, 1995/6[10].

For the most part, atopy and hayfever also arise in childhood although neither may be clearly established before the age of about 8 years.

Sex

As above, asthma and hayfever are each more common in boys than girls. This is unlikely to be entirely explicable by social determinants since there is a similar sex distribution for the more objective measure of atopy. To what extent the remaining determinants are 'genetic' or 'environmental' is unknown.

Place

Nowhere are the difficulties in disease measurement more obvious than in studies that compare the frequencies of asthma between different countries or at different time points. Recent global surveys of asthma and associated diseases, using as far as possible standardized instruments, have confirmed striking geographical patterns of childhood disease prevalence. In Europe, wheeze, seasonal rhinitis ('hayfever'), childhood eczema and atopy are all more common in western and northern countries (particularly the UK) than in those to the south or east of the continent (Fig. 2).

More broadly, the same diseases are more prevalent in New Zealand, Australia and North America than in central or southern Asia or Africa. The apparent predilection for predominantly English speaking countries is remarkable and may reflect complex genetic/cultural/linguistic influences. Atopy and hayfever—and to a lesser extent asthma—may vary at a more local level. In 1989, for example, they were more common among children brought up in Munich than those brought up in Leipzig, the two communities being either side of the old Iron Curtain[12]. Elsewhere, these diseases may be less frequent in rural than urban communities[13,14], although the differences are not consistent. Migrant studies of asthma are relatively few but there is some evidence that movement to a more 'western' environment is associated with an increase in risk[15].

Time

It is widely acknowledged that asthma and hayfever are more common now than previously; the same may be true for atopy although the evidence is more sparse. Studies of this phenomenon have relied exclusively on two-point comparisons of similar populations at different time points. To a large extent, reported increases must reflect changes in ascertainment but repeated surveys using more 'objective' methods suggest that this is

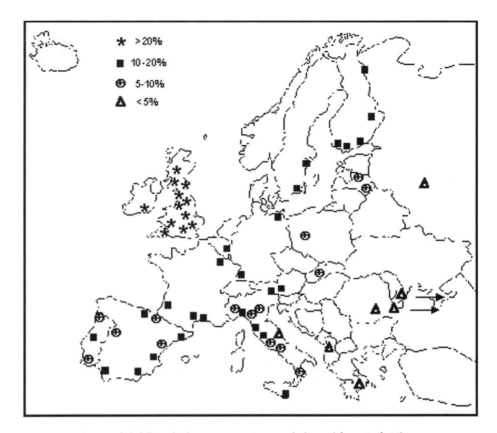

Fig. 2 Prevalence of childhood wheeze across Europe (adapted from Ref. 11).

not entirely the case[16]. Although the magnitude of any increases is fairly small (Fig. 3) the high frequency of asthma in western communities renders them important. It is unclear, however, when they began.

The systematic assessment of recruits to the Finnish army suggests a process beginning in the early 1960s[17], but other data are very sparse. The experience of city populations in the former east Germany, following unification, suggests that once prompted incidence rates of hayfever and atopy can rise very rapidly[18]. There is some evidence that in western Europe the rate of increase may have tailed off if not stopped altogether[19].

Family size

One of the most consistent findings of the descriptive epidemiology of childhood allergies is that relating to family size or birth order. Children with no or few siblings have a strikingly increased risk of developing hayfever, eczema, atopy or—less clearly—asthma[20]. Commonly the

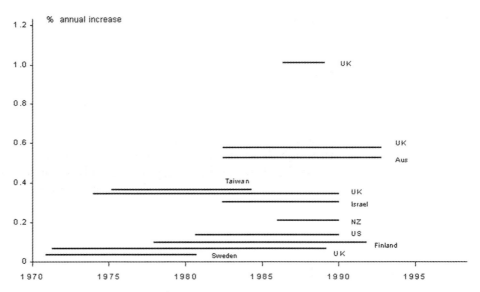

Fig. 3 Annual increases in asthma symptoms in paired cross-sectional surveys. The ends of each line mark the first and second survey years (adapted from Magnus and Jaakkola[16]).

effect is reported to be confined to older siblings but this is not universal. Male siblings may be more 'protective' than female[21,22]. There may also be important interactions with parental allergy with the effect more strongly seen in children with non-allergic parents[22].

Family history

Of similar strength and consistency are the observations that children with allergic or asthmatic parents are more likely to develop the same diseases. This seems to be especially the case for children with allergic mothers. There is little doubt that genetic factors influence the risk of asthma and that there are important interactions between these and relevant environmental aetiologies. A detailed description of asthma genetics is beyond the scope of this chapter—and in any case the subject remains full of uncertainties—but there is no doubt that the disease has multiple, interacting and probably competing genetic influences determining not only susceptibility but also disease expression and response to treatment. With increasing interest in genome scanning there is enormous scope for study, although arguably a more cautious approach might be more fruitful; at this stage very little is known about the interactions between environmental factors and the most basic measures of genetic disposition such as 'maternal asthma', 'parental atopy', *etc*. However, the disease has a relatively low concordance between monozygotic twins, suggesting that genetic factors cannot fully explain its distribution.

Environmental explanations

The variability in asthma frequency both geographically and temporally suggests one or more powerful environmental influences on its aetiology. Its young age of onset indicates that early-life or even pre-natal factors are of especial importance in disease inception if not persistence. Here it is worth emphasizing the distinction between factors that *induce* the disease from those which, once it is present, *provoke* its manifestations. Although these factors may not be wholly distinct, and in some instances cannot be distinguished, they are usefully studied separately and by the use of appropriate study designs. Unfortunately, the epidemiological literature on asthma is replete with hastily designed, generally cross-sectional studies the results of which are not easily interpreted.

Early allergen exposure

A clear example of the scope for confusion lies in the study of the role of allergen exposures. As explained above most asthmatic patients in western countries display sensitization (atopy) to one or more common aeroallergens. This state arises in early childhood. Asthmatics sensitized in this way generally have symptoms after inhalation of the relevant allergen(s) and are more symptomatic in high-exposure environments, with resolution of symptoms and airway hyperresponsiveness when exposure can be avoided. It does not necessarily follow, however, that high allergen exposure—in early life—gives rise to asthma or indeed specific sensitization. Nonetheless, albeit with varying conviction, it has long been held that high exposures, particularly to domestic allergens (house dust mite or cat), are an important risk factor for asthma; and furthermore that exposure thresholds for both sensitization and asthma exist.

The 'allergen hypothesis' was built very largely around information derived from cross-sectional or case-referent studies, supplemented by a single, very small prospective study of highly-selected infants[23]. It is obviously difficult to determine valid estimates of past exposures, particularly using retrospective techniques, and more recent studies have indicated that the relationships between allergen concentrations experienced in infancy and subsequent sensitization and asthma, if they exist, are far from straightforward[24,25]. It has even been suggested that especially high exposures to some allergens exert a tolerizing effect although this requires confirmation[26].

A particular example of the potential role of inhaled allergens is afforded by *occupational asthma*, a form of the disease that arises in adult life from an agent inhaled at work. Some 300 workplace agents have been described as capable of inducing asthma; it is believed that in most cases they act by initiating an immune response very similar to that seen in asthmatic children. Thus the disease is accompanied by evidence of specific sensitization to the causative allergen—universally where the allergen is biological but less commonly where chemical agents are implicated. In western countries, most cases of occupational asthma are attributable to one of around a dozen workplace allergens among which isocyanates (commonly used as hardening agents in spray paints), wheat flour (bakers), proteins excreted by laboratory animals (medical and pharmaceutical research), enzymes (used widely in industry, particularly in detergent manufacture and bakeries) and solder rosin (electronic assembly) are prominent.

Several studies of work populations have demonstrated that the risks of specific sensitization and, probably, occupational asthma rise with increasing levels of exposure to causative allergens[27,28]. Exposure in this sense exerts a more powerful effect than genetic susceptibility[29]. Furthermore, reductions in factory exposure have been shown to lead to reductions in disease incidence[30,31]. The detailed nature of such exposure–response relationships, however, is unknown and it has proved impossible so far to establish threshold levels, below which there is no observable risk, with any certainty.

It is not clear why allergen exposure should apparently be more important in the development of occupational asthma than in the development of childhood asthma. Possibly the discrepancy can be explained by the extremely high exposures frequently encountered by factory employees. Alternatively it may reflect differences in response to allergens between a mature and a relatively immature immune system. More broadly, whereas high rates of occupational asthma in a workforce can generally be attributed to local, high allergen exposures—and whereas outbreaks of community asthma certainly occur for similar reasons[32,33]—it is difficult to explain the wider temporal and geographical patterns of childhood asthma on the basis of an 'allergen hypothesis' alone. Even if exposures of western children to some domestic allergens have risen over recent years, and the evidence for this is weak, it seems improbable that this is the case for all allergens, especially pollen types.

Hygiene hypothesis

If, however, the underlying *susceptibility* to inhaled allergens has increased across a population then higher levels of allergen exposure might not be necessary for a rising incidence of sensitization, asthma and hayfever. A mechanism of broadly increased susceptibility, or loss of some protective factor(s), at a population level and somehow consequential on 'westernization' would be a parsimonious explanation for many of the observed differences in distribution of these diseases.

The 'hygiene hypothesis' posits that the lack, or loss, of early life exposure to one or more microbial agents increases the risk of allergic disease both at an individual and community level. The nature of the protective agent(s) is unspecified except that this should be capable of eliciting a host immune response. Early attempts to provide biological plausibility to the theory were encouraged by the discovery in inbred experimental mice of mutually exclusive maturation of two classes of T-helper (Th) lymphocytes distinguished by the different profiles of their secreted cytokines: Th1 lymphocytes produce IL-2 and γ-interferon in response to microbial stimulation whereas Th2 cells produce cytokines typically associated with allergic responses, IL-5 and IL-4. This, for reasons outlined below, is almost certainly an oversimplification.

The observation that family size or birth order is a determinant of the development of childhood allergies is often cited as evidence in support of the 'hygiene hypothesis'. First-born children or those from small families, it is argued, undergo little microbial exposure in early life and are thus relatively unprotected from allergic responses. That this is actually the case is difficult to demonstrate, at least in contemporary western communities; it may be that the differences between first-born and subsequent children are very subtle. For example, the relationship between family size and atopy among Italian military recruits[34] was lost in those men with serological evidence of prior hepatitis A infection (a 'marker' of oro-faecal infective load in early childhood).

The effects of birth order may, however, be more complex than simply a reflection of differential rates of early infection. There is, for example, some evidence that the apparently protective effects of a large sibship interact with parental allergy[35]. Cord blood IgE concentrations decrease with birth order[36] and cord mononuclear cell proliferative responses to house dust mite are greater in first-born children although it is not clear whether this is related to maternal age[37]. Furthermore, mothers who have had several children appear to have a lower prevalence of atopy than those who have only one[38], a pattern which is independent of maternal age. Other diseases, some clearly not dependent on a Th2-type immune response, are also more common among first-born children[39,40].

Many other sources of evidence in support of an hygiene hypothesis have been sought. These include both direct and indirect measures of early microbial exposure. Single studies of specific agents have examined the role of early measles[41], tuberculosis[42] and schistosomaisis[43], in each case, their findings suggesting a protective role for such infections. These findings have not been replicated and they derive from studies of populations with very different childhood exposures from those in most of Europe. Mycobacterial exposure, both through BCG vaccination and to atypical species, has been further examined in several different settings without any clear evidence of a protective effect[44–46].

Several indirect markers of exposure have been studied. Perhaps the most interesting of these is hepatitis A serology, used in this context as a marker of an 'unhygienic' childhood. The inverse relationship between positive serology and a variety of allergic outcomes among Italian soldiers mentioned above was not replicated in studies of Greek and UK communities[47,48]. A large study of 34,000 US residents, however, showed strongly reduced risks of hayfever, asthma and atopy among subjects seropositive for hepatitis A, *Toxoplasma gondii* or herpes simplex[49]. Analyses by birth cohort suggested that the frequencies of asthma and hayfever had increased between 1920 and 1970 but only among those who had no serological evidence of past infection. However, confounding by birth order may account for all or part of the associations observed.

Other studies have examined the use of daycare[50], anthroposophic lifestyles[51] and antibiotic prescriptions in early life. Studies of this last have indicated, fairly consistently, that antibiotic use in early childhood is associated with an increased risk of asthma[52–54]; it is likely, however, that this is an example of protopathic bias in which the early manifestations of asthma, being indistinguishable from respiratory infection, are themselves the reason for antibiotic prescription.

A childhood spent on a farm may incur unusually high microbial exposures. Children whose parents are farmers tend to have reduced prevalences of atopy (in particular), hayfever and asthma when compared with neighbouring children whose parents have different occupations. These observations have been made largely in communities farming cattle[55–57], and may not be applicable to other types of farming[47]. A 'dose–response' relationship has been reported suggesting that especially intense farm exposure, and perhaps very early exposures, are particularly protective[58]. The precise nature of the protective exposure has not been elucidated but endotoxin, atypical mycobacteria and the drinking of unpasteurized milk have all been suggested. A long-term process of subtle genetic selection, whereby atopic diseases have been excluded from farming families, cannot be entirely ruled out.

If, as is suggested in western communities, there has been a population-wide increase in susceptibility to allergic diseases such as asthma then it might be expected that there would be a similar reduction in diseases with contrary immune mechanisms. Similarly, it might be anticipated that at an individual level, Th1- and Th2-mediated diseases would be inversely correlated. Neither of these appears to be the case. The prevalence of insulin-dependent diabetes, for example, has increased over more or less the same time scale as asthma; indeed the two diseases share several epidemiological distributions and, at an ecological level, are closely correlated both in Europe and elsewhere[59,60]. Across a large US population there were no inverse relationships between atopy or asthma and three Th1-type diseases, thyroid disease, diabetes and rheumatoid arthritis[61].

These findings indicate that explanations based simply around the mutually exclusive maturation of Th1/Th2 lymphocyte responses in early childhood are an oversimplification. Within the immune model, a more plausible mechanism would be that some facet of western development (perhaps related to early hygiene) has resulted in a reduced immunological regulation and the development of inappropriate responses to both external and internal allergens. The mechanisms of this dysregulation have not been studied in any detail although a role for interleukin 10 has been suggested[43].

Respiratory irritants

As a result of their underlying airway hyperresponsiveness, patients with asthma often experience symptoms when exposed to a wide variety of non-allergenic, respiratory irritants. As with allergen exposures, it may be helpful to distinguish the provocation of pre-existing asthma in this way from the induction of new asthma. Such a distinction is, however, to some extent unreal, or at least unhelpful in clinical or public-health terms since the outcome (apparent asthma) is the same and arguably, preventive efforts ought to be aimed at the most important proximal cause of the disease. These issues come to the fore when considering, again, asthma at work and also the relationship between asthma and 'air pollution' in general.

As described above, asthma may arise from an immune response to an agent inhaled at work ('occupational asthma'). Alternatively, asthmatic symptoms may be provoked, in an asthmatic employee, by respiratory irritants encountered at work (often termed 'work-exacerbated asthma'). In clinical practice the distinction can generally be made using tests of an agent-specific immune reaction. Most regulatory bodies and all legislatures consider occupational asthma to be more important in terms of prevention and compensation. However it has been argued that this focus is too narrow[62]. Community surveys of asthma and occupation suggest that approximately 10% of all adult asthma can be attributed to work[63]. Issues of ascertainment aside, this figure is higher than most estimates of occupational asthma frequency indicating that the larger proportion of the apparent asthma-work association is explained by 'work-related' disease. Possibly, the estimate of 10% is too high. Derived from occupation-specific risk estimates for asthma it does not take into account those jobs where the risks of asthma are apparently reduced; nor the distinction between occupations that give rise to asthma from those in which people with asthma find themselves employed. Several of the high-risk categories identified in surveys such as these are not obviously associated with high irritant exposures; and several of the occupations widely recognized as having a high risk of occupational asthma are not represented at all. Nonetheless, exposure to respiratory irritants, even if only at work, is a public health issue of considerable importance.

Very high exposures at work to a respiratory irritant may induce a condition that is physiologically indistinguishable from asthma. First described in 1989 as 'reactive airways dysfunction syndrome'[64], this is now more commonly known as irritant-induced asthma. The term is generally restricted to those without pre-existing asthma who have had a single exposure at toxic levels to an established irritant such as acetic acid. Although useful clinically, this may be too narrow a definition ignoring the likelihood that bronchial hyperreactivity may enhance the response to high-dose irritants; and the possibility that repeated workplace exposures, at lower concentrations, might also induce disease—so-called 'low-dose reactive airways syndrome'[65]. This latter is a difficult area to study and to date there is no convincing evidence that repeated workplace irritant exposures at low doses induce asthma in the absence of any high-dose exposure.

More widespread than exposures in the workplace are, of course, those to potentially irritant material in the general environment. Traffic-derived pollutants have received most attention in this respect and have generated a vast body of published literature. There is little doubt, from both clinical studies and the examination of peak pollution episodes, that at high concentrations, pollutants such as ozone, nitrogen oxides or diesel particulate matter can provoke symptoms in an asthmatic individual; and that they may also increase the bronchial response to inhaled allergen. Analogous to the situation at work, it is more doubtful that they have this effect, in any widespread manner, at the pollution levels generally found in those countries where asthma is endemic. Again this is a difficult area of study and the effects may be too small to be measurable with the small sample sizes available for direct clinical study or with the relatively crude techniques of epidemiology. Even if this is the case, small effects spread widely may produce large attributable risks. Expert consensus suggests that traffic pollutant exposures in general are not an important cause of asthma[66]. In just one study has a traffic-related pollutant (probably ozone) been suggested to induce new cases of the disease[67].

Conclusions

There are clear indications, largely from epidemiological studies, that environmental factors are of high importance in both the development and persistence of asthma. The prevalence, and probably incidence, of the disease have a wide geographical variation, even at a relatively local level, and have increased apparently in accompaniment to the processes of 'western development'. Arguably the most parsimonious explanation lies broadly within the hygiene hypothesis; that the lack or loss of microbial immune stimulation in early childhood leads to the development of poorly regulated T cell responses to both external and internal antigenic stimuli.

However attractive, the goal of parsimony should not deflect from the probability that the aetiology of asthma is complex and it is unlikely that a single explanation will suffice. The determinants of the sibling effect, for example, are probably not confined to differential rates of infection in childhood but may also encompass shifts in maternal immune responses with both age and repeated pregnancies. If it is the case that increases in the frequency of asthma and other allergic diseases have been accompanied by increases in Th1-mediated diseases then a single 'hygiene' explanation implies a major and widespread shift in early childhood immune regulation. If such a shift has already been completed in western communities then it may no longer be possible to identify it, particularly in those with a genetic disposition to asthma. More fruitful may be further study in communities where the epidemic is in its early stages.

References

1. Kuehni CE, Davis A, Brooke AM, Silverman M. Are all wheezing disorders in very young (preschool) children increasing in prevalence? *Lancet* 2001; **357**: 1821–5
2. Siersted HC, Boldsen J, Hansen HS, Mostgaard G, Hyldebrandt N. Population based study of risk factors for underdiagnosis of asthma in adolescence: Odense schoolchild study. *BMJ* 1998; **316**: 651–5
3. Josephs LK, Gregg I, Holgate ST. Does non-specific bronchial responsiveness indicate the severity of asthma? *Eur Respir J* 1990; **3**: 220–7
4. Leung R, Lai CK. The importance of domestic allergens in a tropical environment. *Clin Exp Allergy* 1997; **27**: 856–9
5. Priftanji A, Strachan D, Burr M, Sinamati J, Shkurti A, Grabocka E *et al*. Asthma and allergy in Albania and the UK. *Lancet* 2001; **358**: 1426–7
6. Pearce N, Pekkanen J, Beasley R. How much asthma is really attributable to atopy? *Thorax* 1999; **54**: 268–72
7. Yunginger JW, Reed CE, O'Connell EJ, Melton III LJ, O'Fallon WM, Silverstein MD. A community-based study of the epidemiology of asthma. Incidence rates, 1964–1983. *Am Rev Respir Dis* 1992; **146**: 888–94
8. Martinez FD, Wright AL, Taussig LM, Holberg CJ, Halonen M, Morgan WJ. Asthma and wheezing in the first six years of life. The Group Health Medical Associates. *N Engl J Med* 1995; **332**: 133–8
9. De Marco R, Locatelli F, Cerveri I, Bugiani M, Marinoni A, Giammanco G. Incidence and remission of asthma: a retrospective study on the natural history of asthma in Italy. *J Allergy Clin Immunol* 2002; **110**: 228–35
10. Prescott-Clarke P, Primatesta P (eds) *Health Survey for England 1996*. HMSO, 1998
11. Worldwide variation in prevalence of symptoms of asthma, allergic rhinoconjunctivitis, and atopic eczema: ISAAC. The International Study of Asthma and Allergies in Childhood (ISAAC) Steering Committee. *Lancet* 1998; **351**: 1225–32
12. von Mutius E, Fritzsch C, Weiland SK, Roll G, Magnussen H. Prevalence of asthma and allergic disorders among children in united Germany: a descriptive comparison. *BMJ* 1992; **305**: 1395–9

13 Aberg N. Asthma and allergic rhinitis in Swedish conscripts. *Clin Exp Allergy* 1989; **19**: 59–63
14 Van Niekerk CH, Weinberg EG, Shore SC, Heese HV, Van Schalkwyk J. Prevalence of asthma: a comparative study of urban and rural Xhosa children. *Clin Allergy* 1979; **9**: 319–24
15 Waite DA, Eyles EF, Tonkin SL, O'Donnell TV. Asthma prevalence in Tokelauan children in two environments. *Clin Allergy* 1980; **10**: 71–5
16 Magnus P, Jaakkola JJ. Secular trend in the occurrence of asthma among children and young adults: critical appraisal of repeated cross sectional surveys. *BMJ* 1997; **314**: 1795–9
17 Haahtela T, Lindholm H, Bjorksten F, Koskenvuo K, Laitinen LA. Prevalence of asthma in Finnish young men. *BMJ* 1990; **301**: 266–8
18 von Mutius E, Weiland SK, Fritzsch C, Duhme H, Keil U. Increasing prevalence of hay fever and atopy among children in Leipzig, East Germany. *Lancet* 1998; **351**: 862–6
19 Ronchetti R, Villa MP, Barreto M, Rota R, Pagani J, Martella S *et al.* Is the increase in childhood asthma coming to an end? Findings from three surveys of schoolchildren in Rome, Italy. *Eur Respir J* 2001; **17**: 881–6
20 Karmaus W, Botezan C. Does a higher number of siblings protect against the development of allergy and asthma? A review. *J Epidemiol Community Health* 2002; **56**: 209–17
21 Strachan DP, Harkins LS, Golding J. Sibship size and self-reported inhalant allergy among adult women. ALSPAC Study Team. *Clin Exp Allergy* 1997; **27**: 151–5
22 Svanes C, Jarvis D, Chinn S, Burney P. Childhood environment and adult atopy: results from the European Community Respiratory Health Survey. *J Allergy Clin Immunol* 1999; **103**: 415–20
23 Sporik R, Holgate ST, Platts-Mills TA, Cogswell JJ. Exposure to house-dust mite allergen (Der p I) and the development of asthma in childhood. A prospective study. *N Engl J Med* 1990; **323**: 502–7
24 Lau S, Illi S, Sommerfeld C, Niggemann B, Bergmann R, von Mutius E *et al.* Early exposure to house-dust mite and cat allergens and development of childhood asthma: a cohort study. Multicentre Allergy Study Group. *Lancet* 2000; **356**: 1392–7
25 Pearce N, Douwes J, Beasley R. Is allergen exposure the major primary cause of asthma? *Thorax* 2000; **55**: 424–31
26 Platts-Mills T, Vaughan J, Squillace S, Woodfolk J, Sporik R. Sensitisation, asthma, and a modified Th2 response in children exposed to cat allergen: a population-based cross-sectional study. *Lancet* 2001; **357**: 752–6
27 Cullinan P, Cook A, Gordon S, Nieuwenhuijsen MJ, Tee RD, Venables KM *et al.* Allergen exposure, atopy and smoking as determinants of allergy to rats in a cohort of laboratory employees. *Eur Respir J* 1999; **13**: 1139–43
28 Heederik D, Houba R. An exploratory quantitative risk assessment for high molecular weight sensitizers: wheat flour. *Ann Occup Hyg* 2001; **45**: 175–85
29 Jeal H, Draper A, Jones M, Harris J, Welsh K, Taylor AN *et al.* HLA associations with occupational sensitization to rat lipocalin allergens: a model for other animal allergies? *J Allergy Clin Immunol* 2003; **111**: 795–9
30 Allmers H, Schmengler J, Skudlik C. Primary prevention of natural rubber latex allergy in the German health care system through education and intervention. *J Allergy Clin Immunol* 2002; **110**: 318–23
31 Cathcart M, Nicholson P, Roberts D, Bazley M, Juniper C, Murray P *et al.* Enzyme exposure, smoking and lung function in employees in the detergent industry over 20 years. Medical Subcommittee of the UK Soap and Detergent Industry Association. *Occup Med (Lond)* 1997; **47**: 473–8
32 Anto JM, Sunyer J, Rodriguez-Roisin R, Suarez-Cervera M, Vazquez L. Community outbreaks of asthma associated with inhalation of soybean dust. Toxicoepidemiological Committee. *N Engl J Med* 1989; **320**: 1097–102
33 Li JT, Swanson MC, Rando RJ, Wentz-Murtha P, Ovsyannikova IG, Morell F *et al.* Soybean aeroallergen around the port of New Orleans: a potential cause of asthma. *Aerobiologia* 1996; **12**: 173–6
34 Matricardi PM, Rosmini F, Riondino S, Fortini M, Ferrigno L, Rapicetta M *et al.* Exposure to foodborne and orofecal microbes versus airborne viruses in relation to atopy and allergic asthma: epidemiological study. *BMJ* 2000; **320**: 412–7
35 Mattes J, Karmaus W, Moseler M, Frischer T, Kuehr J. Accumulation of atopic disorders within families: a sibling effect only in the offspring of atopic fathers. *Clin Exp Allergy* 1998; **28**: 1480–6

36 Karmaus W, Arshad H, Mattes J. Does the sibling effect have its origin in utero? Investigating birth order, cord blood immunoglobulin E concentration, and allergic sensitization at age 4 years. *Am J Epidemiol* 2001; **154**: 909–15
37 Devereux G, Barker RN, Seaton A. Antenatal determinants of neonatal immune responses to allergens. *Clin Exp Allergy* 2002; **32**: 43–50
38 Sunyer J, Anto JM, Harris J, Torrent M, Vall O, Cullinan P et al. Maternal atopy and parity. *Clin Exp Allergy* 2001; **31**: 1352–5
39 Bingley PJ, Douek IF, Rogers CA, Gale EA. Influence of maternal age at delivery and birth order on risk of type 1 diabetes in childhood: prospective population based family study. Bart's-Oxford Family Study Group. *BMJ* 2000; **321**: 420–4
40 Gutensohn N, Cole P. Childhood social environment and Hodgkin's disease. *N Engl J Med* 1981; **304**: 135–40
41 Shaheen SO, Aaby P, Hall AJ, Barker DJ, Heyes CB, Shiell AW et al. Measles and atopy in Guinea-Bissau. *Lancet* 1996; **347**: 1792–6
42 Shirakawa T, Enomoto T, Shimazu S, Hopkin JM. The inverse association between tuberculin responses and atopic disorder. *Science* 1997; **275**: 77–9
43 van den Biggelaar AH, van Ree R, Rodrigues LC, Lell B, Deelder AM, Kremsner PG et al. Decreased atopy in children infected with Schistosoma haematobium: a role for parasite-induced interleukin-10. *Lancet* 2000; **356**: 1723–7
44 Alm JS, Lilja G, Pershagen G, Scheynius A. Early BCG vaccination and development of atopy. *Lancet* 1997; **350**: 400–3
45 Omenaas E, Jentoft HF, Vollmer WM, Buist AS, Gulsvik A. Absence of relationship between tuberculin reactivity and atopy in BCG vaccinated young adults. *Thorax* 2000; **55**: 454–8
46 Strannegard IL, Larsson LO, Wennergren G, Strannegard O. Prevalence of allergy in children in relation to prior BCG vaccination and infection with atypical mycobacteria. *Allergy* 1998; **53**: 249–54
47 Barnes M, Cullinan P, Athanasaki P, MacNeill S, Hole AM, Harris J et al. Crete: does farming explain urban and rural differences in atopy? *Clin Exp Allergy* 2001; **31**: 1822–8
48 Bodner C, Anderson WJ, Reid TS, Godden DJ. Childhood exposure to infection and risk of adult onset wheeze and atopy. *Thorax* 2000; **55**: 383–87
49 Matricardi PM, Rosmini F, Panetta V, Ferrigno L, Bonini S. Hay fever and asthma in relation to markers of infection in the United States. *J Allergy Clin Immunol* 2002; **110**: 381–7
50 Nystad W. Daycare attendance, asthma and atopy. *Ann Med* 2000; **32**: 390–6
51 Alm JS, Swartz J, Lilja G, Scheynius A, Pershagen G. Atopy in children of families with an anthroposophic lifestyle. *Lancet* 1999; **353**: 1485–8
52 Droste JH, Wieringa MH, Weyler JJ, Nelen VJ, Vermeire PA, Van Bever HP. Does the use of antibiotics in early childhood increase the risk of asthma and allergic disease? *Clin Exp Allergy* 2000; **30**: 1547–53
53 McKeever TM, Lewis SA, Smith C, Collins J, Heatlie H, Frischer M et al. Early exposure to infections and antibiotics and the incidence of allergic disease: a birth cohort study with the West Midlands General Practice Research Database. *J Allergy Clin Immunol* 2002; **109**: 43–50
54 Wickens K, Pearce N, Crane J, Beasley R. Antibiotic use in early childhood and the development of asthma. *Clin Exp Allergy* 1999; **29**: 766–71
55 Braun-Fahrlander C, Gassner M, Grize L, Neu U, Sennhauser FH, Varonier HS et al. Prevalence of hay fever and allergic sensitization in farmer's children and their peers living in the same rural community. SCARPOL team. Swiss Study on Childhood Allergy and Respiratory Symptoms with Respect to Air Pollution. *Clin Exp Allergy* 1999; **29**: 28–34
56 Ernst P, Cormier Y. Relative scarcity of asthma and atopy among rural adolescents raised on a farm. *Am J Respir Crit Care Med* 2000; **161**: 1563–6
57 Riedler J, Braun-Fahrlander C, Eder W, Schreuer M, Waser M, Maisch S et al. Exposure to farming in early life and development of asthma and allergy: a cross-sectional survey. *Lancet* 2001; **358**: 1129–33
58 Braun-Fahrlander C, Riedler J, Herz U, Eder W, Waser M, Grize L et al. Environmental exposure to endotoxin and its relation to asthma in school-age children. *N Engl J Med* 2002; **347**: 869–77
59 Bach JF. The effect of infections on susceptibility to autoimmune and allergic diseases. *N Engl J Med* 2002; **347**: 911–20

60 Stene LC, Nafstad P. Relation between occurrence of type 1 diabetes and asthma. *Lancet* 2001; **357**: 607–8
61 Sheikh A, Smeeth L, Hubbard R. There is no evidence of an inverse relationship between TH2-mediated atopy and TH1-mediated autoimmune disorders: Lack of support for the hygiene hypothesis. *J Allergy Clin Immunol* 2003; **111**: 131–5
62 Wagner GR, Wegman DH. Occupational asthma: prevention by definition. *Am J Ind Med* 1998; **33**: 427–9
63 Blanc PD, Toren K. How much adult asthma can be attributed to occupational factors? *Am J Med* 1999; **107**: 580–7
64 Brooks SM, Weiss MA, Bernstein IL. Reactive airways dysfunction syndrome (RADS). Persistent asthma syndrome after high level irritant exposures. *Chest* 1985; **88**: 376–84
65 Kipen HM, Blume R, Hutt D. Asthma experience in an occupational and environmental medicine clinic. Low-dose reactive airways dysfunction syndrome. *J Occup Med* 1994; **36**: 1133–7
66 Committee on the Medical Effects of Air Pollution. *Asthma and Outdoor Air Pollution*. London: HMSO, 1995
67 McConnell R, Berhane K, Gilliland F, London SJ, Islam T, Gauderman WJ *et al*. Asthma in exercising children exposed to ozone: a cohort study. *Lancet* 2002; **359**: 386–91

Noise pollution: non-auditory effects on health

Stephen A Stansfeld and **Mark P Matheson**

Department of Psychiatry, Medical Sciences Building, Queen Mary, University of London, London, UK

Noise is a prominent feature of the environment including noise from transport, industry and neighbours. Exposure to transport noise disturbs sleep in the laboratory, but not generally in field studies where adaptation occurs. Noise interferes in complex task performance, modifies social behaviour and causes annoyance. Studies of occupational and environmental noise exposure suggest an association with hypertension, whereas community studies show only weak relationships between noise and cardiovascular disease. Aircraft and road traffic noise exposure are associated with psychological symptoms but not with clinically defined psychiatric disorder. In both industrial studies and community studies, noise exposure is related to raised catecholamine secretion. In children, chronic aircraft noise exposure impairs reading comprehension and long-term memory and may be associated with raised blood pressure. Further research is needed examining coping strategies and the possible health consequences of adaptation to noise.

Introduction

Correspondence to: Professor Stephen A Stansfeld, Department of Psychiatry, Medical Sciences Building, Barts and The London, Queen Mary's School of Medicine and Dentistry, Queen Mary, University of London, Mile End Road, London E1 4NS, UK. E-mail: S.A.Stansfeld@qmul.ac.uk

Noise, defined as 'unwanted sound', is perceived as an environmental stressor and nuisance. Non-auditory effects of noise, as dealt with in this chapter, can be defined as 'all those effects on health and well-being which are caused by exposure to noise, with the exclusion of effects on the hearing organ and the effects which are due to the masking of auditory information (*i.e.* communication problems)[1].

Exposure to continuous noise of 85–90 dBA, particularly over a lifetime in industrial settings, can lead to a progressive loss of hearing, with an increase in the threshold of hearing sensitivity[2]. Hearing impairments due to noise are a direct consequence of the effects of sound energy on the inner ear. However, the levels of environmental noise, as opposed to industrial noise, are much lower and effects on non-auditory health cannot be explained as a consequence of sound energy.

If noise does cause ill-health other than hearing impairment, what might be the mechanism? It is generally believed that noise disturbs activities and communication, causing annoyance. In some cases, annoyance may lead to stress responses, then symptoms and possibly illness[3]. Alternatively, noise may influence health directly and not through annoyance. The response to noise may depend on characteristics of the sound, including intensity, frequency, complexity of sound, duration and the meaning of the noise.

Non-auditory effects of noise on health

Noise and sleep disturbance

There is both objective and subjective evidence for sleep disturbance by noise[4]. Exposure to noise disturbs sleep proportional to the amount of noise experienced in terms of an increased rate of changes in sleep stages and in number of awakenings. Habituation occurs with an increased number of sound exposures by night and across nights. One laboratory study, however, found no habituation during 14 nights of exposure to noise at maximum noise level exposure[5]. Objective sleep disturbance is likely to occur if there are more than 50 noise events per night with a maximum level of 50 dBA indoors or more. In fact, there is a low association between outdoor noise levels and sleep disturbance.

In the Civil Aviation Authority Study[6] around Heathrow and Gatwick airports, the relative proportion of total sleep disturbance attributable to noise increased in noisy areas but not the level of total sleep disturbance. In effect, the work suggested a symptom reporting or attribution effect rather than real noise effects. In a subsequent actigraphy study around four UK airports, sleep disturbance was studied in relation to a wide range of aircraft noise exposure over 15 consecutive nights[7]. Although there was a strong association between sleep EEGs and actigram-measured awakenings and self-reported sleep disturbance, none of the aircraft noise events were associated with awakenings detected by actigram and the chance of sleep disturbance with aircraft noise exposure of <82 dB was insignificant. Although it is likely that the population studied was one already adapted to aircraft noise exposure, this study is also likely to be closer to real life than laboratory studies with subjects newly exposed to noise. However, the actigraph as a sensitive measure of sleep disturbance has been questioned.

Noise exposure during sleep may increase blood pressure, heart rate and finger pulse amplitude as well as body movements. There may also be after-effects during the day following disturbed sleep; perceived sleep quality, mood and performance in terms of reaction time all decreased following sleep disturbed by road traffic noise. Studies on noise abatement show that, by reducing indoor noise level, the amount of REM sleep and slow wave sleep can be increased[8]. It thus seems that, although there may be some adaptation to sleep disturbance by noise, complete habituation does not occur, particularly for heart rate.

Noise exposure and performance

There is good evidence, largely from laboratory studies, that noise exposure impairs performance[9]. Performance may be impaired if speech is played while a subject reads and remembers verbal material, although this effect is not found with non-speech noise[10]. The effects of 'irrelevant speech' are independent of the intensity and meaning of the speech. The susceptibility of complex mental tasks to disruption by 'irrelevant speech' suggests that reading, with its reliance on memory, may also be impaired.

Perceived control over and predictability of noise has been found to be important in determining effects and after-effects of noise exposure. Glass and Singer[11] found that tasks performed during noise were unimpaired but tasks that were carried out after noise had been switched off were impaired, this being reduced when subjects were given perceived control over the noise. Indeed, even anticipation of a loud noise exposure in the absence of real exposure may impair performance and an expectation of control counters this effect. Noise exposure may also slow rehearsal in memory, influence processes of selectivity in memory, and choice of strategies for carrying out tasks[1]. There is also evidence that noise may reduce helping behaviour, increase aggression and reduce the processing of social cues seen as irrelevant to task performance[12].

Noise and cardiovascular disease

Physiological responses to noise exposure

Noise exposure causes a number of predictable short-term physiological responses mediated through the autonomic nervous system. Exposure to noise causes physiological activation including increase in heart rate and blood pressure, peripheral vasoconstriction and thus increased peripheral vascular resistance. There is rapid habituation to brief noise exposure but habituation to prolonged noise is less certain[8].

Occupational studies: noise and high blood pressure

The strongest evidence for the effect of noise on the cardiovascular system comes from studies of blood pressure in occupational settings[13] (Table 1). Many occupational studies have suggested that individuals chronically exposed to continuous noise at levels of at least 85 dB have higher blood pressure than those not exposed to noise[14,15]. In many of these studies, noise exposure has also been an indicator of exposure to other factors, both physical and psychosocial, which are also associated with high blood

Table 1 Occupational studies of noise exposure and blood pressure

Reference	Type of study	Sample	Sample size	Noise intensity	Health measures	Hypertension risk factors controlled for	Findings
Herbold et al, 1989[82]	Cross-sectional	Community sample of men 30–69 years	1046	Self report road traffic noise	SBP >160 mm	Age, BMI, Alcohol consumption	Stratified results indicate noise relates to hypertension. Not confirmed in multivariate analysis
Green et al, 1991[83]	Cross-sectional	Israeli male industrial workers	191	74–102 dBA	Hg DBP >95 mmHg	Age, involvement in physical work, smoking, Quetelets Index, hearing loss, using of hearing protectors	SBP, DBP raised in younger but not older worker
Zhao et al[14]	Cross-sectional	Female Chinese textile mill employees	1101	75–104 dBA	SBP >160 mmHg	Age, years of work, salt intake, family history of hypertension	Dose–response relationships in SBP and DBP
Lang et al[15]	Cross-sectional	Parisian workers	7679	>85 dBA/8 h day	DBP >95 mmHg	Age, BMI, alcohol consumption, occupational category	SBP, DBP related to noise. Not confirmed in multivariate analysis
Fogari et al, 1994[84]	Case control	Workers in a metallurgical factory	8811	>80 dB (n = 8078) versus >80dB (n = 733)	DBP >95 mmHg	Age, BMI, duration of employment	Heart rate, DBP not differ SBP higher in noise
Hessel and Sluis-Cremer, 1994[85]	Cross-sectional prospective	White South African miners	2197	80 dBA	SBP > 140 mmHg; DBP > 90 mmHg	Age, BMI	No noise effects on blood pressure
Kristal-Boneh et al, 1995[86]	Cross-sectional	Blue collar workers from 21 Israeli industrial plants 60% response rate	3105		Means only used	Age, smoking, coffee and cholesterol, industrial sector, physical work load	Noise exposure correlates with resting heart rate (significant in men) and DBP only in women. Intensity of noise exposure significantly associated to resting HR in women

pressure. Unless these other risk factors are controlled, spurious associations between noise and blood pressure may arise. A recent pioneering longitudinal industrial noise study has shown that noise levels predicted raised systolic and diastolic pressure in those doing complex but not simple jobs[16], and predicts increased mortality risk. Occupational noise exposure has also recently been linked to greater risk of death from motor vehicle injury[17]. One possibility is that the effects of noise on blood pressure are mediated through an intermediate psychological response such as noise annoyance[18] although this has not been convincingly proved.

Noise and cardiovascular disease in the community

Aircraft noise exposure around Schiphol Airport, Amsterdam has been related to more medical treatment for heart trouble and hypertension, more cardiovascular drug use and higher blood pressure, even after adjustment for age, sex, smoking, height/weight and socio-economic differences[19]. The evidence of the effects of noise on coronary risk factors has not been especially consistent: effects of noise have been shown on systolic blood pressure (but not diastolic pressure), total cholesterol, total triglycerides[20], blood viscosity, platelet count and glucose level[21]. However, a recent Swedish study found that the prevalence of hypertension was higher among people exposed to time-weighted energy averaged aircraft noise levels of at least 55 dBA or maximum levels above 72 dBA around Arlanda airport, Stockholm[22]. In summary, there is some evidence from community studies that environmental noise is related to hypertension and there is also evidence that environmental noise may be a minor risk factor for coronary heart disease (Relative Risk 1.1–1.5)[22–24].

A sudden intense exposure to noise may stimulate catecholamine secretion and precipitate cardiac dysrhythmias. However, neither studies in coronary care units of the effect of speech noise nor studies of noise from low altitude military flights on patients on continuous cardiac monitoring have detected changes in cardiac rhythm attributable to noise[25].

Endocrine responses to noise

Exposure to high intensity noise in industry has been linked in some studies to raised levels of noradrenaline and adrenaline[26]. In one study, catecholamine secretion decreased when workers wore hearing protection against noise. Some studies, but not all, have shown raised cortisol in relation to noise[27]. The general pattern of endocrine responses to noise is indicative of noise as a stressor, exciting short-term physiological responses, but there are inconsistencies between studies.

Noise and psychiatric disorder

It has been postulated that noise exposure creates annoyance which then leads on to more serious psychological effects. This pathway remains unconfirmed; rather it seems that noise causes annoyance and, independently, mental ill-health also increases annoyance. A more complex model[28] incorporates the interaction between the person and their environment. In this model, the person readjusts their behaviour in noisy conditions to reduce exposure. An important addition is the inclusion of the appraisal of noise (in terms of danger, loss of environmental quality, meaning of the noise, challenges for environmental control, *etc.*) and coping (the ability to alter behaviour to deal with the stressor). This model emphasizes that dealing with noise is not a passive process.

Noise exposure and psychological symptoms

Symptoms reported among industrial workers regularly exposed to high noise levels in settings such as schools[29] and factories[30] include nausea, headaches, argumentativeness and changes in mood and anxiety. Many of these industrial studies are difficult to interpret, however, because workers were exposed to other stressors such as physical danger and heavy work demands, in addition to excessive noise. Community surveys have found that high percentages of people reported 'headaches', 'restless nights', and 'being tense and edgy' in high-noise areas[12,31]. An explicit link between aircraft noise and symptoms emerging in such studies raised the possibility of a bias towards over-reporting of symptoms[32]. Notably, a study around three Swiss airports[33], which did not mention that it was related to aircraft noise, did not find any association between the level of exposure to aircraft noise and symptoms.

Noise and common mental disorder

Early studies found associations between the level of aircraft noise and psychiatric hospital admission rates both in London[34] and Los Angeles[35], but this has not been convincingly confirmed by more recent studies[36]. In community studies such as the West London Survey of Psychiatric Morbidity[37], no overall relationship was found between aircraft noise and the prevalence of psychiatric morbidity using various indices of noise exposure. In longitudinal analyses in the Caerphilly Study, no association was found between road traffic noise and psychiatric disorder, even after adjustment for socio-demographic factors and baseline psychiatric disorder, although there was a small non-linear association of noise with increased anxiety scores[38].

Some studies have found dose–response associations: exposure to higher levels of military aircraft noise around Kadena airport in Japan was related in a dose–response relationship to depressiveness and nervousness[39], and road traffic noise has been weakly associated with mental health symptoms after adjusting for age, sex, income and length of residence[40]. Overall, environmental noise seems to be linked to psychological symptoms but not to clinical psychiatric disorder. However, there may be a link to psychiatric disorder at much higher noise levels.

Noise annoyance

The most widespread and well documented subjective response to noise is annoyance, which may include fear and mild anger, related to a belief that one is being avoidably harmed[41]. Noise is also seen as intrusive into personal privacy, while its meaning for any individual is important in determining whether that person will be annoyed by it[42].

Annoyance reactions are often associated with the degree of interference that any noise causes in everyday activities, which probably precedes and leads on to annoyance[43]. In both traffic and aircraft noise studies, noise levels have been found to be associated with annoyance in a dose–response relationship[44,45]. Overall, it seems that conversation, watching television or listening to the radio (all involving speech communication) are the activities most disturbed by aircraft noise while traffic noise, if present at night, is most disturbing for sleep.

Acoustic predictors of noise annoyance in community surveys

One of the primary characteristics affecting the unwantedness of noise is its loudness or perceived intensity. Loudness comprises the intensity of sound, the tonal distribution of sound and its duration. The evidence is mixed on the importance of both the duration and the frequency components of sound and also the number of events involved in determining annoyance[46]. High frequency noise has been found to be more annoying than low frequency noise[47]. Vibrations are perceived as a complement to loud noise in most community surveys of noise and are found to be important factors in determining annoyance, particularly because they are commonly experienced through other senses as well as hearing. Fields[48] found that, after controlling for noise level, noise annoyance increases with fear of danger from the noise source, sensitivity to noise, the belief that the authorities can control the noise, awareness of the non-noise impacts of the source and the belief that the noise source is not important.

Combined effects of noise exposure and other stressors

Noise effects on health may be augmented by, or in turn may augment, the impact of other stressors on health. Stressors may act synergistically, antagonistically or not at all. Stressors may include physical, chemical, biological, social and work organizational factors[49]. In a laboratory based experiment, an interaction was found between having a cold and noise exposure on simple reaction time[50]. There was little difference between healthy and cold subjects' performance tested in quiet conditions, but for subjects tested in noisy conditions (70 dBA), performance was much slower for the cold subjects. Synergistic effects of exposure to noise and vibration have been demonstrated on diastolic blood pressure, whereas temperature and noise have been shown to affect morning adrenaline secretion[51,52].

There has been much emphasis on laboratory studies without considering that results of such studies may lack external validity. Past research on combined effects has not considered common conditions and levels of stressors across studies, direct and indirect effects, long durations of exposure and complex tasks. Field studies suggest that the effects of multiple stressors have greater combined effects than simply summing individual stressors[53]. Few field studies have examined the effects of multiple environmental stressors. This could be an important new area for the development of noise research.

Noise and non-auditory health effects in children

It is likely that children represent a group which is particularly vulnerable to the non-auditory health effects of noise. They have less cognitive capacity to understand and anticipate stressors and lack well-developed coping strategies[54,55]. Moreover, in view of the fact that children are still developing both physically and cognitively, there is a possible risk that exposure to an environmental stressor such as noise may have irreversible negative consequences for this group.

Cognition

Studies of children exposed to environmental noise have consistently found effects on cognitive performance. The studies which are most informative in terms of the effects of noise on cognition have been field studies focusing on primary school children. This article will focus on these studies. For details of noise effects on pre-school children and of laboratory studies of acute noise exposure, see Ref. 56.

The effects of noise have not been found uniformly across all cognitive functions. The research evidence suggests that chronic exposure to noise affects cognitive functions involving central processing and language comprehension. The effects which have been found can be summarized as follows. Deficits have been found in sustained attention and visual attention[57–62]. Relatedly, according to teachers' reports, noise-exposed children have difficulties in concentrating in comparison with children from quieter schools[2,63]. Children exposed to chronic environmental noise have been found to have poorer auditory discrimination and speech perception[54,60,64–67] as well as poorer memory requiring high processing demands[56,68,69]. Finally, chronically exposed children tend to have poorer reading ability and school performance on national standardized tests[64,65,67,70–76].

The first well-designed naturalistic field study to examine the effects of chronic noise exposure focused on primary school children living in four 32-floor apartment buildings adjacent to a major road[65]. The rationale behind this study was that children in the lower floor of the apartment building would be exposed to higher amounts of noise from the road than those higher up the building. Seventy-three children were tested for auditory discrimination and reading level and the results indicated that children living on the lower floors had greater impairments on these measures than those living higher up the buildings.

A very well controlled study by Bronzaft and McCarthy[71] compared primary school children taught in a classroom which was exposed to high levels of railway noise with children in a quiet classroom in the same school. Significant differences in reading scores were found between children in the two classrooms. In fact, the mean reading age of the noise-exposed children was 3–4 months behind that of the control children.

A series of studies have been carried out in schools around Heathrow Airport in west London. These studies have used repeated-measures designs to compare noise-exposed and control children. In the first of these studies[73], the cognitive performance and stress responses of 9- to 10-year-old children in four high noise schools were compared with those of children in four matched control schools. The results showed that, at baseline, the noise-exposed children had impaired reading comprehension and sustained attention after adjustment for age, main language spoken at home and social deprivation. The results at follow up 1 year later suggest that the children's further development in reading comprehension may be affected.

The second study to be conducted near Heathrow Airport[74] was a multi-level modelling study of national standardized test scores (SATs). The data for 11,000 eleven-year-old children were analysed in relation to aircraft noise exposure contours. The results showed that noise exposure was associated with performance on reading and maths tests in a dose–response function but that this was influenced by socio-economic factors. The most recent study to be carried out at Heathrow[75] compared the cognitive performance and stress responses of children in 10 high-noise schools with those of children in 10 matched control schools. The results indicated that children in the noise-exposed schools experienced greater annoyance and had poorer reading performance on the difficult items of a national standardized reading test.

Perhaps the most important of all the naturalistic field studies to examine the effects of noise exposure on children was that carried out in Munich in the 1990s. This prospective, longitudinal study was able to take advantage of a naturally occurring experiment in which the existing Munich Airport was closed down and a new airport was opened at another location. Data were collected at both sites across three testing waves, one before the closure of the old airport and opening of the new one and two afterwards. The mean age of children was 10.8 years. The cross-sectional results[64] showed that, at Wave 1, children at the old airport displayed effects on long-term episodic memory and reading comprehension. The longitudinal results[77] showed that after three waves of testing, children at the old airport had improvements in long-term memory, suggesting that this effect of noise exposure is reversible. Interestingly, by the third wave of testing children at the new airport were exhibiting deficits in long-term memory and reading comprehension, providing strong evidence for a causal link between noise exposure and cognitive effects.

Motivation

A number of studies have identified an association between chronic exposure to aircraft noise and decreased motivation[54,64,66,78]. The results are however not consistent. In the Los Angeles Airport Study[66,78] children exposed to chronic aircraft noise were less likely to solve a difficult puzzle involving a success or failure experience and were more likely to give up. In a follow-up 1 year later[78] the finding that noise-exposed children were less likely to solve a difficult puzzle was replicated, but the finding that the same children are more likely to give up on a difficult puzzle was not. In the Munich study[64], noise-exposed children gave up on an insoluble puzzle more quickly than their non noise-exposed counterparts.

Cardiovascular effects

In addition to effects on cognitive performance, there is evidence that chronic noise exposure may give rise to physiological effects in terms of raised blood pressure. In the Los Angeles Airport Study[66], chronic exposure to aircraft noise was found to be associated with raised systolic and diastolic blood pressure. These increases, although significant, were within the normal range and were not indicative of hypertension. At follow-up 1 year later[78], the findings were the same, showing that these effects had not habituated. In the Munich study, chronic noise exposure was found to be associated with both baseline systolic blood pressure and lower reactivity of systolic blood pressure to a cognitive task presented under acute noise. After the new airport opened, a significant increase in systolic blood pressure was observed providing evidence for a causal link between chronic noise exposure and raised blood pressure. No association was found between noise and diastolic blood pressure or reactivity.

Endocrine disturbance

The Munich Airport Study[64,79] examined overnight, resting levels of urinary catecholamines (adrenaline and noradrenaline). In the cross-sectional study at the old airport, endocrine levels were significantly higher in the noise-exposed children, indicating raised stress levels. The longitudinal data reveal a sharp increase in catecholamine levels in noise-exposed children following the opening of the new airport. Cortisol levels were also examined but no significant differences were observed in either the cross-sectional or longitudinal data. This latter finding is consistent with that of one of the Heathrow Studies[75].

Noise annoyance

Studies have consistently found evidence that exposure to chronic environmental noise causes annoyance in children, even in young children[64,57,71]. In Munich, noise-exposed children were found to be more annoyed by noise as indexed by a calibrated community measure. In London, child-adapted, standard self-report questions[48,80] were used to assess annoyance and showed higher annoyance levels in noise-exposed children. In a follow-up 1 year later, the same result was found, suggesting that annoyance effects are not subject to habituation.

Conclusions

The evidence for effects of environmental noise on health is strongest for annoyance, sleep and cognitive performance in adults and children. Occupational noise exposure also shows some association with raised blood pressure. Dose–response relationships can be demonstrated for annoyance and, less consistently, for blood pressure. The effects of noise are strongest for those outcomes that, like annoyance, can be classified under 'quality of life' rather than illness. What these effects lack in severity is made up for in numbers of people affected, as these responses are very widespread.

It may be that the risk of developing mental or physical illness attributable to environmental noise is quite small, although it is too soon to be certain of this in terms of the progress of research. Part of the problem is that the interaction between people, noise and ill-health is a complex one. Humans are not usually passive recipients of noise exposure and can develop coping strategies to reduce the impact of noise exposure. If people do not like noise they may take action to avoid it by moving away from noisy environments or, if they are unable to move away, by developing coping strategies. Active coping with noise may be sufficient to mitigate any ill-effects. Perception of control over the noise source may reduce the threat of noise and the belief that it can be harmful. It may also be that noise is more harmful to health in situations where several stressors interact and the overall burden may lead to chronic sympathetic arousal or states of helplessness.

Adaptation to long-term noise exposure needs further study. Most people exposed to chronic noise, for instance from major airports, seem to tolerate it. Yet, questionnaire studies suggest that high levels of annoyance do not decline over time. Another possibility is that adaptation to noise is only achieved with a cost to health. Evans and Johnson[81] found that maintaining task performance in noisy offices was associated with additional physiological effort and hormonal response.

Undoubtedly, there is a need for further research to clarify this complex area, including better measurement of noise exposure and health outcomes. Moreover, there should be a greater emphasis on field studies using longitudinal designs with careful choice of samples to avoid undue bias related to prior noise exposure.

References

1. Smith AP, Broadbent DE. *Non-auditory Effects of Noise at Work: A Review of the Literature*. HSE Contract Research Report No 30, London: HMSO, 1992
2. Kryter KD. *The Effects of Noise on Man*, 2nd edn. Orlando, FL: Academic Press, 1985
3. Van Dijk FJH, Souman AM, de Vries FF. Non-auditory effects of noise in industry. VI. A final field study in industry. *Int Arch Occup Environ Health* 1987; **59**: 55–62
4. Öhrström E, Rylander R, Bjorkman N. Effects of night time road traffic noise—an overview of laboratory and field studies on noise dose and subjective noise sensitivity. *J Sound Vib* 1988; **127**: 441–8
5. Öhrström E. Sleep disturbance, psychosocial and medical symptoms—a pilot survey among persons exposed to high levels of road traffic noise. *J Sound Vib* 1989; **133**: 117–28
6. Civil Aviation Authority. *Aircraft Noise and Sleep Disturbance: Final Report*. DORA Report 8008: London, 1980
7. Horne JA, Pankhurst FL, Reyner LA, Hume K, Diamond ID. A field study of sleep disturbance: effects of aircraft noise and other factors on 5,742 nights of actimetrically monitored sleep in a large subject sample. *Sleep* 1994; **17**: 146–59
8. Vallet M, Gagneux J, Clairet JM *et al*. Heart rate reactivity to aircraft noise after a long-term exposure. In: Rossi G (ed) *Noise as a Public Health Problem*. Milan: Centro Recherche e Studio Amplifon, 1983; 965–75
9. Loeb M. *Noise and Human Efficiency*. Chichester: Wiley, 1986
10. Salame P, Baddeley AD. Disruption of short-term memory by unattended speech: implications for the structure of working memory. *J Verb Learn Verb Behav* 1982; **21**: 150–64
11. Glass DC, Singer JE. *Urban Stress*. New York: Academic Press, 1972
12. Jones DM, Chapman AJ, Auburn TC. Noise in the environment: a social perspective. *J Appl Psychol* 1981; **1**: 43–59
13. Thompson SJ. Non-auditory health effects of noise: an updated review. In *Proceedings of Inter-Noise 1996*, vol. **4**. Liverpool, UK: Institute of Acoustics, 1996; 2177–82
14. Zhao Y, Zhang S, Selin S, Spear RCA. A dose response relation for noise induced hypertension. *Br J Ind Med* 1991; **48**: 179–84
15. Lang T, Fouriaud C, Jacquinet MC. Length of occupational noise exposure and blood pressure. *Int Arch Occup Environ Health*: 1992; **63**: 369–72
16. Melamed S, Kristal-Boneh E, Froom P. Industrial noise exposure and risk factors for cardiovascular disease: findings from the CORDIS Study. *Noise Health* 1999; **4**: 49–56

17 Barreto SM, Swerdlow AJ, Smith PG, Higgins CD. Risk of death from motor-vehicle injury in Brazilian steelworkers: a nested case-control study. *Int J Epidemiol* 1997; **26**: 814–21

18 Lercher P, Hörtnagl J, Kofler WW. Work, noise annoyance and blood pressure: combined effects with stressful working conditions. *Int Arch Occup Environ Health* 1993; **63**: 23–8

19 Knipschild PV. Medical effects of aircraft noise: community cardiovascular survey. *Arch Environ Occup Health* 1977; **40**: 185–90

20 Melamed S, Froom P, Kristal-Boneh E, Gofer D, Ribak J. Industrial noise exposure, noise annoyance, and serum lipid levels in blue-collar workers—the CORDIS Study. *Arch Environ Health* 1997; **52**: 292–8

21 Babisch W, Gallacher JEJ, Elwood PC, Ising H. Traffic noise and cardiovascular risk. The Caerphilly Study, first phase. Outdoor noise levels and risk factors. *Arch Environ Health* 1988; **43**: 407–14

22 Rosenlund M, Berglind N, Pershagen G, Jarup L, Bluhm G. Increased prevalence of hypertension in a population exposed to aircraft noise. *Occup Environ Med* 2001; **58**: 769–73

23 Babisch W, Ising H, Gallacher JE, Sweetnam PM, Elwood PC. Traffic noise and cardiovascular risk: The Caerphilly and Speedwell studies, third phase—10 year follow up. *Arch Environ Health* 1999; **54**: 210–6

24 Babisch W. Traffic noise and cardiovascular disease: Epidemiological review and synthesis. *Noise Health* 2000; **8**: 9–32

25 Brenner H, Oberacker A, Kranig W, Buchwalsky R. A field study on the immediate effects of exposure to low-altitude flights on heart rate and arrhythmia in patients with cardiac diseases. *Int Arch Occup Environ Health* 1993; **65**: 263–8

26 Cavatorta A, Falzoi M, Romanelli A *et al*. Adrenal response in the pathogenesis of arterial hypertension in workers exposed to high noise levels. *J Hypertens* 1987; **5**: 463–6

27 Brandenberger G, Follenius M, Wittersheim G, Salame P. Plasma catecholamines and pituitary adrenal hormones related to mental task demand under quiet and noise conditions. *Biol Psychol* 1980; **10**: 239–52

28 Passchier-Vermeer W. *Noise and Health*. Publication No A93/02E. The Hague: Health Council of the Netherlands, 1993

29 Crook MA, Langdon FJ. The effects of aircraft noise in schools around London Airport. *J Sound Vib* 1974; **34**: 221–32

30 Melamed S, Najenson T, Luz T *et al*. Noise annoyance, industrial noise exposure and psychological stress symptoms among male and female workers. In: Berglund B (ed) *Noise 88: Noise as a Public Health Problem. Vol. 2. Hearing, Communication, Sleep and Non-auditory Physiological Effects*. Swedish Council for Building Research, 1988; 315–20

31 Finke HO, Guski R, Martin R *et al*. Effects of aircraft noise on man. *Proceedings of the Symposium on Noise in Transportation*, Section III, paper 1. Southampton: Institute of Sound and Vibration Research, 1974

32 Barker SM, Tarnopolsky A. Assessing bias in surveys of symptoms attributed to noise. *J Sound Vib* 1978; **59**: 349–54

33 Grandjean E, Graf P, Cauber A *et al*. A survey of aircraft noise in Switzerland. *Proceedings of the International Congress on Noise as a Public Health Problem*. Dubrovnik. US Environmental Protection Agency Publications 500: 1973–008. Washington: US EPA, 1973; 645–59

34 Abey-Wickrama I, A'Brook MF, Gattoni F *et al*. Mental hospital admissions and aircraft noise. *Lancet* 1969; **2**: 633; 1275–7

35 Meecham WC, Smith HG. Effects of jet aircraft noise on mental hospital admissions. *Br J Audiol* 1977; **11**: 81–5

36 Jenkins LM, Tarnopolsky A, Hand DJ. Psychiatric admissions and aircraft noise from London Airport: four-year, three hospitals' study. *Psychol Med* 1981; **11**: 765–82

37 Tarnopolsky A, Morton-Williams J. *Aircraft Noise and Prevalence of Psychiatric Disorders*, Research Report. London: Social and Community Planning Research, 1980

38 Stansfeld SA, Gallacher J, Babisch W, Shipley M. Road traffic noise and psychiatric disorder: Prospective findings from the Caerphilly Study. *BMJ* 1996; **313**: 266–7

39 Hiramatsu K, Yamamoto T, Taira K, Ito A, Nakasone T. A survey on health effects due to aircraft noise on residents living around Kadena airport in the Ryukyus. *J Sound Vib* 1997; **205**: 451–60

40 Halpern D. *Mental Health and the Built Environment. More than Bricks and Mortar?* London: Taylor & Francis Ltd, 1995

41 Cohen S, Weinstein N. Non-auditory effects of noise on behavior and health. *J Social Issues* 1981; **37**: 36–70
42 Gunn WJ. The importance of the measurement of annoyance in prediction of effects of aircraft noise on the health and well-being of noise exposed communities. In: Koelaga HS (ed) *Developments in Toxicology and Environmental Science*. Amsterdam: Elsevier, 1987; 237–55
43 Taylor SM. A path model of aircraft noise annoyance. *J Sound Vib* 1984; **96**: 243–60
44 Schulz TJ. Synthesis of social surveys on noise annoyance. *J Acoust Soc Am* 1978; **64**: 377–405
45 Miedema H. Noise and health: How does noise affect us? *Proceedings of Inter-noise 2001*, The Hague, The Netherlands, vol. **1**; 2001; 3–20
46 Fields JM. The effect of numbers of noise events on people's reactions to noise. An analysis of existing survey data. *J Acoust Soc Am* 1984; **75**: 447–67
47 Bjork EA. Laboratory annoyance and skin conductance responses to some natural sounds. *J Sound Vib* 1986; **109**: 339–45
48 Fields JM. *Effects of Personal and Situational Variables on Noise Annoyance with Special Reference to Implications for En Route Noise*. Report No: FAA-AEE-92-03. Washington, DC: Federal Aviation Administration and NASA, 1992
49 Dormolen van M, Hertog CAWM. Combined workload, methodological considerations on recent research. In: Manninen O (ed) *Recent Advances in Researches on the Combined Effects of Environmental Factors*. Tampere, Finland: Pk-Paino, 1988; 25–39
50 Smith A, Thomas M, Brockman P. Noise respiratory virus infections and performance. In: Vallet M (ed) *Proceedings of the 6th International Congress on Noise as a Public Health Problem*, vol. **2**. Nice, France: INRETS, 1993; 311–4
51 Manninen O. Hormonal, cardiovascular and hearing responses in men due to combined exposures to noise, whole body vibrations and different temperatures. In: Okada A, Manninen O (eds) *Recent Advances in Research on the Combined Effects of Environmental Factors*. Kanazawan, Japan: Kyoei, 1987; 107–30
52 Cnockaert JC, Damongeot A, Floru R. Combined effects of noise and vibrations on performance and physiological activation. In: Manninen O (ed) *Recent Advances in Researches on the Combined Effects of Environmental Factors*. Tampere, Finland: Pk-Paino, 1988; 101–15
53 Rutter ML. Primary prevention of psychopathology. In: Kent MM, Rolf JE (eds) *Primary Prevention of Psychopathology*. Hanover, NH: University Press of New England, 1979; 610–25
54 Cohen S, Evans GW, Stokols D, Krantz DS. *Behavior, Health and Environmental Stress*. New York: Plenum Press, 1986
55 Evans GW, Kielwer W, Martin J. The role of the physical environment in the health and well-being of children. In: Schroeder HE (ed) *New Directions in Health Psychology Assessment*. Series in Applied Psychology: Social Issues and Questions. New York, NY: Hemisphere Publishing Corporation, 1991; 127–57
56 Evans GW, Lepore SJ. Nonauditory effects of noise on children. *Children's Environ* 1993; **10**: 31–51
57 Haines MM, Stansfeld SA, Berglund B, Job RFS. Chronic aircraft noise exposure and child cognitive performance and stress. In: Carter N, Job RFS (eds) *Proceedings of the 7th International Conference on Noise as a Public Health Problem*, vol. **1**. Sydney: Noise Effects '98 Pty, 1998; 329–35
58 Hambrick-Dixon PJ. Effects of experimentally imposed noise on task performance of black children attending day centres near elevated subway trains. *Dev Psychol* 1986; **22**: 259–64
59 Hambrick-Dixon PJ. The effect of elevated subway train noise over time on black children's visual vigilance performance. *J Environ Psychol* 1988; **8**: 299–314
60 Moch-Sibony A. Study of the effects of noise on personality and certain psychomotor and intellectual aspects of children, after a prolonged exposure. *Travail Humane* 1984; **47**: 155–65
61 Muller F, Pfeiffer E, Jilg M, Paulsen R, Ranft U. Effects of acute and chronic traffic noise on attention and concentration of primary school children. In: Carter N, Job RFS (eds) *Proceedings of the 7th International Conference on Noise as a Public Health Problem*, vol. **1**. Sydney: Noise Effects '98 Pty, 1998; 365–8
62 Sanz SA, Garcia AM, Garcia A. Road traffic noise around schools: a risk for pupil's performance. *Int Arch Environ Health* 1993; **65**: 205–7
63 Ko NWM. Responses of teachers to road traffic noise. *J Sound Vib* 1981; **77**: 133–6
64 Evans GW, Hygge S, Bullinger M. Chronic noise and psychological stress. *Psychol Sci* 1995; **6**: 333–8

65 Cohen S, Glass DC, Singer JE. Apartment noise, auditory discrimination, and reading ability in children. *J Exp Soc Psychol* 1973; **9**: 407–22
66 Cohen S, Evans GW, Krantz DS, Stokols D. Physiological, motivational and cognitive effects of aircraft noise on children: Moving from the laboratory to the field. *Am Psychol* 1980; **35**: 231–43
67 Evans GW, Maxwell L. Chronic noise exposure and reading deficits: The mediating effects of language acquisition. *Environ Behav* 1997; **29**: 638–56
68 Hygge S. Classroom experiments on the effects of aircraft, road traffic, train and verbal noise presented at 66dBA Leq, and of aircraft and road traffic presented at 55dBA Leq, on long term recall and recognition in children aged 12–14 years. In: Vallak M (ed) *Noise as a Public Health Problem: Proceedings of the Sixth International Congress*, vol. 2. Arcueil, France: INRETS, 1994; 531–8
69 Hygge S, Evans GW, Bullinger M. The Munich Airport Noise Study: Cognitive effects on children from before to after the change over of airports. In: *Proceedings of Inter-Noise '96*. Book 5. Liverpool, UK: Institute of Acoustics, 1996; 2189–92
70 Bronzaft AL. The effect of a noise abatement program on reading ability. *J Environ Psychol* 1981; **1**: 215–22
71 Bronzaft AL, McCarthy DP. The effects of elevated train noise on reading ability. *Environ Behav* 1975; **7**: 517–27
72 Green KB, Pasternack BS, Shore RE. Effects of aircraft noise on reading ability of school-age children. *Arch Environ Health* 1982; **37**: 24–31
73 Haines MM, Stansfeld SA, Job RFS, Berglund B, Head J. Chronic aircraft noise exposure, stress responses, mental health and cognitive performance in school children. *Psychol Med* 2001; **31**: 265–77
74 Haines MM, Stansfeld SA, Head J, Job RFS. Multi-level modelling of the effects of aircraft noise on national standardised performance tests in primary schools around Heathrow Airport, London. *J Epidemiol Community Health* 2001; **56**: 139–44
75 Haines MM, Stansfeld SA, Brentnall S, Head J, Berry B, Jiggins M, Hygge S. The West London School Study: The effects of chronic aircraft noise exposure on child health. *Psychol Med* 2001; **31**: 1385–96
76 Lukas JS, DuPree RB, Swing JW. *Report of a Study on the Effects of Freeway Noise on Academic Achievement of Elementary School Children, and a Recommendation for a Criterion Level for a School Noise Abatement Program*. Sacramento, CA: California Department of Health Services, 1981
77 Hygge S, Evans GW, Bullinger M. A prospective study of some effects of aircraft noise on cognitive performance in school children. *Psychol Sci* 2002; **13**: 469–74
78 Cohen S, Evans GW, Krantz DS, Stokols D. Aircraft noise and children: Longitudinal and cross-sectional evidence on adaptation to noise and the effectiveness of noise abatement. *J Pers Soc Psychol* 1981; **40**: 331–45
79 Evans GW, Lepore SJ, Shejwal BR, Palsane MN. Chronic residential crowding and children's well-being: an ecological perspective. *Child Dev* 1998; **69**: 1514–23
80 Fields JM, de Jong RG, Brown AL *et al*. Guidelines for reporting core information from community noise reaction surveys. *J Sound Vib* 1997; **206**: 685–95
81 Evans GW, Johnson D. Human response to open office noise. In: Carter N, Job RFS (eds) *Proceedings of the International Congress on Noise as a Public Health Problem*, vol. 1. Sydney: Noise Effects '98 Pty, 1998; 255–8
82 Herbold M, Hense H-W, Keil U (1989). Effects of Road Traffic Noise on Prevalence of Hypertension in Men: Results of the Luebeck Blood Pressure Study. *Soz Praeventivmed*, **34**: 19–23.
83 Green MS, Schwartz K, Harari G, Najenson MD (1991). Industrial Noise Exposure and Ambulatory Blood Pressure and Heart Rate. *Journal of Occupational Medicine*, **33(8)**: 879–883.
84 Fogari R, Zoppi A, Vanasia A, Marasi G, Villa G (1994). Occupational noise exposure and blood pressure. *Journal of Hypertension*, **12**: 475–479
85 Hessel PA & Sluis-Cremer GK (1994). Occupational noise exposure and blood pressure: Longitudinal and cross-sectional observations in a group of underground miners. *Archives of Environmental Health*, **49**: 128–134.
86 Kristal-Boneh E, Melamed S, Harari G. Acute noise exposure on blood pressure and heart rate among industrial employees: The Cordis Study. *Archives of Environmental Health*, **50(4)**: 298–304

Risks associated with ionizing radiation

MP Little

Department of Epidemiology and Public Health, Imperial College Faculty of Medicine, St Mary's Campus, London, UK

This paper reviews current knowledge on the deterministic and stochastic risks (the latter including the risk of cancer and of hereditary disease) associated with exposure to ionizing radiation. Particular attention is paid to cancer risks following exposure to man-made low linear energy transfer radiation. Excess cancer risks have been observed in the Japanese atomic bomb survivors and in many medically and occupationally exposed groups. In general, the relative risks among Japanese survivors of atomic-bomb explosions are greater than those among comparable subsets in studies of medically exposed individuals. Cell sterilization largely accounts for the discrepancy in relative risks between these two populations, although other factors may contribute, such as the generally higher underlying cancer risks in the medical series than in the Japanese atomic bomb survivors. Risks among occupationally exposed groups such as nuclear workforces and underground miners are generally consistent with those observed in the Japanese atomic bomb survivors.

Introduction

Risks associated with ionizing radiation have been known for almost as long as ionizing radiation itself: within a year of the discovery of X-rays by Röntgen, skin burns had been reported[1,2] and within 7 years a case of skin cancer was observed[3], in all cases associated with high dose X-ray exposure. In general, risks associated with ionizing radiation can be divided into the so-called stochastic effects (genetic risks in offspring, somatic effects (cancer) in directly exposed population), and deterministic effects. This review summarizes the stochastic and the deterministic risks associated with exposure to radiation.

Correspondence to:
Dr MP Little, Department of Epidemiology and Public Health, Imperial College Faculty of Medicine, St Mary's Campus, Norfolk Place, London W2 1PG, UK. E-mail: mark.little@imperial.ac.uk

Types of ionizing radiation, sources of exposure, units

Ionizing radiation is any electromagnetic wave or particle that can ionize, that is remove an electron from, an atom or molecule of the medium through which it propagates. The process of ionization in living material necessarily changes atoms and molecules, at least transiently, and may thus damage cells. If cellular damage occurs and is not adequately repaired, the cell may not survive, reproduce or perform its normal function. Alternatively, it may result in a viable but modified cell, which may go on to become cancerous if it is a somatic cell, or lead to inherited disease if it is a germ cell.

The basic quantity used to measure absorbed dose from ionizing radiation is the gray (Gy), defined as 1 J of initial energy (of charged particles released by the ionization events) per kg of tissue[4]. The biological effects per unit of absorbed dose differ with the type of radiation and the part of the body exposed, so that a weighted quantity called the effective dose is used, for which the measure is the sievert (Sv)[4]. Low linear energy transfer (LET) radiation (photons, electrons, muons) is assigned a radiation weighting factor of 1, whereas high LET radiation (neutrons, protons, α-particles) is assigned radiation weighting factors of between 5 and 20, depending on the energy of the particles[5]. A related concept is that of relative biological effectiveness (RBE) of a given dose D_q of some specified type 'q' of radiation, which is defined as $RBE_q(D_q) = D_r/D_q$ where D_r is the dose of the reference radiation (usually X-rays or gamma rays) required to produce the same biological effect[6,7]. Since it is a ratio of doses it is a scalar quantity, without units.

All living organisms are continually exposed to ionizing radiation, for example from cosmic and terrestrial gamma rays, ingestion of potassium-40 and radon exposure. Worldwide, the average human exposure to radiation from natural sources is 2.4 mSv per year, about half of which is due to the effects of radon daughters[4]. Diagnostic medical exposures add about 0.4 mSv per year to this figure, atmospheric nuclear testing about 0.005 mSv per year, the Chernobyl accident 0.002 mSv per year, and nuclear power production about 0.0002 mSv per year[4]. Further details on the typical range of these figures, and how they have changed over time, are given in Table 1. These figures should be compared with the average colon dose of 0.2 Sv, and a maximum in excess of 5 Sv, that the proximally exposed groups received in the Japanese atomic bomb survivors Life Span Study (LSS) cohort[8].

Health effects

Deterministic effects

Deterministic effects generally occur only after high-dose acute exposure (mostly >0.1 Gy, see Table 2) and are characterized by non-linear dose–responses, with a threshold dose below which the effect is not observed.

Table 1 Current sources of exposure, annual effective doses (mSv/year) and typical ranges (taken from Ref. 4)

Source	Source sub-type	Worldwide average (mSv/year)	Typical range (mSv/year), other notes
Natural	External exposure		
	Cosmic rays	0.4	0.3–1.0
	Terrestrial gamma rays	0.5	0.3–0.6
	Internal exposure		
	Inhalation (mostly *via* radon)	1.2	0.2–10
	Ingestion	0.3	0.2–0.8
	Total natural	2.4	1–10
Man-made	Diagnostic medical	0.4	0.04–1.0
	Atmospheric nuclear testing	0.005	Decrease from maximum of 0.15 in 1963
	Chernobyl accident	0.002	Decrease from maximum of 0.04 in 1986
	Nuclear power production	0.0002	Has increased with expansion of programme, but decreased with improved practice

Because of these features, deterministic effects are of most relevance in radiotherapy; normal tissue therapy doses are limited to avoid these effects. Deterministic effects are thought to arise from the killing of large groups of cells in the tissues concerned, leading to functional deterioration in the organs affected. Deterministic effects generally arise within days (*e.g.* prodromal syndrome, gastrointestinal syndrome, central nervous system syndrome) or weeks (*e.g.* haematopoietic syndrome, pulmonary syndrome) of exposure; however, certain deterministic effects (*e.g.* cataracts, hypothyroidism) are manifest only over periods of years or more[9]. Most of the information on deterministic effects of radiation comes from (a) medically exposed groups, (b) the survivors of the atomic bombings of Hiroshima and Nagasaki, (c) radiation accidents and (d) animal experiments[9]. The probability, P, of most deterministic effects following an acute dose, D, is given by a modified Weibull distribution:

$$P = \left\{1 - \exp\left[-\ln[2]\cdot\left(\frac{D}{D_{50}}\right)^V\right]\right\}\cdot 1_{D>T} \quad (1)$$

where V is the shape factor, determining the steepness of the risk function, and T is the threshold dose below which no effect is observed[9,10]. D_{50} (>T) is the risk at which the effect is expected to be observed in half the population, and is a function of the radiation dose rate DR (in Gy h^{-1})[9]:

$$D_{50}(DR) = \theta_\infty + \frac{\theta_1}{DR} \quad (2)$$

Values of the parameters are given in Table 2. The International Commission on Radiological Protection (ICRP)[5] recommends that generally lower

Table 2 Recommended parameters for use in equations (1)–(2) for fatal and non-fatal deterministic effects (taken from Ref. 9)

Symptom		Organ	θ_∞ (Gy)	θ_1 (Gy² h⁻¹)	Shape factor V	RBE	Threshold dose T (Gy)
Mortality							
Bone marrow syndrome	Without medical care	Bone marrow	3	0.07	6	2	1.5
	With medical care	Bone marrow	4.5	0.10	6	2	2.2
Pneumonitis		Lung	10	30	7	7	5.5
Gastrointestinal syndrome	External	Small intestine	15	–ᵃ	10	–ᵃ	9.8
	Internal	Colon	35	–ᵃ	10	–ᵃ	23
Embryonic/fetal death	1–18 days	Fetus	1	0.02	2	2	0.12
	18–150 days		1.5	0.03	3	2	0.37
	>150 days		3	0.07	6	2	1.5
Morbidity							
Prodromal	Vomiting	Abdomen	2	0.2	3	–ᵃ	0.49
	Diarrhoea	Abdomen	3	0.2	2.5	–ᵃ	0.55
Lung fibrosis		Lung	5	15	5	7	2.7
Skin burns		Skin	20	5	5	–ᵃ	8.6
Hypothyroidism		Thyroid	60	30	1.3	–ᵃ	2.3
Thyroiditis		Thyroid	1200	–ᵃ	2	–ᵃ	140
Cataracts		Eye lens	3	0.01	5	–ᵃ	1.3
Suppression of ovulation		Ovary	3.5	0.3	3	–ᵃ	0.85
Suppression of sperm count		Testes	0.7	–ᵃ	10	–ᵃ	0.46

ᵃLack of data (A A Edwards and D C Lloyd, personal communication).

RBEs for high LET radiation be used for most deterministic effects than for stochastic effects, as indicated in Table 2.

Although equation (1) is thought to describe most deterministic effects, there are certain effects for which it does not apply. Particularly problematic in this respect are severe mental retardation and reduction in IQ following irradiation of the fetus. Otake et al[11] and Otake and Schull[12] observed a dose-related increase in severe mental retardation in those exposed *in utero* as a result of the atomic bomb explosions in Hiroshima and Nagasaki. This was particularly marked for those exposed in the period 8–15 weeks post-conception. A threshold in the region of 0.1–0.3 Gy was indicated[11,12]. There was a linear reduction in IQ with increasing uterine dose in the atomic bomb survivors, although thresholds of at least 0.1 Gy are consistent with the data[12].

Stochastic effects

Stochastic effects are the main late health effects that are expected to occur in populations exposed to ionizing radiation; somatic risks dominate the overall estimate of health detriment. For both somatic and genetic effects the probability of their occurrence, but not their severity, is taken to depend on the radiation dose. The dose–response may be non-linear, as for deterministic effects. However, in contrast to the situation for deterministic effects, for most stochastic effects it is generally accepted that at sufficiently low doses there is a non-zero linear component to the dose–response *i.e.* there is no threshold. There is little evidence, epidemiological[13-15] or biological[16], for thresholds for stochastic effects.

Heritable genetic effects

The heritable genetic risks associated with radiation exposure are estimated directly from animal studies in combination with data on baseline incidence of disease in human populations[17]. There are no usable human data on radiation-induced germ cell mutations, let alone induced genetic diseases, and the results of the largest and most comprehensive of human epidemiological studies, namely that carried out on the children of the Japanese atomic bomb survivors, are negative; there are no statistically significant radiation-associated adverse hereditary effects in this cohort[18]. The data on the induction of germline mutations at human minisatellite loci[19,20], although of importance from the standpoint of direct demonstration of radiation-induced heritable genetic changes in humans, are not suitable for risk estimation, because they occur in non-coding DNA and are not associated with heritable disease[17].

Simple linear models of dose are generally employed to model genetic effects. For example, the preferred risk model used in the 2001 report of the United Nations Scientific Committee on the Effects of Atomic Radiation (UNSCEAR)[17] assumes that the excess heritable genetic risk for disease class i associated with a dose D of radiation to the parental gonads is given by:

$$R_i = P_i \cdot \left[\frac{1}{DD_i}\right] \cdot MC_i \cdot PCRF_i \cdot D \qquad (3)$$

where P_i is the baseline incidence of that disease class, MC_i is the mutational component of the disease class [defined as the relative increase in disease frequency (relative to the baseline) per unit relative increase in mutation rate (relative to the spontaneous rate)], $PCRF_i$ is the potential recoverability correction factor for the disease class (the fraction of induced mutations compatible with live births) and DD_i is the mutational doubling dose (*i.e.* the dose required to double the mutational load associated with the disease). Table 3 summarizes estimates of risk for the first- and second-generation progeny of an irradiated population that has sustained radiation exposure in the parental generation and no radiation subsequently. These risk estimates assume a mutational doubling dose, DD_i, of 1 Gy, that is based on human data on spontaneous mutation rates and mouse data on

Table 3 Cancer and heritable genetic risk estimates associated with low dose irradiation (taken from Refs 16 and 17)

		Risk/100 Gy
Somatic (cancer) incidence risks[a]		
Relative risk transport for solid cancer,	Leukaemia	0.5[b]
absolute risk transport for leukaemia	Solid cancer	10–17.5[c]
	Total somatic risk	10.5–18.0
Absolute risk transport for solid	Leukaemia	0.5[b]
cancer and leukaemia	Solid cancer	9–14.5[c]
	Total somatic risk	9.5–15.0
Somatic (cancer) mortality risks[a]		
Relative risk transport for solid cancer,	Leukaemia	0.4[b]
absolute risk transport for leukaemia	Solid cancer	5.5–8.5[c]
	Total somatic risk	5.9–8.9
Absolute risk transport for solid cancer and	Leukaemia	0.4[b]
leukaemia	Solid cancer	4–7[c]
	Total somatic risk	4.4–7.4
Heritable genetic risk[d]		
Autosomal dominant, X-linked	Risk to first generation	0.075–0.15
	Risk to second generation	0.05–0.10
Autosomal recessive	Risk to first generation	≈0
	Risk to second generation	≈0
Chronic multifactorial disease	Risk to first generation	0.025–0.12
	Risk to second generation	0.025–0.12
Developmental abnormalities	Risk to first generation	0.20
	Risk to second generation	0.04–0.10
Total	Risk to first generation	0.30–0.47
	Risk to second generation	0.12–0.32

[a]Risk calculated for a UK population assuming a test dose of 0.1 Gy administered, with a dose and dose-rate effectiveness factor of 2 applied to the solid cancer risk.
[b]Risk calculated using a generalized excess absolute risk model, linear-quadratic in dose, with adjustment to the excess absolute risk for age at exposure, time since exposure and gender.
[c]Risk calculated using a generalized excess relative risk model, linear in dose, with adjustment to the relative risk for either attained age and gender (lower risk) or age at exposure and gender (higher risk).
[d]Risk calculated for radiation exposure only in one generation, with a dose and dose-rate effectiveness factor of 3 applied.

induced mutation rates[17]. Neel et al[18] derived an estimate of the doubling dose in the atomic bomb survivors of 3.4–4.5 Sv, based on examination of a combination of five endpoints. However, as discussed by UNSCEAR[17], in view of the differences between the endpoints considered by Neel et al[18] and those employed by UNSCEAR[17], and the uncertainties in the estimates of Neel et al[18], their estimate of the doubling dose is entirely compatible with that used by UNSCEAR[17].

Somatic effects (cancer)
Most of the information on radiation-induced cancer risk comes from (a) the Japanese atomic bomb survivors, (b) medically exposed populations and (c) occupationally exposed groups[16]. The LSS cohort of Japanese atomic bomb survivors is unusual among exposed populations in that both

genders and a wide range of ages were exposed, comparable with those of a general population[8]. Most medically treated groups are more restricted in the age and gender mix. For example, the International Radiation Study of Cervical Cancer patients (IRSCC), a cohort of women followed up after treatment for cancer of the cervix, were all treated as adults, most above the age of 40[21,22]. Organ doses among those treated with radiotherapy tend to be higher than those received by the Japanese atomic bomb survivors, although there are some exceptions, e.g. breast doses in the IRSCC patients[22]. Occupationally exposed groups are also more restricted in their age and gender mix. For example, the cohorts of workers exposed in the nuclear industry[23,24] are overwhelmingly male and exposed in adulthood, as are the groups of underground miners[25]. For these reasons, most standard setting bodies[5,16,26,27] use the LSS as the basis for estimates of population cancer risk associated with exposure to low LET radiation. For certain cancer sites and types of radiation exposures, other groups are occasionally used; in particular for high LET (α-particle) exposure to the lung, underground miners are used[25], and for breast, bone and liver cancer, certain therapeutically exposed groups are sometimes employed[27]. However, lung cancer risks estimated for the miners by applying those estimated in the LSS in combination with the current ICRP dosimetric model[28] are close to those that can be estimated directly[29,30].

Temporal patterns of risk for radiation-induced cancer

One of the principal uncertainties that surround the calculation of population cancer risks from epidemiological data results from the fact that few radiation-exposed cohorts have been followed up to extinction. For example, 52 years after the atomic bombings of Hiroshima and Nagasaki, about half of the survivors were still alive[8]. In attempting to calculate lifetime population cancer risks, it is therefore important to predict how risks might vary as a function of time after radiation exposure, in particular for that group for whom the uncertainties in projection of risk to the end of life are most uncertain, namely those who were exposed in childhood.

Analyses of solid cancers in the LSS and other exposed groups have found that the radiation-induced excess risk can be approximately described by a constant relative risk model[5]. The time-constant excess relative risk (ERR) model assumes that if a dose of radiation is administered to a population, then, after some latent period, there is an increase in the cancer rate, the excess rate being proportional to the underlying cancer rate in an unirradiated population. For leukaemia, this model provides an unsatisfactory fit, and consequently a number of other models have been used for this group of malignancies, including one in which the excess cancer rate resulting from exposure is assumed to be constant rather than proportional to the underlying rate, i.e. the time-constant excess absolute risk (EAR) model[10].

It is well known that for all cancer subtypes (including leukaemia) the ERR diminishes with increasing age at exposure[16] (Figs 1 and 2). For those irradiated in childhood, there is evidence of a reduction in the ERR of solid cancer 25 or more years after exposure[8,31–33] (Fig. 1). Therefore, even for solid cancers, various factors have to be employed to modify the ERR. For many solid cancers, a generalized relative risk model is commonly used, in which the cancer rate t years after exposure for gender s following exposure at age e to a dose D of radiation is given by:

$$r_0(a, s) [1 + F(D) \times \phi(t, e, s)] \qquad (4)$$

where $r_0(a, s)$ is the cancer rate in the absence of irradiation, $i.e.$ the baseline cancer rate, $a = t + e$ is the age at observation (attained age) of the person and $F(D)$ is the function determining the dose dependency of the cancer risk, to be discussed below. The expression $\phi(t, e, s)$ describes the modification to the ERR, $F(D)$, as a function of time since exposure t, age at exposure e, and gender s.

For leukaemia, neither the time-constant EAR model nor the time-constant ERR model fits well (Fig. 2). For reasons largely of ease of interpretation,

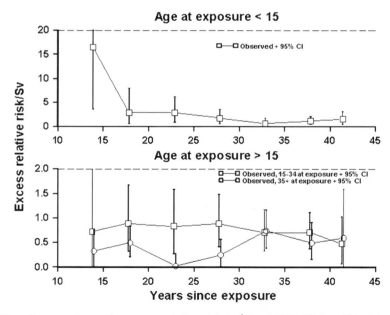

Fig. 1 The diagram shows the excess relative risk Sv^{-1} (and 95% CI) for all incident solid tumour cases in Japanese atomic bomb survivors (excluding survivors with >4 Gy shielded kerma) by age at exposure group (top panel: age at exposure < 15 years; bottom panel: 15 ≤ age at exposure < 35years and age at exposure ≥ 35) as a function of time since exposure (taken from Ref. 51). In particular, this shows a reduction of excess relative risk with increasing age after exposure, and a reduction of excess relative risk with increasing time after exposure for those exposed in childhood (age at exposure < 15).

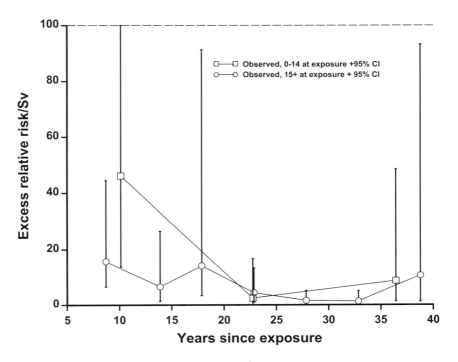

Fig. 2 The diagram shows the excess relative risk Sv^{-1} (and 95% CI) for all radiogenic leukaemias (acute myeloid leukaemia, chronic myeloid leukaemia, acute lymphocytic leukaemia) in Japanese atomic bomb survivors (excluding survivors with >4 Gy shielded kerma) by age at exposure group and as a function of time since exposure. In particular, this shows a reduction of excess relative risk with increasing age after exposure, and a reduction of excess relative risk with increasing time after exposure.

Preston *et al*[34] present most of their analyses of the LSS leukaemia incidence dataset using a generalized absolute risk model, in which the cancer rate t years after exposure for gender s following exposure at age e to a dose D of radiation is given by:

$$r_0(a, s) + F(D) \cdot \psi(t, e, s) \tag{5}$$

The expression $\psi(t, e, s)$ describes the modification to the EAR, $F(D)$, as a function of time since exposure t, age at exposure e and gender s.

Given appropriate forms of the modifying functions $\phi(t, e, s)$ and $\psi(t, e, s)$ of the relative and absolute risk, respectively, equivalently good fits to the leukaemia incidence dataset were achieved using both generalized ERR and generalized EAR models[34]. It is to some extent arbitrary which of these two models one uses. However, as can be seen from Table 3, models with equivalent fit can yield very different risks. The reason for this is that, as noted above, about half the LSS cohort are still alive[8], so that population risk calculations based on this dataset (used by many scientific committees[5,16,26,27]) crucially depend on extrapolating the current mortality and incidence follow-up of this group to the end of life. Uncertainties due to risk projection are greatest for solid cancers, because the radiation-associated excess risk in the LSS is still increasing[8,33]. For leukaemia, the excess risk is reducing over time[34] and most models used predict very few radiation-associated leukaemia deaths or cases from the current follow-up point in the LSS to extinction.

UNSCEAR[16] used a variety of generalized ERR models, linear in dose, for solid cancer risk calculations, including one in which the ERR varied with age at exposure and gender, and one in which the ERR varied with attained age and gender (Table 3). As shown in Table 3, the model with the ERR varying according to attained age yields slightly lower risks than the model in which the ERR varies with age at exposure. For leukaemia, UNSCEAR[16] used a generalized EAR model, linear-quadratic in dose, in which the EAR varied with age at exposure, time since exposure and gender (Table 3).

Forms of cancer dose–response

It has been customary to model the dose–response function $F(D)$ that appears in Expressions (4) and (5) in fits to biological[35] and epidemiological data[16] by the linear-quadratic expression:

$$F(D) = \alpha D + \beta D^2 \qquad (6)$$

There is significant curvilinearity in the dose–response for leukaemia in the LSS[13,14,36], although for solid cancers, apart from non-melanoma skin cancer[33,37] and bone cancer[38,39], there is little evidence for anything other than a linear dose–response in the Japanese cohort[13–15] or in any other group[16]. It should be noted that as well as modifications in effectiveness (per unit dose) relating to alterations in the total dose, there are also possible variations of effectiveness as a result of dose fractionation (the process of splitting a given dose into a number of smaller doses suitably separated in time) and dose-rate[35]. This is not surprising radiobiologically; by administering a given dose at progressively lower dose rates (*i.e.* giving the same total dose over longer periods of time), or by splitting it into many fractions, the biological system has time to repair the damage, so that the total damage induced will be less[35]. Therefore, although for cancers other than leukaemia there is generally little justification for assuming anything other than a linear dose–response, *i.e.* $\beta = 0$, it may nevertheless be justifiable to employ a dose and dose-rate effectiveness factor (DDREF) other than 1. The DDREF is the factor by which one divides risks for high dose and high dose-rate exposure to obtain risks for low doses and low dose-rates. The ICRP[5] recommended that a DDREF of 2 be used together with models linear in dose for all cancer sites, on the basis largely of the observations in various epidemiological datasets. UNSCEAR[35] recommended that a DDREF of no more than 3 be used in conjunction with these linear models.

Another form of dose–response, perhaps less commonly used, slightly generalizes Expression (6):

$$F(D) = (\alpha D + \beta D^2) \exp(\gamma \cdot D) \tag{7}$$

and this has been employed in fits to biological data[35] and epidemiological data[13,21,37,40–42]. The $\alpha D + \beta D^2$ component represents the effect of (carcinogenic) mutation induction, while the $\exp(\gamma \cdot D)$ term represents the effect of cell sterilization or killing. In general, the cell sterilization coefficient γ is <0. Essentially, this is saying that there is a competing effect due to cell killing which is greater at higher radiation doses. A dead cell cannot proliferate and become the focus of a malignant clone. Variant forms of the cell-sterilization term $\exp(\gamma \cdot D)$, incorporating higher powers of dose D, i.e. $\exp(\gamma \cdot D^k)$ for $k > 1$, are sometimes employed[35,37].

Although it is generally assumed that protraction of radiation dose results in a reduction of effect (i.e. DDREF > 1), largely as a result of the extra time this allows for cellular repair processes to operate, there are biological mechanisms that could result in health effects increasing when dose is protracted (i.e. DDREF < 1). Bystander effects, whereby cells that are not directly exposed to radiation exhibit adverse biological effects, have been observed in a number of experimental systems[16]. The bystander effect implies that the dose–response after broad-beam irradiation could be highly concave at low doses because of saturation of the bystander effect at high doses, so that predictions of low-dose effects obtained by linear extrapolation from data for high-dose exposures would be substantial underestimates. Recently, Brenner et al[43] proposed a model for the bystander effect based on the oncogenic transformation data of Sawant et al[44] and Miller et al[45] for in vitro exposure of C3H 10T$^1/_2$ cells to α-particles. Brenner et al[43] discussed evidence from experimental systems that would be consistent with the linear extrapolation of high-dose effects to low doses underestimating oncogenic transformation rates by a factor of between 60 and 3000. However, Little and Wakeford[46] assess the ratio of the lung cancer risk among persons exposed to low (residential) doses of radon daughters to that among persons (underground miners) exposed to high doses of radon daughters to lie in the range 2–4, with an upper 95% confidence limit of about 14, and a lower 95% confidence limit of less than 1. This implies that low dose-rate lung cancer risks associated with α-particle exposure are not seriously underestimated by extrapolation from the high-dose miner data, and that the bystander effect observed in the C3H 10T$^1/_2$ cell system cannot play a large part in the process of radon-induced lung carcinogenesis in humans[46].

Projection of cancer risk across populations

Associated with the issue of projection of cancer risk over time is that of projection of cancer risk between two populations with differing underlying susceptibilities to cancer. Analogous to the constant relative-risk time projection model is the multiplicative transfer of risks, in which the ratio of the radiation-induced excess cancer rates to the underlying cancer rates in the two populations is assumed to be identical. Similarly, akin to the absolute-risk time projection model is the additive transfer of risks, in which the radiation-induced excess cancer rates in the two populations are assumed to be identical. The data that are available suggest that there is no simple solution to the problem[47]. For example, there are weak indications that the ERRs of stomach cancer following radiation exposure may be more comparable than the EARs in populations with different background stomach cancer rates[47]. Comparison of breast cancer risks observed in the Japanese atomic bomb survivor incidence data and those in various medically exposed populations, many from North America and Europe, where underlying breast cancer rates are higher than in Japan, suggests that ERRs are rather higher in the LSS than those in the medically irradiated groups, but (time- and age-adjusted) EARs are more similar[48,49]. The observation that gender differences in solid tumour ERR are generally offset by differences in gender-specific background cancer rates[47] might suggest that EARs are more alike than ERRs. Taken together, these considerations suggest that in various circumstances, relative or absolute transfers of risk between populations may be advocated or, indeed, the use of some sort of hybrid approach such as that employed by Muirhead and Darby[50] and Little et al[51]. Table 3 illustrates the effect of transporting either the absolute risk or the relative risk for solid cancers on the population risk for a UK population. As can be seen, for solid tumours in aggregate the transfer of absolute risk results in slightly lower risks than the transfer of relative risk. However, the differences can be much greater (and in different directions) for specific types of solid tumour.

ERRs in medically exposed groups are generally lower than those in age-, gender- and interval-of-follow-up-matched subsets of the Japanese atomic bomb survivors[52,53]. The most striking discrepancies between the ERRs in the medical series and in the LSS were for leukaemia[52,53]. The discrepancy between the LSS and radiotherapy ERRs for leukaemia and various other sites can be largely explained by cell-sterilization effects[52,53]. This finding is supported by a joint analysis of leukaemia in the LSS[34], in a group of women treated for cervical cancer[21], and in a group treated for ankylosing spondylitis[41], which found evidence of a quadratic-exponential dose–response for all radiogenic leukaemia subtypes, with no significant differences between the dose–response in the LSS and the two medical series[42].

Radiogenic cancer risks derived from groups of nuclear workers are generally consistent with those obtained from the LSS[23,24]. For example

Muirhead et al[24] estimate that the ratio of the leukaemia ERR coefficient in the UK nuclear workers to that in the LSS is 1.18 (90% CI <0, 3.73), and the corresponding ratio for all malignant neoplasms excluding lung cancer and leukaemia is 0.89 (90% CI <0, 3.65). The ratio of lung cancer risk coefficients in the LSS and the Colorado Plateau uranium miners is very close to the value suggested by the latest ICRP[28] model of lung dosimetry[30].

Interaction of radiation-associated cancer risk with chemotherapy

Factors confounded with radiation dose are present in many treatment regimens in medically exposed groups, and in the most recent cases treated occur as a clinically necessary requirement. The most important confounding factor is chemotherapy at some time before, during, or after radiotherapy. This has recently been reviewed elsewhere[52,54]. There is limited quantitative information to assess this interaction, and no marked differences in ERR have been noted between patients receiving and those not receiving adjuvant chemotherapy[52,54]. However, radiogenic EARs are higher in some chemotherapy-treated groups than those among patients not so treated[54].

Interaction of radiation-associated cancer risk with smoking and other lifestyle factors

There is limited information on interaction of ionizing radiation with lifestyle factors. In particular, few studies of medically-exposed groups have high-quality information on smoking. Inskip et al[55] found that, after adjustment for smoking status, the ERR associated with radiotherapy changed only marginally. Davis et al[56] observed a statistically significant excess risk of lung cancer only among radiation-unexposed patients with an unknown smoking history. The smoking information in both of these studies was limited to knowledge of whether or not the person had ever been a smoker. Much better information on smoking status was collected by van Leeuwen et al[57], who observed a stronger trend of lung-cancer risk with radiotherapy dose among people who had more than 1 pack-year of cigarette exposure than in those smoking less than this amount; the interaction between the effects of radiation and cigarette smoke on lung-cancer risk was roughly multiplicative. In the LSS smoking and radiation interact additively on the ERR[58]. Analysis of lung cancer in relation to radon daughter and cigarette-smoke exposure in various cohorts of underground miners suggests that the interaction of radon daughters and cigarette smoke is intermediate between additive and multiplicative in the effects on relative risk of lung cancer[25,30], although the quality of the smoking information is poor. Evidence for interaction of ionizing radiation with other lifestyle factors has been reviewed by UNSCEAR[16].

Interaction of radiation-associated cancer risk with cancer-prone disorders

There is limited information on interactions of radiotherapy and cancer-prone conditions in relation to second-cancer risk in studies of people treated for cancer, and this has been reviewed elsewhere[52,54]. The published data indicate that second-cancer ERR was somewhat lower among patients with a familial cancer syndrome than among those patients without[52,54]. Although there are indications of rather lower radiogenic ERRs among people with cancer-prone disorders, the radiogenic EARs can be higher[52,54]. As discussed above, the fact that ERRs are lower in people with cancer-prone disorders is consistent with a more general pattern observed in epidemiological data, whereby higher underlying cancer risks are to some extent offset by lower radiogenic cancer ERRs[16].

Non-cancer late health effects

There is emerging evidence of excess risks of non-cancer late health effects in the LSS[8,59]. In particular, excess radiation-associated mortality due to circulatory, digestive and respiratory diseases has been observed in this cohort[8,59]. However, the form of the dose–response is uncertain[8,59], and there is little evidence of elevated non-cancer risks in other exposed groups[16], so that it may be premature to use these data to estimate low dose risks for a general population.

Key points for clinical practice

- Except in radiotherapy, where deterministic effects may arise, stochastic effects, and particularly cancer in the directly exposed population, are the principal adverse health effects of exposure to ionizing radiation. Most information on cancer risk is derived from the survivors of the Japanese atomic bomb explosions.

- In general, the radiation-associated excess relative risks among the Japanese atomic bomb survivors are greater than those among medically-exposed individuals, although excess absolute risks can be higher in the medically-exposed groups.

- There is no evidence of higher radiation-associated excess relative risk among those with cancer-prone disorders, although excess absolute risks can be higher in these cancer-prone groups.

- Risks among occupationally exposed groups such as nuclear workforces and underground miners are generally consistent with those observed in the Japanese atomic bomb survivors.

Acknowledgements

The author is grateful for the detailed and helpful remarks of Dr Richard Wakeford, Dr David Lloyd, Mr Alan Edwards, Dr Monty Charles, Dr E. Janet Tawn, Professor Krishnaswami Sankaranarayanan, Dr Colin Muirhead, Dr Mike Joffe and a referee. This work was funded partially by the European Commission under contract FIGD-CT-2000–0079. This paper makes use of data obtained from the Radiation Effects Research Foundation (RERF) in Hiroshima, Japan. RERF is a private foundation funded equally by the Japanese Ministry of Health and Welfare and the US Department of Energy through the US National Academy of Sciences. The conclusions in this paper are those of the author and do not necessarily reflect the scientific judgement of RERF or its funding agencies.

References

1 Stevens LG. Injurious effects on the skin. *BMJ* 1896; **1**: 998
2 Gilchrist TC. A case of dermatitis due to the x rays. *Bull Johns Hopkins Hosp* 1897; **8(71)**: 17–22
3 Frieben A. Demonstration eines Cancroids des rechten Handrückens, das sich nach langdauernder Einwirkung von Röntgenstrahlen bei einem 33 jährigen Mann entwickelt hatte. *Fortschr Röntgenstr* 1902; **6**: 106
4 United Nations Scientific Committee on the Effects of Atomic Radiation (UNSCEAR). *Sources and Effects of Ionizing Radiation. UNSCEAR 2000 Report to the General Assembly, with Scientific Annexes. Volume I: Sources.* New York: United Nations, 2000
5 International Commission on Radiological Protection (ICRP). *Publication 60. 1990 Recommendations of the International Commission on Radiological Protection. Ann ICRP* 1991; **21(1–3)**. Oxford: Pergamon
6 International Commission on Radiation Units and Measurements (ICRU). *The Quality Factor in Radiation Protection.* ICRU Report 40. Bethesda, MD: ICRU, 1986
7 National Council on Radiation Protection and Measurements (NCRP). *The Relative Biological Effectiveness of Radiations of Different Quality.* NCRP Report No. 104. Bethesda, MD: NCRP, 1990
8 Preston DL, Shimizn Y, Pierce DA, Suyama A, Kabuchi K. Studies of the mortality of atomic bomb survivors. Report 13: Solid cancer and noncancer disease mortality: 1950–1997. *Radiat Res* 2003; **160**: 381–407
9 Edwards AA, Lloyd DC. Risk from deterministic effects of ionising radiation. *Docs NRPB* 1996; **7(3)**: 1–31
10 United Nations Scientific Committee on the Effects of Atomic Radiation (UNSCEAR). *Sources, Effects and Risks of Ionizing Radiation. UNSCEAR 1988 Report to the General Assembly, with Annexes.* New York: United Nations, 1988
11 Otake M, Schull WJ, Lee S. Threshold for radiation-related severe mental retardation in prenatally exposed A-bomb survivors: a re-analysis. *Int J Radiat Biol* 1996; **70**: 755–63
12 Otake M, Schull WJ. Review: Radiation-related brain damage and growth retardation among the prenatally exposed atomic bomb survivors. *Int J Radiat Biol* 1998; **74**: 159–71
13 Little MP, Muirhead CR. Evidence for curvilinearity in the cancer incidence dose–response in the Japanese atomic bomb survivors. *Int J Radiat Biol* 1996; **70**: 83–94
14 Little MP, Muirhead CR. Curvature in the cancer mortality dose response in Japanese atomic bomb survivors: absence of evidence of threshold. *Int J Radiat Biol* 1998; **74**: 471–80
15 Pierce DA, Preston DL. Radiation-related cancer risks at low doses among atomic bomb survivors. *Radiat Res* 2000; **154**: 178–86

16   United Nations Scientific Committee on the Effects of Atomic Radiation (UNSCEAR). *Sources and Effects of Ionizing Radiation. UNSCEAR 2000 Report to the General Assembly, with Scientific Annexes. Volume II: Effects.* New York: United Nations, 2000
17   United Nations Scientific Committee on the Effects of Atomic Radiation (UNSCEAR). *Hereditary Effects of Radiation. UNSCEAR 2001 Report to the General Assembly, with Scientific Annex.* New York: United Nations, 2001
18   Neel JV, Schull WJ, Awa AA et al. The children of parents exposed to atomic bombs: estimates of the genetic doubling dose of radiation for humans. *Am J Hum Genet* 1990; **46**: 1053–72
19   Dubrova YE, Nesterov VN, Krouchinsky NG et al. Human minisatellite mutation rate after the Chernobyl accident. *Nature* 1996; **380**: 683–6
20   Dubrova YE, Bersimbaev RI, Djansugurova LB et al. Nuclear weapons tests and human germline mutation rate. *Science* 2002; **295**: 1037
21   Boice JD Jr, Blettner M, Kleinerman RA et al. Radiation dose and leukemia risk in patients treated for cancer of the cervix. *J Natl Cancer Inst* 1987; **79**: 1295–311
22   Boice JD Jr, Engholm G, Kleinerman RA et al. Radiation dose and second cancer risk in patients treated for cancer of the cervix. *Radiat Res* 1988; **116**: 3–55; **127**: 118
23   Cardis E, Gilbert ES, Carpenter L et al. Effects of low doses and low dose rates of external ionizing radiation: cancer mortality among nuclear industry workers in three countries. *Radiat Res* 1995; **142**: 117–32
24   Muirhead CR, Goodill AA, Haylock RGE et al. Occupational radiation exposure and mortality: second analysis of the National Registry for Radiation Workers. *J Radiol Prot* 1999; **19**: 3–26
25   US National Academy of Sciences, National Research Council, Committee on Health Risks of Exposure to Radon (BEIR VI). *Health Effects of Exposure to Radon*. Washington, DC: National Academy Press, 1999
26   US National Academy of Sciences, National Research Council, Committee on the Biological Effects of Ionizing Radiations, *Health Effects of Exposure to Low Levels of Ionizing Radiation (BEIR V)*. Washington, DC: National Academy Press, 1990
27   Muirhead CR, Cox R, Stather JW, MacGibbon BH, Edwards AA, Haylock RGE. Estimates of late radiation risks to the UK population. *Docs NRPB* 1993; **4(4)**: 15–157
28   International Commission on Radiological Protection (ICRP). *Human Respiratory Tract Model for Radiological Protection. A Report of a Task Group of the International Commission on Radiological Protection. Ann ICRP* 1994; **24(1–3)**: 1–482. Oxford: Pergamon
29   Birchall A, James AC. Uncertainty analysis of the effective dose per unit exposure from radon progeny and implications for ICRP risk-weighting factors. *Radiat Prot Dosim* 1994; **53**: 133–40
30   Little MP. Comparisons of lung tumour mortality risk in the Japanese A-bomb survivors and in the Colorado Plateau uranium miners: support for the ICRP lung model. *Int J Radiat Biol* 2002; **78**: 145–63
31   Little MP, Hawkins MM, Shore RE, Charles MW, Hildreth NG. Time variations in the risk of cancer following irradiation in childhood. *Radiat Res* 1991; **126**: 304–16; **132**: 126
32   Little MP, de Vathaire F, Charles MW, Hawkins MM, Muirhead CR. Variations with time and age in the risks of solid cancer incidence after radiation exposure in childhood. *Stat Med* 1998; **17**: 1341–55
33   Thompson DE, Mabuchi K, Ron E et al. Cancer incidence in atomic bomb survivors. Part II: solid tumors, 1958–1987. *Radiat Res* 1994; **137**: S17–S67; **139**: 129
34   Preston DL, Kusumi S, Tomonaga M et al. Cancer incidence in atomic bomb survivors. Part III: leukemia, lymphoma and multiple myeloma, 1950–1987. *Radiat Res* 1994; **137**: S68–S97; **139**: 129
35   United Nations Scientific Committee on the Effects of Atomic Radiation (UNSCEAR). *Sources and Effects of Ionizing Radiation. UNSCEAR 1993 Report to the General Assembly, with Scientific Annexes.* New York: United Nations, 1993
36   Pierce DA, Vaeth M. The shape of the cancer mortality dose–response curve for the A-bomb survivors. *Radiat Res* 1991; **126**: 36–42
37   Little MP, Charles MW. The risk of non-melanoma skin cancer incidence in the Japanese atomic bomb survivors. *Int J Radiat Biol* 1997; **71**: 589–602
38   Rowland RE, Stehney AF, Lucas HF Jr. Dose–response relationships for female radium dial workers. *Radiat Res* 1978; **76**: 368–83

39 Thomas RG. The US radium luminisers: a case for a policy of 'below regulatory concern'. *J Radiol Prot* 1994; **14**: 141–53
40 Thomas DC, Blettner M, Day NE. Use of external rates in nested case-control studies with application to the International Radiation Study of Cervical Cancer patients. *Biometrics* 1992; **48**: 781–94
41 Weiss HA, Darby SC, Fearn T, Doll R. Leukemia mortality after X-ray treatment for ankylosing spondylitis. *Radiat Res* 1995; **142**: 1–11
42 Little MP, Weiss HA, Boice JD Jr, Darby SC, Day NE, Muirhead CR. Risks of leukemia in Japanese atomic bomb survivors, in women treated for cervical cancer, and in patients treated for ankylosing spondylitis. *Radiat Res* 1999; **152**: 280–92; **153**: 243
43 Brenner DJ, Little JB, Sachs RK. The bystander effect in radiation oncogenesis: II. A quantitative model. *Radiat Res* 2001; **155**: 402–8
44 Sawant SG, Randers-Pehrson G, Geard CR, Brenner DJ, Hall EJ. The bystander effect in radiation oncogenesis: I. Transformation in C3H 10T$^{1}/_{2}$ cells *in vitro* can be initiated in the unirradiated neighbors of irradiated cells. *Radiat Res* 2001; **155**: 397–401
45 Miller RC, Randers-Pehrson G, Geard CR, Hall EJ, Brenner DJ. The oncogenic transforming potential of the passage of single α particles through mammalian cell nuclei. *Proc Natl Acad Sci USA* 1999; **96**: 19–22
46 Little MP, Wakeford R. The bystander effect in C3H 10T$^{1}/_{2}$ cells and radon-induced lung cancer. *Radiat Res* 2001; **156**: 695–9
47 United Nations Scientific Committee on the Effects of Atomic Radiation (UNSCEAR). *Sources and Effects of Ionizing Radiation. UNSCEAR 1994 Report to the General Assembly, with Scientific Annexes*. New York: United Nations, 1994
48 Little MP, Boice JD Jr. Comparison of breast cancer incidence in the Massachusetts tuberculosis fluoroscopy cohort and in the Japanese atomic bomb survivors. *Radiat Res* 1999; **151**: 218–24
49 Preston DL, Mattsson A, Holmberg E, Shore R, Hildreth NG, Boice JD Jr. Radiation effects on breast cancer risk: a pooled analysis of eight cohorts. *Radiat Res* 2002; **158**: 220–35; 666
50 Muirhead CR, Darby SC. Modelling the relative and absolute risks of radiation-induced cancers. *J R Stat Soc A* 1987; **150**: 83–118
51 Little MP, Muirhead CR, Charles MW. Describing time and age variations in the risk of radiation-induced solid tumour incidence in the Japanese atomic bomb survivors using generalized relative and absolute risk models. *Stat Med* 1999; **18**: 17–33
52 Little MP, Muirhead CR, Haylock RGE, Thomas JM. Relative risks of radiation-associated cancer: comparison of second cancer in therapeutically irradiated populations with the Japanese atomic bomb survivors. *Radiat Environ Biophys* 1999; **38**: 267–83
53 Little MP. Comparison of the risks of cancer incidence and mortality following radiation therapy for benign and malignant disease with the cancer risks observed in the Japanese A-bomb survivors. *Int J Radiat Biol* 2001; **77**: 431–64; 745–60
54 Little MP. Cancer after exposure to radiation in the course of treatment for benign and malignant disease. *Lancet Oncol* 2001; **2**: 212–20
55 Inskip PD, Stovall M, Flannery JT. Lung cancer risk and radiation dose among women treated for breast cancer. *J Natl Cancer Inst* 1994; **86**: 983–8
56 Davis FG, Boice JD Jr, Hrubec Z, Monson RR. Cancer mortality in a radiation-exposed cohort of Massachusetts tuberculosis patients. *Cancer Res* 1989; **49**: 6130–6
57 van Leeuwen FE, Klokman WJ, Stovall M *et al*. Roles of radiotherapy and smoking in lung cancer following Hodgkin's disease. *J Natl Cancer Inst* 1995; **87**: 1530–7
58 Pierce DA, Sharp GB, Mabuchi K. Joint effects of radiation and smoking on lung cancer risk among atomic bomb survivors. *Radiat Res* 2003; **159**: 511–520
59 Wong FL, Yamada M, Sasaki H, Kodama K, Akiba S, Shimaoka K, Hosoda Y. Noncancer disease incidence in the atomic bomb survivors: 1958–1986. *Radiat Res* 1993; **135**: 418–430

Index

2,3,7,8-tetrachlorodibenzo-p-dioxin (TCDD), 39, 115, 132, 133
N-Acetyl-β-D-glucosaminidase (NAG), 171, 176

Achilles heel, 124
Acute respiratory infections (ARI), 96, 97, 212
Agricultural chemicals, 204–205
Ahlbom A & Feychting M:
  Electromagnetic radiation, 157–165
Air Pollution and Health: a European Approach (APHEA), 98, 101, 146, 147, 148, 150
Air pollution and industrial pollution sources, 37–38
Air pollution and infection in respiratory illness: **Chauhan AJ & Johnston SJ**, 95–112
  combustion of biomass fuels and infections, 97–98
  contemporary indoor and outdoor pollutants and infections, 98–99
  controlled exposure and infection studies, 102–104
  future, 109
  improving air quality improve health, 108–109
  lesson from history, 95–96
  link with infection, 99–102
  mechanisms of interaction between infection and air pollution, 104–108
Air pollution risk transition, 222
Airway hyperresponsiveness, 227, 228, 233, 237
Allergen
  exposure, 233–234
  hypothesis, 233, 234
Allergic contact dermatitis (ACD), 130
Allergic skin reactions, 130
Alveolar macrophages (AMs), 106
Amalgam disease, 173
Ambient air pollution and health: **Katsouyanni K**, 143–156
Ambient particles, 144
  current guidelines and regulations for, 145–146
  measurement, sources, distribution and relevant components of the particle mix, 144–145
Animal infectivity models, 104–105
Anomalies of the male genitalia, 55–56
  genetic factors, 55–56
  trends and spatial/ethnic variation, 55
Arsenic, 178–179, 203
  health effects, 179
  occurrence, exposure and dose, 178–179
Asthma
  definitions of, 227–228
  measuring, 228–229
Asthma: environmental and occupational factors: **Cullinan P & Taylor AN**, 227–238
Atmospheric emissions by source in the European Union, 6
Atopic dermatitis (AD), 129, 130–131

Benzo(a)pyrene, 136
Biological process, 59–70
Black Smoke (BS), 144
Boffetta P & Nyberg F: Contribution of environmental factors to cancer risk, 71–94
Briggs, D: Environmental pollution and the global burden of disease, 1–24
By-products of water treatment, 205

Cadmium, 170–172
  health effects, 171–172
  occurrence, exposure and dose, 170–171
Cancer, 172
Cancer dose–response, forms of, 268–269
Cancer risk across populations, projection of, 270–271
Carbon monoxide (CO), 97, 152–153, 210
  health effects, 153
  measurement, sources, and distribution, 152
  regulations, 152
Carboxyhaemoglobin (COHb), 152, 153

Chauhan AJ & Johnston SJ: Air pollution and infection in respiratory illness, 95–112
Chemical
 contaminants, 202–206
 depigmentation, 133–134
Chernobyl incident, 8
Childhood wheeze, 231
Chloracne (halogen acne), 131–133
Chlorofluorocarbons (CFCs), 139, 140
Chronic obstructive pulmonary disease (COPD), 97, 101, 147, 212
Clinical and public health practice, implications for, 41–42
Clinical practice, key points for, 140, 272–273
Clover Disease, 49
Combined effects of noise exposure and other stressors, 249–250
Confidence interval (CI), 192
Confounder, 30
Congenital anomalies, 27
 sources of pollution and their impact on, 31–40
Contaminants in drinking water: **Fawell J & Nieuwenhuijsen MJ**, 199–208
Contamination of food, 38
Contribution of environmental factors to cancer risk: **Boffetta P & Nyberg F**, 71–94
 environmental exposure to asbestos, 72–74
 exposure to environmental tobacco smoke, 80–81
 inorganic arsenic in drinking water, 83–85
 other drinking water pollutants, 86–88
 other sources of indoor air pollution, 82–83
 outdoor air pollution, 74–80
 residential radon exposure, 81–82
 water chlorination by-products, 85–86
Cryptorchidism, 32, 55, 58, 59, 60, 65, 206
Cullinan P & Taylor AN: Asthma: environmental and occupational factors, 227–238

Dawe RS *see* English JSC
Dermatitis, 129
Deterministic effects, 260–262
Dibromochloropropane (DBCP), 51, 56

Diethylstilboestrol (DES), 49, 50, 60
Dioxin, 62
Disability-adjusted life years (DALYs), 1, 17, 18, 19, 20, 21
Distributions, 229–232
 age, 229–230
 family history, 232
 family size, 231–232
 place, 230
 sex, 230
 time, 230–231
Dolk H & Vrijheid M: The impact of environmental pollution on congenital anomalies, 25–45
Dose–rate effectiveness factor (DDREF), 268, 269
Dose–response relationship, 11, 12, 13, 40, 100, 117, 118, 144, 172, 246, 248, 249
Down Syndrome, 28, 30, 33, 34, 40
Drinking water contamination, 33–34

Effects of known agents, 48–51
 infertility in females, 48–51
 infertility in males, 51
Electromagnetic fields (EMF), 157, 158, 163
Electromagnetic radiation: **Ahlbom A & Feychting M**, 157–165
 balancing risk, 163–164
 electromagnetic spectrum and low frequency fields, 158–159
 established mechanisms of interaction, 159
 potential health risks from weak long-term exposure: ELF, 160–161
 potential health risks from weak long-term exposure: RF, 161–163
Elliott P *see* **Rushton L**
Emission processes and exposure pathways, model of, 15
End stage renal disease (ESRD), 171
Endocrine disrupters, 206
Endocrine disruption, 59–61
 anti-androgens and other types of endocrine disruption, 60–61
 expected spectrum of effects, 61
 oestrogens, 59–60
Endocrine disruptors, 49
Endocrine responses to noise, 247
English JSC, Dawe RS & Ferguson J: Environmental effects and skin disease, 129–142

Index

Environmental effects and skin disease: **English JSC, Dawe RS & Ferguson J**, 129–142
Environmental explanations, 233
Environmental health risks, challenges and future developments of, 123–127
   data availability and quality, 123–124
   data protection and confidentiality, 124–125
   dealing with uncertainties, 125–126
   molecular epidemiology, 125
   role of systematic review and meta-analysis, 126–127
Environmental media, sources and pathways of emission, 5
Environmental pollution
   and health, 2–14
   disasters involving, 39–40
Environmental pollution and the global burden of disease: **Briggs, D**, 1–24
Environmental Protection Agency (EPA), 217
Environmental tobacco smoke (ETS), 72, 80, 81, 152, 213, 214, 215
Epidemiological characteristics of the responsible agent(s), 57–59
   heritability, 58–59
   question of linkage, 57–58
Epidemiological literature, limitations of the, 26–31
Epidemiological study designs, 120
Epidemiology: conditions affecting the male reproductive system, 51–59
Epithelial interaction, 106–108
Essential elements of assessing risk, 118
EUROCAT data, 27
Evaluating evidence on environmental health risks: **Rushton L & Elliott P**, 113–128
   general considerations, 114–117
   human study methods, 119–123
   risk assessment methodology, 118–119
Excess relative risk (ERR), 264, 265
Exposure–response relationships, 11, 168, 179, 180, 206, 234
Exposure–side indicators, 16

Farmer's lung, 13
Fawell J & Nieuwenhuijsen MJ: Contaminants in drinking water, 199–208
Ferguson J see **English JSC**
Fertility and semen quality, 51–53
   genetic factors, 53
   spatial variation, 53
   trends, 52–53
Feychting M see **Ahlbom A**
Fluoride, 203

Gene for infertility, 58
Genetic damage, 62–66
   hypothesis, 62–64
   mechanism and genomic localization, 64–66
Germ line genetic damage, 63
Global burden of disease, 14–22
   environmental burden of disease, 18–22
   estimation of, 14–18
Global Burden of Disease, 147
Global health concern in the future, 220–223
Global production and consumption of selected toxic metals, 169
Glomerular filtration rate (GFR), 171
Great Smog, 95

Harvesting effects, 12
Hazards of heavy metal contamination: **Järup L**, 167–182
Health effects, 260–272
   ambient particle concentrations, 146–147
Health hazards and waste management: **Rushton L**, 183–197
Health outcomes, 115–116
Henderson's hypothesis, 59, 60
Heritable genetic effects, 263–264
High atmospheric ozone and lower atmospheric pollution, 140
Hygiene hypothesis, 234–237
Hypospadias, 27, 32, 36, 55, 56, 59, 60, 65, 189, 206

Impaired bronchial immunity, 105
Incineration, 190–193
   health effects in communities, 191–192
   individual pollutants, 190–191
   worker populations, 192–193
Indoor
   inorganic contaminants, 215–218
   organic contaminants, 218–220
   smoke from burning solid fuels, 211–213

Indoor air pollution: a global health concern: **Zhang J & Smith KR**, 209–225
Infertility and environmental pollutants: **Joffe M**, 47–70
Inorganic mercury, 173–174
Intake fraction, 221
Intercellular adhesion molecule 1 (ICAM-1), 107
International Agency for Research on Cancer (IARC), 161, 163, 172, 177, 218
International Commission on Radiological Protection (ICRP), 261, 265, 268, 271
International Radiation Study of Cervical Cancer patients (IRSCC), 265
Interquartile range (IQR), 101
Ionizing radiation, types, sources, units of, 259–260
Iron and manganese, 204
Irritant contact dermatitis (ICD), 129
Itai-itai (ouch-ouch) disease, 171

Järup L: Hazards of heavy metal contamination, 167–182
Joffe M: Infertility and environmental pollutants, 47–70
Johnston SJ *see* **Chauhan AJ**

Katsouyanni K: Ambient air pollution and health, 143–156

Lead, 174–177
  concentrations in petrol and children's blood, 175
  health effects, 176–177
  occurrence, exposure and dose, 174
Life Span Study (LSS), 260
Lipopolysaccharide (LPS), 106
Little MP: Risks associated with ionizing radiation, 259–275
Low-dose reactive airways syndrome, 238
Low linear energy transfer (LET), 260

Magnetic resonance imaging (MRI), 159
Matheson MP *see* **Stansfeld SA**
Mercury, 172–174
  occurrence, exposure and dose, 172–173

Meteorological Synthesizing Centre-East (MSC-E), 169
Microbial contamination, 200–202
Monocytes (Mo), 106
Municipal solid waste (MSW), 184, 192
Murine cytomegalovirus (MCMV) infection, 105

National Mortality, Morbidity and Air Pollution Study (NMMAPS), 146, 147, 148
Natural killer (NK) cells, 105
Nature of exposure, 116–117
Nieuwenhuijsen MJ *see* **Fawell J**
Nitrogen dioxide (NO_2), 97, 98, 99, 149–151, 210
  health effects, 150–151
  measurement, sources and distribution, 149
  regulations, 149–150
No Observed Adverse Effect Level (NOAEL), 118, 119
Noise and cardiovascular disease, 245
  noise and cardiovascular disease in the community, 247
  occupational studies: noise and high blood pressure, 245–247
  physiological responses to noise exposure, 245
Noise and non-auditory health effects in children, 250–253
  cardiovascular effects, 252
  cognition, 250–252
  endocrine disturbance, 253
  motivation, 252
  noise annoyance, 253
Noise and psychiatric disorder, 248
  acoustic predictors of noise annoyance in community surveys, 249
  and common mental disorder, 248–249
  and sleep disturbance, 244–245
  annoyance, 249
  exposure and psychological symptoms, 248
Noise exposure and performance, 245
Noise pollution: non-auditory effects on health: **Stansfeld SA & Matheson MP**, 243–257
Non-auditory effects of noise on health, 244–247
Non-cancer late health effects, 272

Non-methane volatile organic compounds (NMVOCs), 6
Non-specific-building-related illness (NSBRI), 219, 220
Nyberg F *see* **Boffetta P**

Obstructive bronchitis, 99
Occupational asthma, 234, 237, 238
Oestrogen hypothesis, 60, 66
Organic mercury, 174
Ozone (O_3), 147–149, 215
   health effects, 148–149
   measurement, sources and distribution, 147–148
   regulations, 148

Peak expiratory flow (PEF), 102
Pesticide
   aggregate exposure assessment of, 117
   in agricultural areas, 37
Pollutants: effects on high atmospheric ozone, 139–140
Polychlorinated biphenyls (PCBs), 26, 34, 36, 38, 39, 49, 51, 62, 131, 132, 186, 191
Polycyclic aromatic hydrocarbons (PAHs), 36, 145, 186, 187, 212, 218
Polymorphic light eruption (PLE), 140
Polyvinyl chloride (PVC), 216

Radiation-associated cancer risk, interaction of
   with cancer-prone disorders, 272
   with chemotherapy, 271
   with smoking and other lifestyle factors, 271
Radiation Effects Research Foundation (RERF), 273
Radiation-induced cancer, temporal patterns of risk for, 265–268
Radiofrequency (RF), 158, 159, 161, 162, 163
Reactive airways dysfunction syndrome, 238
Relative biological effectiveness (RBE), 260
Relative risk (RR), 74, 121, 265, 270, 272
Respiratory irritants, 237–238

Respiratory syncytial (RS), 101, 102, 107
Risks associated with ionizing radiation: **Little MP**, 259–275
Rushton L & Elliott P: Evaluating evidence on environmental health risks, 113–128
Rushton L: Health hazards and waste management, 183–197

Scleroderma-like disease, 134
   Quartz induced, 134–135
Selenium and uranium, 203
Semivolatile organic compounds (SVOCs), 211
Sensitive period, 26
Sequences, 27
Sex ratio, 56
   spatial and other variation, 56
   trends, 56
Sick building syndrome, 211
Single-nucleotide polymorphisms (SNPs), 64
Skin cancer, 136–138
Smith KR *see* **Zhang J**
Smoky coal, 212
Somatic effects (cancer), 264–265
Source–effect chain, 3–14
   atmospheric emissions, 3–7
   emissions to surface water, groundwater and soil, 7
   environmental fate, 7–9
   exposure and dose, 9–10
   health effects: dose–response relationships, latency and attributable risk, 11–13
   models of pollutant pathways, 13–14
Specific absorption rate (SAR), 159
Sperm chromatin stability assay (SCSA), 65
Standardized Mortality Ratio (SMR), 192
Standardized test scores (SATs), 251
Stansfeld SA & Matheson MP: Noise pollution: non-auditory effects on health, 243–257
Stochastic effects, 262–272
Sulphur dioxide (SO_2), 96, 97, 143, 191, 213
   health effects, 151–152
   measurement, sources, and distribution, 151

regulations, 151
Sunlight effects on the skin, 138–139

T helper 1 (Th1), 130
T helper 2 (Th2), 130
Taylor AN *see* **Cullinan P**
Testicular cancer, 53–54
 genetic factors, 54
 spatial and ethnic variation, 54
 trends, 54
Testicular dysgenesis syndrome, 55
Texan sharpshooter, 31
The impact of environmental pollution on congenital anomalies: **Dolk H & Vrijheid M**, 25–45
Total particle mass (TSP) concentration, 145
Total volatile organic compounds (TVOC), 219
Trihalomethanes (THM), 34
Tumour necrosis factor-α (TNF-α), 106

Ultraviolet and visible rays, 138
United Nations Scientific Committee on the Effects of Atomic Radiation (UNSCEAR), 263, 264, 268, 269, 271
Urban pollution, 205
US Centre for Disease Control and Prevention (CDC), 214

Vibration white finger (VWF), 135, 136
Volatile organic compounds (VOCs), 186, 211, 219, 220
Vrijheid M *see* **Dolk H**

Waste disposal (landfill sites and incinerators) and contaminated land, 34–37
Waste management, 183
 hazardous substances associated with, 185–187
 methods of, 184–185
Waste management practices on health, 187
 birth defects and reproductive disorders, 188–189
 cancer, 189–190
 landfill sites, 187–188
 studies of self-reported health symptoms, 190
Waste production in England and Wales, 185
World Health Organization (WHO), 211

Years of life lost (YLL), 17
Yusho disease (oil disease), 39

Zhang J & Smith KR: Indoor air pollution: a global health concern, 209–225

OXFORD UNIVERSITY PRESS
Great Clarendon Street, Oxford OX2 6DP, UK

© The British Council 2003

Apart from any fair dealing for the purposes of research or private study, or criticism or review, as permitted under the UK Copyright, Designs and Patents Act, 1988, no part of this publication may be reproduced, stored or transmitted, in any form or by any means, without the prior permission in writing of the publishers, or in the case of reprographic reproduction in accordance with the terms of licences issued by the Copyright Licensing Agency in the UK, or in accordance with the terms of licences issued by the appropriate Reproduction Rights Organisation outside the UK. Enquiries concerning reproduction outside the terms stated here should be sent to the publishers at the UK address printed on this page.

British Library Cataloguing in Publication Data
A catalogue record for this book is available from the British Library
ISBN 0-19-852698-9
ISSN 0007-1420

Subscription information *British Medical Bulletin* is published quarterly on behalf of The British Council. Subscription rates for 2003 are £185/$315 for four volumes, each of one issue. Prices include distribution; the British Medical Bulletin is distributed by surface mail within Europe, by air freight and second class post within the USA*, and by various methods of air-speeded delivery to all other countries. Subscription orders, single issue orders and enquiries regarding volumes from 2001 onwards should be sent to:

Oxford University Press, Great Clarendon Street, Oxford OX2 6DP, UK
(Tel +44 (0)1865 353907; Fax +44(0)1865 353485; E-mail: jnl.orders@oup.co.uk

*Periodicals postage paid at Rahway, NJ. US Postmaster: Send address changes to *British Medical Bulletin*, c/o Mercury Airfreight International Ltd, 365 Blair Road, Avenel, NJ 07001, USA.

Back numbers of titles published 1996–2000 are available from The Royal Society of Medicine Press Limited, 1 Wimpole St, London W1G 0AE, UK. (Tel. +44 (0)20 7290 2921; Fax +44 (0)20 7290 2929); www.rsm.ac.uk/pub/bmb/htm).

Pre-1996 back numbers: Contact Jill Kettley, Subscriptions Manager, Harcourt Brace, Foots Cray, Sidcup, Kent DA14 5HP (Tel +44 (0)20 8308 5700; Fax +44 (0)20 8309 0807).

This journal is indexed, abstracted and/or published online in the following media: Adonis, Biosis, BRS Colleague (full text), Chemical Abstracts, Colleague (Online), Current Contents/Clinical Medicine, Current Contents/Life Sciences, Elsevier BIOBASE/Current Awareness in Biological Sciences, EMBASE/Excerpta Medica, Index Medicus/Medline, Medical Documentation Service, Reference Update, Research Alert, Science Citation Index, Scisearch, SIIC-Database Argentina, UMI (Microfilms)

Printed in Great Britain by Bell & Bain Ltd, Glasgow, Scotland.

BRITISH MEDICAL BULLETIN
VOLUME 68   2003